An Invitation to Applied Category Theory

Category theory is unmatched in its ability to organize a ___ ___ to find commonalities between structures of all sorts. No longer the exclusive preserve of pure mathematicians, it is now proving itself to be a powerful tool in science, informatics, and industry. By facilitating communication between communities and building rigorous bridges between disparate worlds, applied category theory has the potential to be a major organizing force.

This book offers a self-contained tour of applied category theory. Each chapter follows a single thread motivated by a real-world application and discussed with category-theoretic tools. We see data migration as an adjoint functor, electrical circuits in terms of monoidal categories and operads, and collaborative design via enriched profunctors. All the relevant category theory, from simple to sophisticated, is introduced in an accessible way with many examples and exercises, making this an ideal guide even for those without experience of university-level mathematics.

Brendan Fong is a postdoctoral associate in the Department of Mathematics at the Massachusetts Institute of Technology. His research explores how we use pictures to represent and reason about the systems around us, and how to understand the world from a relational point of view. These topics find their intersection in applied category theory.

David I. Spivak is a research scientist in the Department of Mathematics at the Massachusetts Institute of Technology. He has found applications of category theory ranging from database integration to knowledge representation, from materials science to dynamical systems and behavior. He is the author of two other books in category theory.

An Invitation to Applied Category Theory

Seven Sketches in Compositionality

BRENDAN FONG

Massachusetts Institute of Technology

DAVID I. SPIVAK

Massachusetts Institute of Technology

CAMBRIDGE
UNIVERSITY PRESS

CAMBRIDGE
UNIVERSITY PRESS

University Printing House, Cambridge CB2 8BS, United Kingdom

One Liberty Plaza, 20th Floor, New York, NY 10006, USA

477 Williamstown Road, Port Melbourne, VIC 3207, Australia

314–321, 3rd Floor, Plot 3, Splendor Forum, Jasola District Centre, New Delhi – 110025, India

79 Anson Road, #06–04/06, Singapore 079906

Cambridge University Press is part of the University of Cambridge.

It furthers the University's mission by disseminating knowledge in the pursuit of education, learning, and research at the highest international levels of excellence.

www.cambridge.org
Information on this title: www.cambridge.org/9781108482295
DOI: 10.1017/9781108668804

First published 2019
3rd printing 2021

Printed in Singapore by Markono Print Media Pte Ltd

A catalogue record for this publication is available from the British Library.

Library of Congress Cataloging-in-Publication Data
Names: Fong, Brendan, 1988– author. | Spivak, David I., 1978– author.
Title: An invitation to applied category theory : seven sketches in compositionality / Brendan Fong (Massachusetts Institute of Technology), David I. Spivak (Massachusetts Institute of Technology).
Description: Cambridge ; New York, NY : Cambridge University Press, 2019. | Includes bibliographical references and index.
Identifiers: LCCN 2018058456 | ISBN 9781108482295
Subjects: LCSH: Categories (Mathematics) | Computable functions. | Logic, Symbolic and mathematical. | Mathematical analysis.
Classification: LCC QA9.25 .F66 2019 | DDC 512/.62–dc23
LC record available at https://lccn.loc.gov/2018058456

ISBN 978-1-108-48229-5 Hardback
ISBN 978-1-108-71182-1 Paperback

Cambridge University Press has no responsibility for the persistence or accuracy of URLs for external or third-party internet websites referred to in this publication and does not guarantee that any content on such websites is, or will remain, accurate or appropriate.

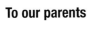

To our parents

Contents

Preface

Category theory is becoming a central hub for all of pure mathematics. It is unmatched in its ability to organize and layer abstractions, to find commonalities between structures of all sorts, and to facilitate communication between different mathematical communities.

But it has also been branching out into science, informatics, and industry. We believe that it has the potential to be a major cohesive force in the world, building rigorous bridges between disparate worlds, both theoretical and practical. The motto at MIT is *mens et manus*, Latin for mind and hand. We believe that category theory – and pure math in general – has stayed in the realm of mind for too long; it is ripe to be brought to hand.

Purpose and audience

The purpose of this book is to offer a self-contained tour of applied category theory. It is an invitation to discover advanced topics in category theory through concrete real-world examples. Rather than try to give a comprehensive treatment of these topics – which include adjoint functors, enriched categories, proarrow equipments, toposes, and much more – we merely provide a taste of each. We want to give readers some insight into how it feels to work with these structures as well as some ideas about how they might show up in practice.

The audience for this book is quite diverse: anyone who finds the above description intriguing. This could include a motivated high school student who hasn't seen calculus yet but has loved reading a weird book on mathematical logic they found at the library. Or a machine-learning researcher who wants to understand what vector spaces, design theory, and dynamical systems could possibly have in common. Or a pure mathematician who wants to imagine what sorts of applications their work might have. Or a recently retired programmer who's always had an eerie feeling that category theory is what they've been looking for to tie it all together, but who's found the usual books on the subject impenetrable.

For example, we find it something of a travesty that at the time of publication there is almost no introductory material available on monoidal categories. Even beautiful modern introductions to category theory, e.g. by Riehl [Rie17] or Leinster [Lei14], do not include anything on this rather central topic. The only exceptions we can think of are [CK17, Chapter 3] and [CP10], each of which has a very user-friendly introduction to monoidal categories; however, readers who are not drawn to physics may not think to look there.

The basic idea of monoidal categories is certainly not too abstract; modern human intuition seems to include a pre-theoretical understanding of monoidal categories that is just waiting to be formalized. Is there anyone who wouldn't correctly understand the basic idea being communicated in the following diagram?

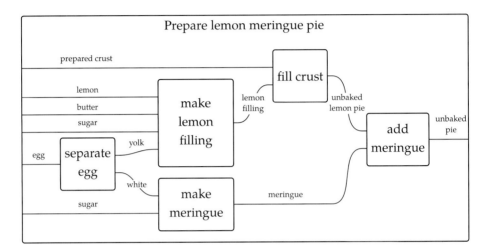

Many applied category theory topics seem to take monoidal categories as their jumping-off point. So one aim of this book is to provide a reference – even if unconventional – for this important topic.

We hope this book inspires both new visions and new questions. We intend it to be self-contained in the sense that it is approachable with minimal prerequisites, but not in the sense that the complete story is told here. On the contrary, we hope that readers use this as an invitation to further reading, to orient themselves in what is becoming a large literature, and to discover new applications for themselves.

This book is, unashamedly, our take on the subject. While the abstract structures we explore are important to any category theorist, the specific topics have simply been chosen to our personal taste. Our examples are ones that we find simple but powerful, concrete but representative, entertaining but in a way that feels important and expansive at the same time. We hope our readers will enjoy themselves and learn a lot in the process.

How to read this book

The basic idea of category theory – which threads through every chapter – is that if one pays careful attention to structures and coherence, the resulting systems will be extremely reliable and interoperable. For example, a category involves several structures: a collection of objects, a collection of morphisms relating objects, and a formula for combining any chain of morphisms into a morphism. But these structures need to *cohere* or work together in a simple commonsense way: a chain of chains is itself a long chain, so combining a chain of chains should be the same as combining the long chain. That's it!

We shall see structures and coherence come up in pretty much every definition we give: "here are some things and here are how they fit together." We ask the reader to be on the lookout for structures and coherence as they read the book, and to realize that as we layer abstraction upon abstraction, it is the coherence that makes all the parts work together harmoniously in concert.

Each chapter in this book is motivated by a real-world topic, such as electrical circuits, control theory, cascade failures, information integration, and hybrid systems. These motivations lead us into and through various sorts of category-theoretic concepts. We generally have one motivating idea and one category-theoretic purpose per chapter, and this forms the title of the chapter, e.g. Chapter 4 is "Co-design: Profunctors, Categorification, and Monoidal Categories."

In many math books, the difficulty is roughly a monotonically increasing function of the page number. In this book, this occurs in each chapter, but not so much in the book as a whole. The chapters start out fairly easy and progress in difficulty.

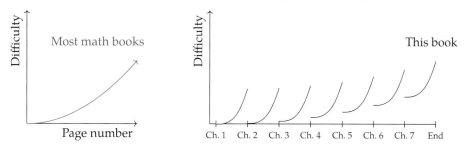

The upshot is that if you find the end of a chapter very difficult, hope is certainly not lost: you can start on the next one and make good progress. This format lends itself to giving you a first taste now, but also leaving open the opportunity for you to come back to the book at a later date and get more deeply into it. But by all means, if you have the gumption to work through each chapter to its end, we very much encourage that!

We include about 240 exercises throughout the text, with solutions in the Appendix at the end of the book. Usually these exercises are fairly straightforward; the only thing they demand is that readers change their mental state from passive to active, reread the previous paragraphs with intent, and put the pieces together. Readers become *students* when they work through the exercises; until then they are more tourists, riding on a bus and listening off and on to the tour guide. Hey, there's nothing wrong with that, but we do encourage you to get off the bus and make direct contact with the native population and local architecture as often as you can.

Acknowledgments

Thanks to Jared Briskman, James Brock, Ronnie Brown, Thrina Burana, David Chudzicki, Jonathan Castello, Margo Crawford, Fred Eisele, David Ellerman, Cam Fulton,

Bruno Gavranović, Sebastian Galkin, John Garvin, Peter Gates, Juan Manuel Gimeno, Alfredo Gómez, Leo Gorodinski, Jason Grossman, Jason Hooper, Yuxi Liu, Jesús López, MTM, Nicolò Martini, Martin MacKerel, Pete Morcos, Nelson Niu, James Nolan, Dan Oneata, Paolo Perrone, Thomas Read, Rif A. Saurous, Dan Schmidt, Samantha Seaman, Marcello Seri, Robert Smart, Valter Sorana, Adam Theriault-Shay, Emmy Trewartha, Sergey Tselovalnikov, Andrew Turner, Joan Vazquez, Daniel Wang, Jerry Wedekind for helpful comments and conversations.

We also thank our sponsors at the AFOSR; this work was supported by grants FA9550–14–1–0031 and FA9550–17–1–0058.

Finally, we extend a very special thanks to John Baez for running an online course (https://forum.azimuthproject.org) on this material and generating tons of great feedback.

Personal note

Our motivations to apply category theory outside of math are, perhaps naively, grounded in the hope it can help bring humanity together to solve our big problems. But category theory is a tool for thinking, and like any tool it can be used for purposes we align with and those we don't.

In this personal note, we ask that readers try to use what they learn in this book to do something they would call "good," in terms of contributing to the society they'd want to live in. For example, if you're planning to study this material with others, consider specifically inviting someone from an underrepresented minority – a group that is more highly represented in society than in upper-level math classes – to your study group. As another example, perhaps you can use the material in this book to design software that helps people relate to and align with each other. What is the mathematics of a well-functioning society?

The way we use our tools affects all our lives. Our society has seen the results – both the wonders and the waste – resulting from rampant selfishness. We would be honored if readers found ways to use category theory as part of an effort to connect people, to create common ground, to explore the cross-cutting categories in which life, society, and environment can be represented, and to end the ignorance entailed by limiting ourselves to a singular ontological perspective on anything.

If you do something of the sort, please let us and the community know about it.

Brendan Fong and David I. Spivak

1 Generative Effects: Orders and Galois Connections

In this book, we explore a wide variety of situations – in the world of science, engineering, and commerce – where we see something we might call *compositionality*. These are cases in which systems or relationships can be combined to form new systems or relationships. In each case we find category-theoretic constructs – developed for their use in pure math – which beautifully describe the compositionality of the situation.

This chapter, being the first of the book, must serve this goal in two capacities. First, it must provide motivating examples of compositionality, as well as the relevant categorical formulations. Second, it must provide the mathematical foundation for the rest of the book. Since we are starting with minimal assumptions about the reader's background, we must begin slowly and build up throughout the book. As a result, examples in the early chapters are necessarily simplified. However, we hope the reader will already begin to see the sort of structural approach to modeling that category theory brings to the fore.

1.1 More Than the Sum of Their Parts

We motivate this first chapter by noticing that while many real-world structures are compositional, the results of observing them are often not. The reason is that observation is inherently "lossy": in order to extract information from something, one must drop the details. For example, one stores a real number by rounding it to some precision. But if the details are actually relevant in a given system operation, then the observed result of that operation will not be as expected. This is clear in the case of roundoff error, but it also shows up in non-numerical domains: observing a complex system is rarely enough to predict its behavior because the observation is lossy.

A central theme in category theory is the study of structures and structure-preserving maps. A map $f : X \to Y$ is a kind of observation of object X via a specified relationship it has with another object, Y. For example, think of X as the subject of an experiment and Y as a meter connected to X, which allows us to extract certain features of X by looking at the reaction of Y.

Asking which aspects of X one wants to preserve under the observation f becomes the question "what category are you working in?" As an example, there are many functions f from $\mathbb{R}$ to $\mathbb{R}$ (where $\mathbb{R}$ is the set of real numbers), and we can think of them as observations: rather than view x "directly," we only observe $f(x)$. Out of all the functions $f: \mathbb{R} \to \mathbb{R}$, only some of them preserve the order of numbers, only some of them preserve the distance between numbers, only some of them preserve the sum of numbers, etc. Let's check in with an exercise; a solution can be found in the Appendix.

Exercise 1.1. Some terminology: a function $f: \mathbb{R} \to \mathbb{R}$ is said to be

(a) *order-preserving* if $x \le y$ implies $f(x) \le f(y)$, for all $x, y \in \mathbb{R}$;[1]
(b) *metric-preserving* if $|x - y| = |f(x) - f(y)|$;
(c) *addition-preserving* if $f(x + y) = f(x) + f(y)$.

For each of the three properties defined above – call it *foo* – find an f that is *foo*-preserving and an example of an f that is not *foo*-preserving.   ◇

In category theory we want to keep control over which aspects of our systems are being preserved under various observations. As we said above, the less structure is preserved by our observation of a system, the more "surprises" occur when we observe its operations. One might call these surprises *generative effects*.

In using category theory to explore generative effects, we follow the basic ideas from work by Adam [Ada17]. He goes much more deeply into the issue than we can here; see Section 1.5. But as mentioned above, we must also use this chapter to give an order-theoretic warm-up for the full-fledged category theory to come.

1.1.1   A First Look at Generative Effects

To explore the notion of a generative effect we need a sort of system, a sort of observation, and a system-level operation that is not preserved by the observation. Let's start with a simple example.

A simple system
Consider three points; we'll call them •, ∘, and ∗. In this example, a *system* will simply be a way of connecting these points together. We might think of our points as sites on a power grid, with a system describing connection by power lines, or as people susceptible to some disease, with a system describing interactions that can lead to contagion. As an abstract example of a system, there is a system where • and ∘ are connected, but neither is connected to ∗. We shall draw this like so:

---

[1] We are often taught to view functions $f: \mathbb{R} \to \mathbb{R}$ as plots in the (x, y)-coordinate system, where x is the domain (independent) variable and y is the codomain (dependent) variable. In this book, we do not adhere to that naming convention; e.g. in Example 1.1, both x and y are being "plugged in" as input to f. As an example consider the function $f(x) = x^2$. Then f being order-preserving would say that, for any $x, y \in \mathbb{R}$, if $x \le y$ then $x^2 \le y^2$; is that true?

Connections are symmetric, so if a is connected to b, then b is connected to a. Connections are also transitive, meaning that if a is connected to b, and b is connected to c, then a is connected to c; that is, all a, b, and c are connected. Friendship is not transitive – my friend's friend is not necessarily my friend – but possible communication of a concept or a disease is.

Here we depict two more systems, one in which none of the points are connected, and one in which all three points are connected.

There are five systems in all, and we depict them below.

Now that we have defined the sort of system we want to discuss, suppose that Alice is observing this system. Her observation of interest, which we call Φ, extracts a single feature from a system, namely whether the point $\bullet$ is connected to the point $*$; this is what she wants to know. Her observation of the system will be an assignment of either `true` or `false`; she assigns `true` if $\bullet$ is connected to $*$, and `false` otherwise. So Φ assigns the value `true` to the following two systems:

and Φ assigns the value `false` to the three remaining systems:

(1.1)

The last piece of setup is to give a sort of operation that Alice wants to perform on the systems themselves. It's a very common operation – one that will come up many times throughout the book – called *join*. If the reader has been following the story arc, the expectation here is that Alice's connectivity observation will not be compositional with respect to the operation of system joining; that is, there will be generative effects. Let's see what this means.

Joining our simple systems

Joining two systems A and B is performed simply by combining their connections. That is, we shall say the *join* of systems A and B, denoted $A \vee B$, has a connection between points x and y if there are some points $z_1, \ldots, z_n$ such that each of the following is true in at least one of A or B: x is connected to z_1, z_i is connected to z_{i+1}, and z_n is connected to y. In a three-point system, the above definition is overkill, but we want to say something that works for systems with any number of elements. The high-level way to say it is "take the transitive closure of the union of the connections in A and B." In our three-element system, it means for example that

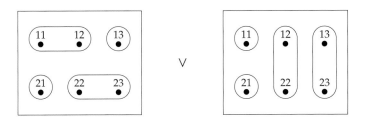

and   (1.2)

Exercise 1.2.   What is the result of joining the following two systems?

We are now ready to see the generative effect. We don't want to build it up too much – this example has been made as simple as possible – but we shall see that Alice's observation fails to preserve the join operation. We've been denoting her observation – measuring whether • and ∗ are connected – by the symbol Φ; it returns a boolean result, either true or false.

We see above in Eq. (1.1) that $\Phi(\text{⊘}) = \Phi(\text{⊘}) = $ false: in both cases • is not connected to ∗. On the other hand, when we join these two systems as in Eq. (1.2), we see that $\Phi(\text{⊘} \vee \text{⊘}) = \Phi(\text{⊘}) = $ true: in the joined system, • *is* connected to ∗. The question that Alice is interested in, that of Φ, is inherently lossy with respect to join, and there is no way to fix it without a more detailed observation, one that includes not only ∗ and • but also ○.

While this was a simple example, it should be noted that whether the potential for such effects exist – i.e. determining whether an observation is operation-preserving – can be incredibly important information to know. For example, Alice could be in charge of putting together the views of two local authorities regarding possible contagion between an infected person • and a vulnerable person ∗. Alice has noticed that if they separately extract information from their raw data and combine the results, it gives a different answer than if they combine their raw data and extract information from it.

1.1.2 Ordering Systems

Category theory is all about organizing and layering structures. In this section we will explain how the operation of joining systems can be derived from a more basic structure: order. We shall see that while joining is not preserved by Alice's connectivity observation Φ, order is.

To begin, we note that the systems themselves are ordered in a hierarchy. Given systems A and B, we say that $A \leq B$ if, whenever x is connected to y in A, then x is connected to y in B. For example,

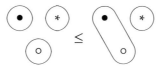

This notion of $\leq$ leads to the following diagram:

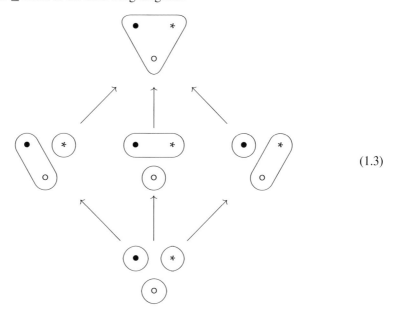

(1.3)

where an arrow from system A to system B means $A \leq B$. Such diagrams are known as *Hasse diagrams*.

As we were saying above, the notion of join is derived from this order. Indeed, for any two systems A and B in the Hasse diagram (1.3), the joined system $A \vee B$ is the smallest system that is bigger than both A and B. That is, $A \leq (A \vee B)$ and $B \leq (A \vee B)$, and for any C, if $A \leq C$ and $B \leq C$ then $(A \vee B) \leq C$. Let's walk through this with an exercise.

Exercise 1.3.

1. Write down all the partitions of a two-element set $\{\bullet, *\}$, order them as above, and draw the Hasse diagram.

2. Now do the same thing for a four-element set, say $\{1, 2, 3, 4\}$. There should be 15 partitions.

Choose any two systems in your 15-element Hasse diagram, call them A and B.

3. What is $A \vee B$, using the definition given in the paragraph above Eq. (1.2)?
4. Is it true that $A \leq (A \vee B)$ and $B \leq (A \vee B)$?
5. What are all the systems C for which both $A \leq C$ and $B \leq C$?
6. Is it true that in each case $(A \vee B) \leq C$?   ◇

The set $\mathbb{B} = \{\texttt{true}, \texttt{false}\}$ of booleans also has an order, $\texttt{false} \leq \texttt{true}$:

Thus $\texttt{false} \leq \texttt{false}$, $\texttt{false} \leq \texttt{true}$, and $\texttt{true} \leq \texttt{true}$, but $\texttt{true} \not\leq \texttt{false}$. In other words, $A \leq B$ if A implies B.[2]

For any A, B in $\mathbb{B}$, we can again write $A \vee B$ to mean the least element that is greater than both A and B.

Exercise 1.4.   Using the order $\texttt{false} \leq \texttt{true}$ on $\mathbb{B} = \{\texttt{true}, \texttt{false}\}$, what is:

1. $\texttt{true} \vee \texttt{false}$?
2. $\texttt{false} \vee \texttt{true}$?
3. $\texttt{true} \vee \texttt{true}$?
4. $\texttt{false} \vee \texttt{false}$?   ◇

Let's return to our systems with •, ∘, and ∗, and Alice's "• is connected to ∗" function, which we called Φ. It takes any such system and returns either $\texttt{true}$ or $\texttt{false}$. Note that the map Φ preserves the $\leq$ order: if $A \leq B$ and there is a connection between • and ∗ in A, then there is such a connection in B too. The possibility of a generative effect is captured in the inequality

$$\Phi(A) \vee \Phi(B) \leq \Phi(A \vee B). \tag{1.4}$$

We saw on page 4 that this can be a strict inequality: we showed two systems A and B with $\Phi(A) = \Phi(B) = \texttt{false}$, so $\Phi(A) \vee \Phi(B) = \texttt{false}$, but where $\Phi(A \vee B) = \texttt{true}$. In this case, a generative effect exists.

These ideas capture the most basic ideas in category theory. Most directly, we have seen that the map Φ preserves some structure but not others: it preserves order but not join. In fact, we have seen here hints of more complex notions from category theory, without making them explicit; these include the notions of category, functor, colimit, and adjunction. In this chapter we will explore these ideas in the elementary setting of ordered sets.

---

[2] In mathematical logic, $\texttt{false}$ implies $\texttt{true}$ but $\texttt{true}$ does not imply $\texttt{false}$. That is "P implies Q" means, "if P is true, then Q is true too, but if P is not true, I'm making no claims."

1.2 What is Order?

Above we informally spoke of two different ordered sets: the order on system connectivity and the order on booleans $\text{false} \leq \text{true}$. Then we related these two ordered sets by means of Alice's observation Φ. Before continuing, we need to make such ideas more precise. We begin in Section 1.2.1 with a review of sets and relations. In Section 1.2.2 we will give the definition of a *preorder* – short for preordered set – and a good number of examples.

1.2.1 Review of Sets, Relations, and Functions

We will not give a definition of *set* here, but informally we will think of a set as a collection of things, known as elements. These things could be all the leaves on a certain tree, or the names of your favorite fruits, or simply some symbols a, b, c. For example, we write $A = \{h, 1\}$ to denote the set, called A, that contains exactly two elements, one called h and one called 1. The set $\{h, h, 1, h, 1\}$ is exactly the same as A because they both contain the same elements, h and 1, and repeating an element more than once in the notation doesn't change the set.[3] For an arbitrary set X, we write $x \in X$ if x is an element of X; so we have $h \in A$ and $1 \in A$, but $0 \notin A$.

Example 1.5. Here are some important sets from mathematics – and the notation we will use – that will appear again in this book.

- $\varnothing$ denotes the empty set; it has no elements.
- $\{1\}$ denotes a set with one element; it has one element, 1.
- $\mathbb{B}$ denotes the set of *booleans*; it has two elements, true and false.
- $\mathbb{N}$ denotes the set of *natural numbers*; it has elements $0, 1, 2, 3, \ldots, 90^{717}, \ldots$.
- $\underline{n}$, for any $n \in \mathbb{N}$, denotes the nth *ordinal*; it has n elements $1, 2, \ldots, n$. For example, $\underline{0} = \varnothing, \underline{1} = \{1\}$, and $\underline{5} = \{1, 2, 3, 4, 5\}$.
- $\mathbb{Z}$, the set of *integers*; it has elements $\ldots, -2, -1, 0, 1, 2, \ldots, 90^{717}, \ldots$.
- $\mathbb{R}$, the set of *real numbers*; it has elements like $\pi, 3.14, 5 * \sqrt{2}, e, e^2, -1457, 90^{717}$, etc.

Given sets X and Y, we say that X is a *subset* of Y, and write $X \subseteq Y$, if every element in X is also in Y. For example $\{h\} \subseteq A$. Note that the empty set $\varnothing := \{\}$ is a subset of every other set.[4] Given a set Y and a property P that is either true or false for each element of Y, we write $\{y \in Y \mid P(y)\}$ to mean the subset of those y's that satisfy P.

Exercise 1.6.

1. Is it true that $\mathbb{N} = \{n \in \mathbb{Z} \mid n \geq 0\}$?

---

[3] If you want a notion where "$h, 1$" is different from "$h, h, 1, h, 1$," you can use something called *bags*, where the number of times an element is listed matters, or *lists*, where order also matters. All of these are important concepts in applied category theory, but sets will come up the most for us.

[4] When we write $Z := \text{foo}$, it means "assign the meaning foo to variable Z," whereas $Z = \text{foo}$ means simply that Z is equal to foo, perhaps as discovered via some calculation. In particular, $Z := \text{foo}$ implies $Z = \text{foo}$ but not vice versa; indeed it *would not* be proper to write $3 + 2 := 5$ or $\{\} := \varnothing$.

2. Is it true that $\mathbb{N} = \{n \in \mathbb{Z} \mid n \geq 1\}$?

3. Is it true that $\varnothing = \{n \in \mathbb{Z} \mid 1 < n < 2\}$?   ◇

If both X_1 and X_2 are subsets of Y, their *union*, denoted $X_1 \cup X_2$, is also a subset of Y, namely the one containing the elements in X_1 and the elements in X_2 but no more. For example if $Y = \{1, 2, 3, 4\}$ and $X_1 = \{1, 2\}$ and $X_2 = \{2, 4\}$, then $X_1 \cup X_2 = \{1, 2, 4\}$. Note that $\varnothing \cup X = X$ for any $X \subseteq Y$.

Similarly, if both X_1 and X_2 are subsets of Y, then their *intersection*, denoted $X_1 \cap X_2$, is also a subset of Y, namely the one containing all the elements of Y that are both in X_1 and in X_2, and no others. So $\{1, 2, 3\} \cap \{2, 5\} = \{2\}$.

What if we need to union together or intersect a lot of subsets? For example, consider the sets $X_0 = \varnothing$, $X_1 = \{1\}$, $X_2 = \{1, 2\}$, etc. as subsets of $\mathbb{N}$, and we want to know what the union of all of them is. This union is written $\bigcup_{n \in \mathbb{N}} X_n$, and it is the subset of $\mathbb{N}$ that contains every element of every X_n, but no others. Namely, $\bigcup_{n \in \mathbb{N}} X_n = \{n \in \mathbb{N} \mid n \geq 1\}$. Similarly one can write $\bigcap_{n \in \mathbb{N}} X_n$ for the intersection of all of them, which will be empty in the above case.

Given two sets X and Y, the *product* $X \times Y$ of X and Y is the set of pairs (x, y), where $x \in X$ and $y \in Y$.

Finally, we may want to take a *disjoint* union of two sets, even if they have elements in common. Given two sets X and Y, their *disjoint union* $X \sqcup Y$ is the set of pairs of the form $(x, 1)$ or $(y, 2)$, where $x \in X$ and $y \in Y$.

Exercise 1.7.   Let $A := \{h, 1\}$ and $B := \{1, 2, 3\}$.

1. There are eight subsets of B; write them out.
2. Take any two nonempty subsets of B and write out their union.
3. There are six elements in $A \times B$; write them out.
4. There are five elements of $A \sqcup B$; write them out.
5. If we consider A and B as subsets of the set $\{h, 1, 2, 3\}$, there are four elements of $A \cup B$; write them out.   ◇

Relationships between different sets – for example between the set of trees in your neighborhood and the set of your favorite fruits – are captured using subsets and product sets.

Definition 1.8.   Let X and Y be sets. A *relation between X and Y* is a subset $R \subseteq X \times Y$. A *binary relation on X* is a relation between X and X, i.e. a subset $R \subseteq X \times X$.

It is convenient to use something called *infix notation* for binary relations $R \subseteq A \times A$. This means one picks a symbol, say $\star$, and writes $a \star b$ to mean $(a, b) \in R$.

Example 1.9.   There is a binary relation on $\mathbb{R}$ with infix notation $\leq$. Rather than writing $(5, 6) \in R$, we write $5 \leq 6$.

Other examples of infix notation for relations are $=, \approx, <, >$. In number theory, we are interested in whether one number divides without remainder into another number; this relation is denoted with infix notation $|$, so $5|10$.

Partitions and equivalence relations

We can now define partitions more formally.

Definition 1.10. If A is a set, a *partition* of A consists of a set P and, for each $p \in P$, a nonempty subset $A_p \subseteq A$, such that

$$A = \bigcup_{p \in P} A_p \qquad \text{and} \qquad \text{if } p \neq q \text{ then } A_p \cap A_q = \varnothing. \qquad (1.5)$$

We may denote the partition by $\{A_p\}_{p \in P}$. We refer to P as the set of *part labels* and if $p \in P$ is a part label, we refer to A_p as the pth *part*. The condition (1.5) says that each element $a \in A$ is in exactly one part.

We consider two different partitions $\{A_p\}_{p \in P}$ and $\{A'_{p'}\}_{p' \in P'}$ of A to be the same if for each $p \in P$ there exists a $p' \in P'$ with $A_p = A'_{p'}$. In other words, if two ways to divide A into parts are exactly the same – the only change is in the labels – then we don't make a distinction between them.

Exercise 1.11. Suppose that A is a set and $\{A_p\}_{p \in P}$ and $\{A'_{p'}\}_{p' \in P'}$ are two partitions of A such that for each $p \in P$ there exists a $p' \in P'$ with $A_p = A'_{p'}$.

1. Show that for each $p \in P$ there is at most one $p' \in P'$ such that $A_p = A'_{p'}$.
2. Show that for each $p' \in P'$ there is a $p \in P$ such that $A_p = A'_{p'}$.   ◇

Exercise 1.12. Consider the partition shown below:

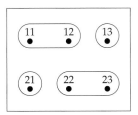

For any two elements $a, b \in \{11, 12, 13, 21, 22, 23\}$, let's allow ourselves to write a twiddle (tilde) symbol $a \sim b$ between them if a and b are both in the same part. Write down every pair of elements (a, b) that are in the same part. There should be 10.[5]   ◇

We shall see in Proposition 1.14 that there is a strong relationship between partitions and something called equivalence relations, which we define next.

---

[5] Hint: whenever someone speaks of "two elements a, b in a set A," the two elements may be the same!

Definition 1.13. Let A be a set. An *equivalence relation* on A is a binary relation, let's give it infix notation $\sim$, satisfying the following three properties:

(a) $a \sim a$, for all $a \in A$,
(b) $a \sim b$ iff[6] $b \sim a$, for all $a, b \in A$,
(c) if $a \sim b$ and $b \sim c$ then $a \sim c$, for all $a, b, c \in A$.

These properties are called *reflexivity*, *symmetry*, and *transitivity*, respectively.

Proposition 1.14. Let A be a set. There is a one-to-one correspondence between the ways to partition A and the equivalence relations on A.

Proof. We first show that every partition gives rise to an equivalence relation, and then that every equivalence relation gives rise to a partition. Our two constructions will be mutually inverse, proving the proposition.

Suppose we are given a partition $\{A_p\}_{p \in P}$; we define a relation $\sim$ and show it is an equivalence relation. Define $a \sim b$ to mean that a and b are in the same part: there is some $p \in P$ such that $a \in A_p$ and $b \in A_p$. It is obvious that a is in the same part as itself. Similarly, it is obvious that if a is in the same part as b then b is in the same part as a, and that if further b is in the same part as c then a is in the same part as c. Thus $\sim$ is an equivalence relation as defined in Definition 1.13.

Suppose we are given an equivalence relation $\sim$; we will form a partition on A by saying what the parts are. Say that a subset $X \subseteq A$ is ($\sim$)-closed if, for every $x \in X$ and $x' \sim x$, we have $x' \in X$. Say that a subset $X \subseteq A$ is ($\sim$)-connected if it is nonempty and $x \sim y$ for every $x, y \in X$. Then the parts corresponding to $\sim$ are exactly the ($\sim$)-closed, ($\sim$)-connected subsets. It is not hard to check that these indeed form a partition. $\square$

Exercise 1.15. Let's complete the "it's not hard to check" part in the proof of Proposition 1.14. Suppose that $\sim$ is an equivalence relation on a set A, and let P be the set of ($\sim$)-closed and ($\sim$)-connected subsets $\{A_p\}_{p \in P}$.

1. Show that each part A_p is nonempty.
2. Show that if $p \neq q$, i.e. if A_p and A_q are not exactly the same set, then $A_p \cap A_q = \varnothing$.
3. Show that $A = \bigcup_{p \in P} A_p$. $\diamond$

Definition 1.16. Given a set A and an equivalence relation $\sim$ on A, we say that the *quotient* $A / \sim$ of A under $\sim$ is the set of parts of the corresponding partition.

Functions
The most frequently used sort of relation between sets is that of functions.

---

[6] "Iff" is short for "if and only if."

Definition 1.17. Let S and T be sets. A *function from S to T* is a subset $F \subseteq S \times T$ such that for all $s \in S$, there exists a unique $t \in T$ with $(s, t) \in F$.

The function F is often denoted $F : S \to T$. From now on, we write $F(s) = t$, or sometimes $s \mapsto t$, to mean $(s, t) \in F$. For any $t \in T$, the *preimage of t along F* is the subset $f^{-1}(t) := \{s \in S \mid F(s) = t\}$.

A function is called *surjective*, or a *surjection*, if for all $t \in T$, there exists $s \in S$ with $F(s) = t$. A function is called *injective*, or an *injection*, if for all $t \in T$ and $s_1, s_2 \in S$ with $F(s_1) = t$ and $F(s_2) = t$, we have $s_1 = s_2$. A function is called *bijective* if it is both surjective and injective.

We use various decorations on arrows, $\to$, $\twoheadrightarrow$, $\rightarrowtail$, $\xrightarrow{\cong}$ to denote these special sorts of functions. Here is a table with the name, arrow decoration, and an example of each sort of function:

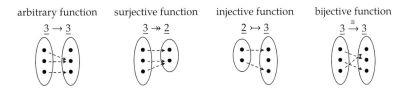

Example 1.18. An important but very simple sort of function is the *identity function* on a set X, denoted id_X. It is the bijective function $\mathrm{id}_X(x) = x$.

For notational consistency with Definition 1.17, the arrows in Example 1.18 might be drawn as $\mapsto$ rather than $\dashrightarrow$. The $\dashrightarrow$-style arrows were drawn because we thought it was prettier, i.e. easier on the eye. Beauty is important too; an imbalanced preference for strict correctness over beauty becomes *pedantry*. But, outside of pictures, we will be careful.

Exercise 1.19. In the following, do not use any examples already drawn above.

1. Find two sets A and B and a function $f : A \to B$ that is injective but not surjective.
2. Find two sets A and B and a function $f : A \to B$ that is surjective but not injective.

Now consider the four relations shown here:

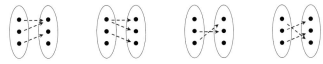

For each relation, answer the following two questions.

3. Is it a function?
4. If not, why not? If so, is it injective, surjective, both (i.e. bijective), or neither?   ◇

Exercise 1.20. Suppose that A is a set and $f : A \to \varnothing$ is a function to the empty set. Show that A is empty.   ◇

Example 1.21. A partition on a set A can also be understood in terms of surjective functions out of A. Given a surjective function $f : A \twoheadrightarrow P$, where P is any other set, the preimages $f^{-1}(p) \subseteq A$, one for each element $p \in P$, form a partition of A. Here is an example.

Consider the partition of $S := \{11, 12, 13, 21, 22, 23\}$ shown below:

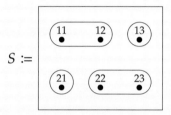

It has been partitioned into four parts, so let $P = \{a, b, c, d\}$ and let $f : S \twoheadrightarrow P$ be given by

$$f(11) = a, \quad f(12) = a, \quad f(13) = b, \quad f(21) = c, \quad f(22) = d, \quad f(23) = d.$$

Exercise 1.22. Write down a surjection corresponding to each of the five partitions in Eq. (1.3).   ◇

Definition 1.23. If $F : X \to Y$ is a function and $G : Y \to Z$ is a function, their *composite* is the function $X \to Z$ defined to be $G(F(x))$ for any $x \in X$. It is often denoted $G \circ F$, but we prefer to denote it $F \,\fatsemi\, G$. It takes any element $x \in X$, evaluates F to get an element $F(x) \in Y$ and then evaluates G to get an element $G(F(x))$.

Example 1.24. If X is any set and $x \in X$ is any element, we can think of x as a function $\{1\} \to X$, namely the function sending 1 to x. For example, the three functions $\{1\} \to \{1, 2, 3\}$ shown below correspond to the three elements of $\{1, 2, 3\}$:

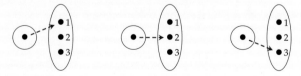

Suppose we are given a function $F: X \to Y$ and an element of X, thought of as a function $x: \{1\} \to X$. Then evaluating F at x is given by the composite $F(x) = x \mathbin{\fatsemi} F$.

1.2.2 Preorders

In Section 1.1, we often used the symbol $\leq$ to denote a sort of order. Here is a formal definition of what it means for a set to have an order.

Definition 1.25. A *preorder relation* on a set X is a binary relation on X, here denoted with infix notation $\leq$, such that

(a) $x \leq x$; and
(b) if $x \leq y$ and $y \leq z$, then $x \leq z$.

The first condition is called *reflexivity* and the second is called *transitivity*. If $x \leq y$ and $y \leq x$, we write $x \cong y$ and say x and y are *equivalent*. We call a pair $(X, \leq)$ consisting of a set equipped with a preorder relation a *preorder*.

Remark 1.26. Observe that reflexivity and transitivity are familiar from Definition 1.13: equivalence relations are preorders with an additional symmetry condition.

Example 1.27 (Discrete preorders). Every set X can be considered as a discrete preorder $(X, =)$. This means that the only order relationships on X are of the form $x \leq x$; if $x \neq y$ then neither $x \leq y$ nor $y \leq x$ hold.

We depict discrete preorders as simply a collection of points:

Example 1.28 (Codiscrete preorders). From every set we may also construct its codiscrete preorder $(X, \leq)$ by equipping it with the total binary relation $X \times X \subseteq X \times X$. This is a very trivial structure: it means that for *all* x and y in X we have $x \leq y$ (and hence also $y \leq x$).

Example 1.29 (Booleans). The booleans $\mathbb{B} = \{\texttt{false}, \texttt{true}\}$ form a preorder with $\texttt{false} \leq \texttt{true}$.

Remark 1.30 (Partial orders are skeletal preorders).   A preorder is a *partial order* if we additionally have that

(c) $x \cong y$ implies $x = y$.

In category theory terminology, the requirement that $x \cong y$ implies $x = y$ is known as *skeletality*, so partial orders are *skeletal preorders*. For short, we also use the term *poset*, a contraction of partially ordered set.

The difference between preorders and partial orders is rather minor. A partial order already is a preorder, and every preorder can be made into a partial order by equating any two elements x, y for which $x \cong y$, i.e. for which $x \leq y$ and $y \leq x$.

For example, any discrete preorder is already a partial order, while any codiscrete preorder simply becomes the unique partial order on a one-element set.

We have already introduced a few examples of preorders using Hasse diagrams. It will be convenient to continue to do this, so let us be a bit more formal about what we mean. First, we need to define a graph.

Definition 1.31.   A *graph* $G = (V, A, s, t)$ consists of a set V whose elements are called *vertices*, a set A whose elements are called *arrows*, and two functions $s, t \colon A \to V$ known as the *source* and *target* functions respectively. Given $a \in A$ with $s(a) = v$ and $t(a) = w$, we say that a is an arrow from v to w.

By a *path* in G we mean any sequence of arrows such that the target of one arrow is the source of the next. This includes sequences of length 1, which are just arrows $a \in A$ in G, and sequences of length 0, which just start and end at the same vertex v, without traversing any arrows.

Example 1.32.   Here is a picture of a graph:

It has $V = \{1, 2, 3, 4\}$ and $A = \{a, b, c, d, e\}$. The source and target functions, $s, t \colon A \to V$ are given by the following partially filled-in tables (see Exercise 1.33):

Arrow a	source $s(a) \in V$	target $t(a) \in V$
a	1	?
b	1	3
c	?	?
d	?	?
e	?	?

There are no paths from 4 to 3, but there is one path from 4 to 4, namely the path of length 0. There are infinitely many paths $1 \to 2$ because one can loop and loop and loop through d as many times as one pleases.

Exercise 1.33. Copy and complete the table from Example 1.32. ◇

Remark 1.34. From every graph we can get a preorder. Indeed, a Hasse diagram is a graph $G = (V, A, s, t)$ that gives a *presentation* of a preorder $(P, \leq)$. The elements of P are the vertices V in G, and the order $\leq$ is given by $v \leq w$ iff there is a path $v \to w$. For any vertex v, there is always a path $v \to v$, and this translates into the reflexivity law from Definition 1.25. The fact that paths $u \to v$ and $v \to w$ can be concatenated to a path $u \to w$ translates into the transitivity law.

Exercise 1.35. What preorder relation $(P, \leq)$ is depicted by the graph G in Example 1.32? That is, write down the elements of P and write down every pair (p_1, p_2) for which $p_1 \leq p_2$. ◇

Exercise 1.36. Does a collection of points, like the one in Example 1.27, count as a Hasse diagram? ◇

Exercise 1.37. Let X be the set of partitions of $\{\bullet, \circ, *\}$; it has five elements and an order by coarseness, as shown in the Hasse diagram Eq. (1.3). Write down every pair (x, y) of elements in X such that $x \leq y$. There should be 12. ◇

Remark 1.38. In Example 1.30 we discussed partial orders – preorders with the property that whenever two elements are equivalent, they are the same – and then said that this property is fairly inconsequential: any preorder can be converted to a partial order that's "equivalent" category-theoretically. A partial order is like a preorder with a fancy haircut: some mathematicians might not even notice it.

However, there are other types of preorders that are more special and noticeable. For example, a *total order* has the following additional property:

(d) for all x, y, either $x \leq y$ or $y \leq x$.

We say two elements x, y of a preorder are *comparable* if either $x \leq y$ or $y \leq x$, so a total order is a preorder where *every* two elements are comparable.

Exercise 1.39. Is it correct to say that a discrete preorder is one where *no* two elements are comparable? ◇

Example 1.40 (Natural numbers). The natural numbers $\mathbb{N} := \{0, 1, 2, 3, \ldots\}$ are a preorder with the order given by the usual size ordering, e.g. $0 \leq 1$ and $5 \leq 100$. This is a total order: either $m \leq n$ or $n \leq m$ for all m, n. One can see that its Hasse diagram looks like a line:

$$\underset{0}{\bullet} \longrightarrow \underset{1}{\bullet} \longrightarrow \underset{2}{\bullet} \longrightarrow \underset{3}{\bullet} \longrightarrow \cdots$$

What made Eq. (1.3) not look like a line is that there are non-comparable elements a and b – namely all those in the middle row – which satisfy neither $a \leq b$ nor $b \leq a$.

Note that, for any set S, there are many different ways of assigning an order to S. Indeed, for the set $\mathbb{N}$, we could also use the discrete ordering: only write $n \leq m$ if $n = m$. Another ordering is the reverse ordering, such as $5 \leq 3$ and $3 \leq 2$, like how golf is scored (5 is worse than 3).

Yet another ordering on $\mathbb{N}$ is given by division: we say that $n \leq m$ if n divides into m without remainder. In this ordering $2 \leq 4$, for example, but $2 \not\leq 3$, since there is a remainder when 2 is divided into 3.

Exercise 1.41. Write down the numbers $1, 2, \ldots, 10$ and draw an arrow $a \to b$ if a divides perfectly into b. Is it a total order?   ◇

Example 1.42 (Real numbers). The real numbers $\mathbb{R}$ also form a preorder with the "usual ordering," e.g. $-500 \leq -499 \leq 0 \leq \sqrt{2} \leq 100/3$.

Exercise 1.43. Is the usual $\leq$ ordering on the set $\mathbb{R}$ of real numbers a total order?   ◇

Example 1.44 (Partition from preorder). Given a preorder, i.e. a preordered set $(P, \leq)$, we defined the notion of equivalence of elements, denoted $x \cong y$, to mean $x \leq y$ and $y \leq x$. This is an equivalence relation, so it induces a partition on P. (The phrase "A induces B" means that we have an automatic way to turn an A into a B. In this case, we're saying that we have an automatic way to turn equivalence relations into partitions, which we do; see Proposition 1.14.)

For example, the preorder whose Hasse diagram is drawn on the left corresponds to the partition drawn on the right.

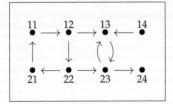   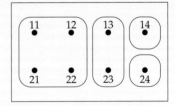

Example 1.45 (Power set). Given a set X, the set of subsets of X is known as the *power set* of X; we denote it $\mathsf{P}(X)$. The power set can naturally be given an order by inclusion of subsets (and from now on, whenever we speak of the power set as an ordered set, this is the order we mean).

For example, taking $X = \{0, 1, 2\}$, we depict P(X) as

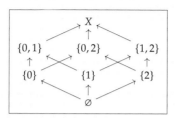

See the cube? The Hasse diagram for the power set of a finite set, say P$\{1, 2, \dots, n\}$,[7] always looks like a cube of dimension n.

Exercise 1.46. Draw the Hasse diagrams for P($\varnothing$), P$\{1\}$, and P$\{1, 2\}$. ◊

Example 1.47 (Partitions). We talked about getting a partition from a preorder; now let's think about how we might order the set Prt(A) of *all partitions* of A, for some set A. In fact, we have done this before in Eq. (1.3). Namely, we order partitions by fineness: a partition P is *finer* than a partition Q if, for every part $p \in P$, there is a part $q \in Q$ such that $A_p \subseteq A_q$. We could also say that Q is *coarser* than P.

Recall from Example 1.21 that partitions on A can be thought of as surjective functions out of A. Then $f : A \twoheadrightarrow P$ is finer than $g : A \twoheadrightarrow Q$ if there is a function $h : P \to Q$ such that $f \, \mathring{,} \, h = g$.

Exercise 1.48. For any set S there is a coarsest partition, having just one part. What surjective function does it correspond to?

There is also a finest partition, where everything is in its own part. What surjective function does it correspond to? ◊

Example 1.49 (Upper sets). Given a preorder $(P, \leq)$, an *upper set* in P is a subset U of P satisfying the condition that if $p \in U$ and $p \leq q$, then $q \in U$. "If p is an element then so is anything bigger." Write U(P) for the set of upper sets in P. We can give the set U an order by letting $U \leq V$ if U is contained in V.

For example, if $(\mathbb{B}, \leq)$ is the booleans (Example 1.29), then its preorder of upper sets U($\mathbb{B}$) is

---

[7] Note that we omit the parentheses here, writing PX instead of P(X); throughout this book we will omit parentheses if we judge the presentation is cleaner and it is unlikely to cause confusion.

The subset {false} ⊆ 𝔹 is not an upper set, because false ≤ true and true ∉ {false}.

Exercise 1.50.   Prove that the preorder of upper sets on a discrete preorder (see Example 1.27) on a set X is simply the power set $P(X)$.   ◇

Example 1.51 (Product preorder). Given preorders $(P, \leq)$ and $(Q, \leq)$, we may define a preorder structure on the product set $P \times Q$ by setting $(p, q) \leq (p', q')$ if and only if $p \leq p'$ and $q \leq q'$. We call this the *product preorder*. This is a basic example of a more general construction known as the product of categories.

Exercise 1.52.   Draw the Hasse diagram for the product of the two preorders drawn below:

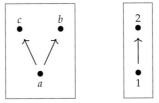

For bonus points, compute the upper set preorder on the result.   ◇

Example 1.53 (Opposite preorder). Given a preorder $(P, \leq)$, we may define the opposite preorder $(P, \leq^{op})$ to have the same set of elements, but with $p \leq^{op} q$ if and only if $q \leq p$.

1.2.3   Monotone Maps

We have said that the categorical perspective emphasizes relationships between things. For example, a preorder is a setting – or world – in which we have one sort of relationship, $\leq$, and any two objects may be, or may not be, so-related. Jumping up a level, the categorical perspective emphasizes that preorders themselves – each a miniature world composed of many relationships – can be related to one another.

The most important sort of relationship between preorders is called a *monotone map*. These are functions that preserve preorder relations – in some sense mappings that respect $\leq$ – and are hence considered the right notion of *structure-preserving map* for preorders.

Definition 1.54.   A *monotone map* between preorders $(A, \leq_A)$ and $(B, \leq_B)$ is a function $f : A \to B$ such that, for all elements $x, y \in A$, if $x \leq_A y$ then $f(x) \leq_B f(y)$.

A monotone map $A \to B$ between two preorders associates to each element of preorder A an element of the preorder B. We depict this by drawing a dotted arrow from each element $x \in A$ to its image $f(x) \in B$. Note that the order must be preserved in order to count as a valid monotone map, so if element x is above element y in the left-hand preorder A, then the image $f(x)$ will be above the image $f(y)$ in the right-hand preorder.

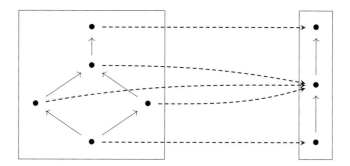

Example 1.55. Let $\mathbb{B}$ and $\mathbb{N}$ be the preorders of booleans from Example 1.29 and $\mathbb{N}$ be the preorder of natural numbers from Example 1.40. The map $\mathbb{B} \to \mathbb{N}$ sending `false` to 17 and `true` to 24 is a monotone map, because it preserves order.

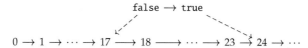

Example 1.56 (The tree of life). Consider the set of all animal classifications, for example "tiger," "mammal," "sapiens," "carnivore," etc. These are ordered by specificity: since "tiger" is a type of "mammal," we write tiger $\leq$ mammal. The result is a preorder, which in fact forms a tree, often called the tree of life. At the top of the following diagram we see a small part of it:

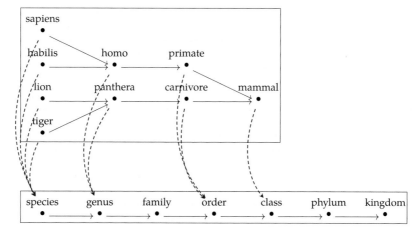

At the bottom we see the hierarchical structure as a preorder. The dashed arrows show a monotone map, call it F, from the classifications to the hierarchy. It is monotone because it preserves order: whenever there is a path $x \to y$ upstairs, there is a path $F(x) \to F(y)$ downstairs.

Example 1.57. Given a finite set X, recall the power set $\mathsf{P}(X)$ and its natural order relation from Example 1.45. The map $|\cdot| : \mathsf{P}(X) \to \mathbb{N}$ sending each subset S to its number of elements $|S|$, also called its *cardinality*, is a monotone map.

Exercise 1.58. Let $X = \{0, 1, 2\}$.

1. Draw the Hasse diagram for $\mathsf{P}(X)$.
2. Draw the Hasse diagram for the preorder $0 \le 1 \le 2 \le 3$.
3. Draw the cardinality map $|\cdot|$ from Example 1.57 as dashed lines between them.   ◇

Example 1.59. Recall the notion of upper set from Example 1.49. Given a preorder $(P, \le)$, the map $\mathsf{U}(P) \to \mathsf{P}(P)$ sending each upper set of $(P, \le)$ to itself – considered as a subset of P – is a monotone map.

Exercise 1.60. Consider the preorder $\mathbb{B}$. The Hasse diagram for $\mathsf{U}(\mathbb{B})$ was drawn in Example 1.49, and you drew the Hasse diagram for $\mathsf{P}(\mathbb{B})$ in Exercise 1.46. Now draw the monotone map between them, as described in Example 1.59.   ◇

Exercise 1.61. Let $(P, \le)$ be a preorder, and recall the notion of opposite preorder from Example 1.53.

1. Show that the set $\uparrow p := \{p' \in P \mid p \le p'\}$ is an upper set, for any $p \in P$.
2. Show that this construction defines a monotone map $\uparrow : P^{\mathrm{op}} \to \mathsf{U}(P)$.
3. Show that $p \le p'$ in P if and only if $\uparrow(p') \subseteq \uparrow(p)$.
4. Draw a picture of the map $\uparrow$ in the case where P is the preorder $(b \ge a \le c)$ from Exercise 1.52.

This is known as the *Yoneda lemma* for preorders. The if and only if condition proved in part 3 implies that, up to equivalence, knowing an element p is the same as knowing its upper set P – that is, knowing its web of relationships with the other elements of the preorder. The general Yoneda lemma is a powerful tool in category theory, and a fascinating philosophical idea besides.   ◇

Exercise 1.62. As you know, a monotone map $f : (P, \le_P) \to (Q, \le_Q)$ consists of a function $f : P \to Q$ that satisfies a "monotonicity" property. Show that when $(P, \le_P)$ is a discrete preorder, then *every* function $P \to Q$ satisfies the monotonicity property, regardless of the order $\le_Q$.   ◇

Example 1.63. Recall from Example 1.47 that given a set X we define $\mathsf{Prt}(X)$ to be the set of partitions on X, and that a partition may be defined using a surjective function $s \colon X \twoheadrightarrow P$ for some set P.

Any surjective function $f \colon X \twoheadrightarrow Y$ induces a monotone map $f^* \colon \mathsf{Prt}(Y) \to \mathsf{Prt}(X)$, going "backwards." It is defined by sending a partition $s \colon Y \twoheadrightarrow P$ to the composite $f \mathbin{\mathring{,}} s \colon X \twoheadrightarrow P$.[8]

Exercise 1.64. Choose two sets X and Y with at least three elements each and choose a surjective, non-identity function $f \colon X \twoheadrightarrow Y$ between them. Write down two different partitions P and Q of Y, and then find $f^*(P)$ and $f^*(Q)$. ⬦

The following proposition, Proposition 1.65, is straightforward to check. Recall the definition of the identity function from Example 1.18 and the definition of composition from Definition 1.23.

Proposition 1.65. For any preorder $(P, \leq_P)$, the identity function is monotone.

If $(Q, \leq_Q)$ and $(R, \leq_R)$ are preorders and $f \colon P \to Q$ and $g \colon Q \to R$ are monotone, then $(f \mathbin{\mathring{,}} g) \colon P \to R$ is also monotone.

Exercise 1.66. Check the two claims made in Proposition 1.65. ⬦

Example 1.67. Recall again the definition of opposite preorder from Example 1.53. The identity function $\mathrm{id}_P \colon P \to P$ is a monotone map $(P, \leq) \to (P, \leq^{\mathrm{op}})$ if and only if for all $p, q \in P$ we have $q \leq p$ whenever $p \leq q$. For historical reasons connected to linear algebra, when this is true, we call $(P, \leq)$ a *dagger preorder*.

But in fact, we have seen dagger preorders before in another guise. Indeed, if $(P, \leq)$ is a dagger preorder, then the relation $\leq$ is symmetric: $p \leq q$ if and only if $q \leq p$, and it is also reflexive and transitive by definition of preorder. So in fact $\leq$ is an equivalence relation (Definition 1.13).

Exercise 1.68. Recall the notion of skeletal preorders (Remark 1.30) and discrete preorders (Example 1.27). Show that a skeletal dagger preorder is just a discrete preorder, and hence can be identified with a set. ⬦

Remark 1.69. We say that an A "can be identified with" a B when any A gives us a unique B and any B gives us a unique A, and both round-trips – from an A to a B and back to an A, or from a B to an A and back to a B – return us where we started. For example, any discrete preorder $(P, \leq)$ has an underlying set P, and any set P can be made into a discrete preorder ($p_1 \leq p_2$ iff $p_1 = p_2$), and the round-trips return us where we started. So what's the difference? It's like the notion of *object-permanence*

---

[8] We shall later see that any function $f \colon X \to Y$, not necessarily surjective, induces a monotone map $f^* \colon \mathsf{Prt}(Y) \to \mathsf{Prt}(X)$, but it involves an extra step. See Section 1.4.2.

from child development jargon: we can recognize "the same chair, just moved from one room to another." A chair in the room of sets can be moved to a chair in the room of preorders. The lighting is different but the chair is the same.

Eventually, we will be able to understand this notion in terms of *equivalence of categories*, which are related to isomorphisms, which we will explore next in Definition 1.70.

Definition 1.70. Let $(P, \leq_P)$ and $(Q, \leq_Q)$ be preorders. A monotone function $f \colon P \to Q$ is called an *isomorphism* if there exists a monotone function $g \colon Q \to P$ such that $f \,\mathbin{\fatsemi}\, g = \mathrm{id}_P$ and $g \,\mathbin{\fatsemi}\, f = \mathrm{id}_Q$. This means that, for any $p \in P$ and $q \in Q$, we have

$$p = g(f(p)) \quad \text{and} \quad q = f(g(q)).$$

We refer to g as the *inverse* of f, and vice versa: f is the inverse of g.

If there is an isomorphism $P \to Q$, we say that P and Q are *isomorphic*.

An isomorphism between preorders is basically just a relabeling of the elements.

Example 1.71. Here are the Hasse diagrams for three preorders P, Q, and R, all of which are isomorphic:

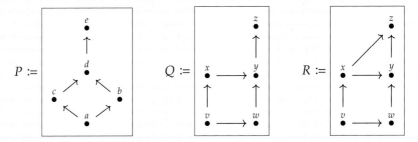

The map $f \colon P \to Q$ given by $f(a) = v$, $f(b) = w$, $f(c) = x$, $f(d) = y$, and $f(e) = z$ has an inverse.

In fact Q and R are the same preorder. One may be confused by the fact that there is an arrow $x \to z$ in the Hasse diagram for R and not one in Q, but in fact this arrow is superfluous. By the transitivity property of preorders (Definition 1.25), since $x \leq y$ and $y \leq z$, we must have $x \leq z$, whether it is drawn or not. Similarly, we could have drawn an arrow $v \to y$ in either Q or R and it would not have changed the preorder.

Recall the preorder $\mathbb{B} = \{\mathtt{false}, \mathtt{true}\}$, where $\mathtt{false} \leq \mathtt{true}$. As simple as this preorder is, it is also one of the most important.

Exercise 1.72. Show that the map Φ from Section 1.1.1, which was roughly given by "Is • connected to ∗?" is a monotone map $\mathrm{Prt}(\{∗, •, \circ\}) \to \mathbb{B}$; see also Eq. (1.3). $\diamond$

Proposition 1.73. Let P be a preorder. Monotone maps $P \to \mathbb{B}$ are in one-to-one correspondence with upper sets of P.

Proof. Let $f: P \rightarrow \mathbb{B}$ be a monotone map. We will show that the subset $f^{-1}(\texttt{true}) \subseteq P$ is an upper set. Suppose $p \in f^{-1}(\texttt{true})$ and $p \leq q$; then $\texttt{true} = f(p) \leq f(q)$. But in $\mathbb{B}$, if $\texttt{true} \leq f(q)$ then $\texttt{true} = f(q)$. This implies $q \in f^{-1}(\texttt{true})$ and thus shows that $f^{-1}(\texttt{true})$ is an upper set.

Conversely, if U is an upper set in P, define $f_U: P \rightarrow \mathbb{B}$ such that $f_U(p) = \texttt{true}$ when $p \in U$, and $f_U(p) = \texttt{false}$ when $p \notin U$. This is a monotone map, because if $p \leq q$, then either $p \in U$, so $q \in U$ and $f_U(p) = \texttt{true} = f(q)$, or $p \notin U$, so $f_U(p) = \texttt{false} \leq f_U(q)$.

These two constructions are mutually inverse, and hence prove the proposition. $\square$

Exercise 1.74 (Pullback map). Let P and Q be preorders, and $f: P \rightarrow Q$ be a monotone map. Then we can define a monotone map $f^*: \mathsf{U}(Q) \rightarrow \mathsf{U}(P)$ sending an upper set $U \subseteq Q$ to the upper set $f^{-1}(U) \subseteq P$. We call this the *pullback along f*.

Viewing upper sets as monotone maps to $\mathbb{B}$ as in Proposition 1.73, the pullback can be understood in terms of composition. Indeed, show that f^* is defined by taking $u: Q \rightarrow \mathbb{B}$ to $(f \,\mathring{\,}\, u): P \rightarrow \mathbb{B}$. $\diamond$

1.3   Meets and Joins

As we have said, a preorder is a set P endowed with an order $\leq$ relating the elements. With respect to this order, certain elements of P may have distinctive characterizations, either absolutely or in relation to other elements. We have discussed joins before, but we discuss them again now that we have built up some formalism.

1.3.1   Definition and Basic Examples

Consider the preorder $(\mathbb{R}, \leq)$ of real numbers ordered in the usual way. The subset $\mathbb{N} \subseteq \mathbb{R}$ has many lower bounds, namely -1.5 is a lower bound: every element of $\mathbb{N}$ is bigger than -1.5. But within all lower bounds for $\mathbb{N} \subseteq \mathbb{R}$, one is distinctive: a *greatest lower bound* – also called a *meet* – namely 0. It is a lower bound, and there is no lower bound for $\mathbb{N}$ that is above it. However, the set $\mathbb{N} \subseteq \mathbb{R}$ has no upper bound, and certainly no least upper bound – which would be called a *join*. On the other hand, the set

$$\left\{ \frac{1}{n+1} \,\middle|\, n \in \mathbb{N} \right\} = \left\{ 1, \frac{1}{2}, \frac{1}{3}, \frac{1}{4}, \dots \right\} \subseteq \mathbb{R}$$

has both a greatest lower bound (meet), namely 0, and a least upper bound (join), namely 1.

These notions will have correlates in category theory, called limits and colimits, which we will discuss in Chapter 3. More generally, we say these distinctive characterizations are *universal properties*, since, for example, a greatest lower bound is greatest among *all* lower bounds. For now, however, we simply want to make the definition of greatest lower bounds and least upper bounds, called meets and joins, precise.

Exercise 1.75.

1. Why is 0 a lower bound for $\{\frac{1}{n+1} \mid n \in \mathbb{N}\} \subseteq \mathbb{R}$?
2. Why is 0 a *greatest* lower bound (meet)? $\diamond$

Definition 1.76.   Let $(P, \leq)$ be a preorder, and let $A \subseteq P$ be a subset. We say that an element $p \in P$ is a *meet* of A if

(a) for all $a \in A$, we have $p \leq a$,
(b) for all q such that $q \leq a$ for all $a \in A$, we have that $q \leq p$.

We write $p = \bigwedge A$, $p = \bigwedge_{a \in A} a$, or, if the dummy variable a is clear from context, just $p = \bigwedge_A a$. If A just consists of two elements, say $A = \{a, b\}$, we can denote $\bigwedge A$ simply by $a \wedge b$.

Similarly, we say that p is a *join* of A if

(a) for all $a \in A$ we have $a \leq p$,
(b) for all q such that $a \leq q$ for all $a \in A$, we have that $p \leq q$.

We write $p = \bigvee A$ or $p = \bigvee_{a \in A} a$, or when $A = \{a, b\}$ we may simply write $p = a \vee b$.

Remark 1.77.   In Definition 1.76, we committed a seemingly egregious abuse of notation. We shall see next in Example 1.79 that there could be two different meets of $A \subseteq P$, say $p = \bigwedge A$ and $q = \bigwedge A$ with $p \neq q$, which does not make sense if $p \neq q$!

But in fact, as we use the symbol $\bigwedge A$, this abuse won't matter because any two meets p, q are automatically isomorphic: the very definition of meet forces both $p \leq q$ and $q \leq p$, and thus we have $p \cong q$. So, for any $x \in P$, we have $p \leq x$ iff $q \leq x$ and $x \leq p$ iff $x \leq q$. Thus as long as we are only interested in elements of P based on their relationships to other elements (and in category theory, this is the case: we should only care about things based on how they interact with other things, rather than on some sort of "internal essence"), the distinction between p and q will never matter.

This foreshadows a major theme of – as well as standard abuse of notation in – category theory, where any two things defined by the same universal property are automatically equivalent in a way known as "unique up to unique isomorphism"; this means that we generally do not run into trouble if we pretend they are equal. We'll pick up this theme of "the" vs. "a" again in Remark 3.70.

Example 1.78 (Meets or joins may not exist).   Note that, in an arbitrary preorder $(P, \leq)$, a subset A need not have a meet or a join. Consider the three-element set $P = \{p, q, r\}$ with the discrete ordering. The set $A = \{p, q\}$ does not have a join in P because if x was a join, we would need $p \leq x$ and $q \leq x$, and there is no such element x.

Example 1.79 (Multiple meets or joins may exist).   It may also be the case that a subset A has more than one meet or join. Here is an example.

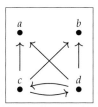

Let A be the subset $\{a, b\}$ in the preorder specified by this Hasse diagram. Then both c and d are meets of A: any element less than both a and b is also less than c, and also less than d. Note that, as in Remark 1.77, $c \leq d$ and $d \leq c$, so $c \cong d$. Such will always the case when there is more than one meet: any two meets of the same subset will be isomorphic.

Exercise 1.80.   Let $(P, \leq)$ be a preorder and $p \in P$ an element. Consider the set $A = \{p\}$ with one element.

1. Show that $\bigwedge A \cong p$.
2. Show that if P is in fact a partial order, then $\bigwedge A = p$.
3. Are the analogous facts true when $\bigwedge$ is replaced by $\bigvee$?   ◇

Example 1.81.   In any partial order P, we have $p \vee p = p \wedge p = p$. The reason is that our notation says $p \vee p$ means $\bigvee\{p, p\}$. But $\{p, p\} = \{p\}$ (see Section 1.2.1), so $p \vee p = p$ by Exercise 1.80.

Example 1.82.   In a power set $\mathsf{P}(X)$, the meet of a collection of subsets, say $A, B \subseteq X$, is their intersection $A \wedge B = A \cap B$, while the join is their union, $A \vee B = A \cup B$.

Perhaps this justifies the terminology: the joining of two sets is their union, the meeting of two sets is their intersection.

Example 1.83.   In the booleans $\mathbb{B} = \{\text{false}, \text{true}\}$ (Example 1.29), the meet of any two elements is given by AND and the join of any two elements is given by OR (recall Exercise 1.4).

Example 1.84. In a total order, the meet of a set is its infimum, while the join of a set is its supremum. Note that $\mathbb{B}$ is a total order, and this generalizes Example 1.83.

Exercise 1.85. Recall the division ordering on $\mathbb{N}$ from Example 1.40: we write $n|m$ if n divides perfectly into m. The meet of any two numbers in this preorder has a common name, which you may have learned when you were around 10 years old; what is it? Similarly the join of any two numbers has a common name; what is it?   ◇

Proposition 1.86. Suppose $(P, \leq)$ is a preorder and $A \subseteq B \subseteq P$ are subsets that have meets. Then $\bigwedge B \leq \bigwedge A$.
  Similarly, if A and B have joins, then $\bigvee A \leq \bigvee B$.

Proof. Let $m = \bigwedge A$ and $n = \bigwedge B$. Then for any $a \in A$ we also have $a \in B$, so $n \leq a$ because n is a lower bound for B. Thus n is also a lower bound for A and hence $n \leq m$, because m is A's greatest lower bound. The second claim is proved similarly.   □

1.3.2   Back to Observations and Generative Effects

In his thesis [Ada17], Adam thinks of monotone maps as observations. A monotone map $\Phi \colon P \to Q$ is a phenomenon (we might say "feature") of P as observed by Q. Adam defines the *generative effect* of such a map Φ to be its failure to preserve joins (or more generally, for categories, its failure to preserve colimits).

Definition 1.87. We say that a monotone map $f \colon P \to Q$ *preserves meets* if $f(a \wedge b) \cong f(a) \wedge f(b)$ for all $a, b \in P$. We similarly say f *preserves joins* if $f(a \vee b) \cong f(a) \vee f(b)$ for all $a, b \in P$.

Definition 1.88. We say that a monotone map $f \colon P \to Q$ *has a generative effect* if there exist elements $a, b \in P$ such that

$$f(a) \vee f(b) \not\cong f(a \vee b).$$

In Definition 1.88, if we think of Φ as an observation or measurement of the systems a and b, then the left-hand side $f(a) \vee f(b)$ may be interpreted as the combination of the observation of a with the observation of b. On the other hand, the right-hand side $f(a \vee b)$ is the observation of the combined system $a \vee b$. The inequality implies that we see something when we observe the combined system that we could not expect by merely combining our observations of the pieces. That is, that there are generative effects from the interconnection of the two systems.

Exercise 1.89. In Definition 1.88, we defined generativity of f as the inequality $f(a \vee b) \neq f(a) \vee f(b)$, but in the subsequent text we seemed to imply there would be not just a difference, but *more stuff* in $f(a \vee b)$ than in $f(a) \vee f(b)$.

Prove that for any monotone map $f : P \to Q$, if $a, b \in P$ have a join and $f(a), f(b) \in Q$ have a join, then indeed $f(a) \vee f(b) \leq f(a \vee b)$.   ◇

In his work on generative effects, Adam restricts his attention to generative maps that preserve meets (but do not preserve joins). The preservation of meets implies that the map Φ behaves well when restricting to subsystems, even though it can throw up surprises when joining systems.

This discussion naturally leads into Galois connections, which are pairs of monotone maps between preorders, one of which preserves all joins and the other of which preserves all meets.

1.4   Galois Connections

The preservation of meets and joins, and in particular issues concerning generative effects, is tightly related to the theory of *Galois connections*, which is a special case of a more general theory we will discuss later, namely that of *adjunctions*. We will use some adjunction terminology when describing Galois connections.

1.4.1   Definition and Examples of Galois Connections

Galois connections between preorders were first considered by Évariste Galois – who didn't call them by that name – in the context of a connection he found between "field extensions" and "automorphism groups." We will not discuss this further, but the idea is that, given two preorders P and Q, a Galois connection is a pair of maps back and forth – from P to Q and from Q to P – with certain properties, which make it like a relaxed version of isomorphisms. To be a bit more precise, preorder isomorphisms are examples of Galois connections, but Galois connections need not be preorder isomorphisms.

Definition 1.90. A *Galois connection* between preorders P and Q is a pair of monotone maps $f : P \to Q$ and $g : Q \to P$ such that

$$f(p) \leq q \quad \text{if and only if} \quad p \leq g(q). \tag{1.6}$$

We say that f is the *left adjoint* and g is the *right adjoint* of the Galois connection.

Example 1.91. Consider the map $(3 \times -) : \mathbb{Z} \to \mathbb{R}$ which sends $x \in \mathbb{Z}$ to $3x$, which we can consider as a real number $3x \in \mathbb{Z} \subseteq \mathbb{R}$. Let's find a left adjoint for the map $(3 \times -)$.

Write $\lceil z \rceil$ for the smallest natural number above $z \in \mathbb{R}$, and write $\lfloor z \rfloor$ for the largest integer below $z \in \mathbb{R}$, e.g. $\lceil 3.14 \rceil = 4$ and $\lfloor 3.14 \rfloor = 3$.[9] As the left adjoint $\mathbb{R} \to \mathbb{Z}$, let's see if $\lceil -/3 \rceil$ works.

It is easily checked that

$$\lceil x/3 \rceil \leq y \text{ if and only if } x \leq 3y.$$

Success! Thus we have a Galois connection between $\lceil -/3 \rceil$ and $(3 \times -)$.

Exercise 1.92. In Example 1.91 we found a left adjoint for the monotone map $(3 \times -) \colon \mathbb{Z} \to \mathbb{R}$. Now find a right adjoint for the same map, and show it is correct. ◇

Exercise 1.93. Consider the preorder $P = Q = \underline{3}$.

1. Let f, g be the monotone maps shown below:

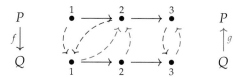

   Is it the case that f is left adjoint to g? Check that for each $1 \leq p, q \leq 3$, one has $f(p) \leq q$ iff $p \leq g(q)$.
2. Let f, g be the monotone maps shown below:

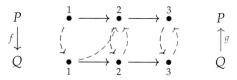

   Is it the case that f is left adjoint to g? ◇

Remark 1.94. The diagrams in Exercise 1.93 suggest the following idea. If P and Q are total orders and $f \colon P \to Q$ and $g \colon Q \to P$ are drawn with arrows bending counterclockwise, then f is left adjoint to g iff the arrows do not cross. With a little bit of thought, this can be formalized. We think this is a pretty neat way of visualizing Galois connections between total orders!

Exercise 1.95.

1. Does $\lceil -/3 \rceil$ have a left adjoint $L \colon \mathbb{Z} \to \mathbb{R}$?
2. If not, why? If so, does its left adjoint have a left adjoint? ◇

---

[9] By "above" and "below," we mean *greater than or equal to* or *less than or equal to*; the latter being a mouthful. Anyway, $\lfloor 3 \rfloor = 3 = \lceil 3 \rceil$.

1.4.2   Back to Partitions

Recall from Example 1.47 that we can understand the set $\mathsf{Prt}(S)$ of partitions on a set S in terms of surjective functions out of S.

Suppose we are given any function $g \colon S \to T$. We will show that this function g induces a Galois connection $g_! \colon \mathsf{Prt}(S) \leftrightarrows \mathsf{Prt}(T) \colon g^*$ between the preorder of S-partitions and the preorder of T-partitions. The way you might explain it to a seasoned category theorist is:

The left adjoint is given by taking any surjection out of S and pushing out along g to get a surjection out of T. The right adjoint is given by taking any surjection out of T, composing with g to get a function out of S, and then taking the epi-mono factorization to get a surjection out of S.

By the end of this book, the reader will understand pushouts and epi-mono factorizations, so he or she will be able to make sense of the above statement. But for now we will explain the process in more down-to-earth terms.

Start with $g \colon S \to T$; we first want to understand $g_! \colon \mathsf{Prt}(S) \to \mathsf{Prt}(T)$. So start with a partition $\sim_S$ of S. To begin the process of obtaining a partition $\sim_T$ on T, say that two elements $t_1, t_2 \in T$ are in the same part, $t_1 \sim_T t_2$, if there exist $s_1, s_2 \in S$ such that $s_1 \sim_S s_2$ and $g(s_1) = t_1$ and $g(s_2) = t_2$. However, the result of doing so will not necessarily be transitive – you may get $t_1 \sim_T t_2$ and $t_2 \sim_T t_3$ without $t_1 \overset{?}{\sim_T} t_3$ – and partitions must be transitive. So complete the process by just adding in the missing pieces (take the transitive closure). The result is $g_!(\sim_S) := \sim_T$.

Again starting with g, we want to get the right adjoint $g^* \colon \mathsf{Prt}(T) \to \mathsf{Prt}(S)$. So start with a partition $\sim_T$ of T. Get a partition $\sim_S$ on S by saying that $s_1 \sim_S s_2$ iff $g(s_1) \sim_T g(s_2)$. The result is $g^*(\sim_T) := \sim_S$.

Example 1.96. Let $S = \{1, 2, 3, 4\}$, $T = \{12, 3, 4\}$, and define $g \colon S \to T$ by $g(1) := g(2) := 12$, $g(3) := 3$, and $g(4) := 4$. The partition shown left below is translated by $g_!$ to the partition shown on the right.

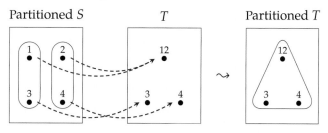

Exercise 1.97. There are 15 different partitions of a set with four elements. Choose four different ones and, for each one, call it $c\colon S \twoheadrightarrow P$, find $g_!(c)$, where S, T, and $g\colon S \to T$ are the same as they were in Example 1.96.   ◇

Example 1.98. Let S, T be as below, and let $g\colon S \to T$ be the function shown in blue. Here is a picture of how g^* takes a partition on T and "pulls it back" to a partition on S:

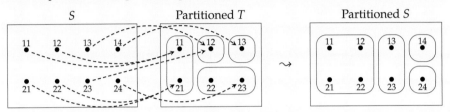

Exercise 1.99. There are five partitions possible on a set with three elements, say $T = \{12, 3, 4\}$. Using the same S and $g\colon S \to T$ as in Example 1.96, determine the partition $g^*(c)$ on S for each of the five partitions $c\colon T \twoheadrightarrow P$.   ◇

To check that, for any function $g\colon S \to T$, the monotone map $g_!\colon \mathrm{Prt}(S) \to \mathrm{Prt}(T)$ really is left adjoint to $g^*\colon \mathrm{Prt}(T) \to \mathrm{Prt}(S)$ would take too much time for this sketch. But the following exercise gives some evidence.

Exercise 1.100. Let S, T, and $g\colon S \to T$ be as in Example 1.96.

1. Choose a nontrivial partition $c\colon S \twoheadrightarrow P$ and let $g_!(c)$ be its push forward partition on T.
2. Choose any coarser partition $d\colon T \twoheadrightarrow P'$, i.e. where $g_!(c) \leq d$.
3. Choose any non-coarser partition $e\colon T \twoheadrightarrow Q$, i.e. where $g_!(c) \not\leq e$. (If you can't do this, revise your answer for #1.)
4. Find $g^*(d)$ and $g^*(e)$.
5. The adjunction formula Eq. (1.6) in this case says that since $g_!(c) \leq d$ and $g_!(c) \not\leq e$, we should have $c \leq g^*(d)$ and $c \not\leq g^*(e)$. Show that this is true.   ◇

1.4.3   Basic Theory of Galois Connections

Proposition 1.101. Suppose that $f\colon P \to Q$ and $g\colon Q \to P$ are monotone maps. The following are equivalent:

(a) f and g form a Galois connection where f is left adjoint to g,
(b) for every $p \in P$ and $q \in Q$ we have

$$p \leq g(f(p)) \qquad \text{and} \qquad f(g(q)) \leq q. \tag{1.7}$$

Proof. Suppose f is left adjoint to g. Take any $p \in P$, and let $q := f(p)$. By reflexivity, we have $f(p) \leq q$, so by Definition 1.90 of Galois connection we have $p \leq g(q)$, but this means $p \leq g(f(p))$. The proof that $f(g(q)) \leq q$ is similar.

Now suppose that Eq. (1.7) holds for all $p \in P$ and $q \in Q$. We want to show that $f(p) \leq q$ iff $p \leq g(q)$. Suppose $f(p) \leq q$; then since g is monotonic, $g(f(p)) \leq g(q)$, but $p \leq g(f(p))$ so $p \leq g(q)$. The proof that $p \leq g(q)$ implies $f(p) \leq q$ is similar. $\square$

Exercise 1.102. Complete the proof of Proposition 1.101 by showing that

1. if f is left adjoint to g then for any $q \in Q$, we have $f(g(q)) \leq q$,
2. if Eq. (1.7) holds, then $p \leq g(q)$ iff $f(p) \leq q$ holds, for all $p \in P$ and $q \in Q$. $\diamond$

If we replace $\leq$ with $=$ in Eq. (1.7), we get back the definition of isomorphism (Definition 1.70); this is why we said at the beginning of Section 1.4.1 that Galois connections are a kind of relaxed version of isomorphisms.

Exercise 1.103.

1. Show that if $f : P \to Q$ has a right adjoint g, then it is unique up to isomorphism. That means, for any other right adjoint g', we have $g(q) \cong g'(q)$ for all $q \in Q$.
2. Is the same true for left adjoints? That is, if $h : P \to Q$ has a left adjoint, is it necessarily unique up to isomorphism? $\diamond$

Proposition 1.104 (Right adjoints preserve meets). Let $f : P \to Q$ be left adjoint to $g : Q \to P$. Suppose that $A \subseteq Q$ is any subset, and let $g(A) := \{g(a) \mid a \in A\}$ be its image. Then if A has a meet $\bigwedge A \in Q$ then $g(A)$ has a meet $\bigwedge g(A)$ in P, and we have

$$g\left(\bigwedge A\right) \cong \bigwedge g(A).$$

That is, right adjoints preserve meets. Similarly, left adjoints preserve joins: if $A \subseteq P$ is any subset that has a join $\bigvee A \in P$, then $f(A)$ has a join $\bigvee f(A)$ in Q, and we have

$$f\left(\bigvee A\right) \cong \bigvee f(A).$$

Proof. Let $f : P \to Q$ and $g : Q \to P$ be adjoint monotone maps, with g right adjoint to f. Let $A \subseteq Q$ be any subset and let $m := \bigwedge A$ be its meet. Then since g is monotone $g(m) \leq g(a)$ for all $a \in A$, so $g(m)$ is a lower bound for the set $g(A)$. We will be done if we can show $g(m)$ is a greatest lower bound.

So take any other lower bound b for $g(A)$; that is, suppose that for all $a \in A$ we have $b \leq g(a)$ and we want to show that $b \leq g(m)$. Then by definition of g being a right adjoint (Definition 1.90), we also have $f(b) \leq a$. This means that $f(b)$ is a lower bound for A in Q. Since the meet m is the greatest lower bound, we have $f(b) \leq m$.

Once again using the Galois connection, $b \le g(m)$, proving that $g(m)$ is indeed the greatest lower bound for $g(A)$, as desired.

The second claim is proved similarly; see Exercise 1.105. $\qquad\square$

Exercise 1.105. Complete the proof of Proposition 1.104 by showing that left adjoints preserve joins. $\qquad\diamond$

Since left adjoints preserve joins, we know that they cannot have generative effects. In fact, we shall see in Theorem 1.108 that a monotone map does not have generative effects – i.e. it preserves joins – if and only if it is a left adjoint to some other monotone.

Example 1.106. Right adjoints need not preserve joins. Here is an example:

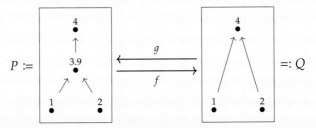

Let g be the map that preserves labels, and let f be the map that preserves labels as far as possible but with $f(3.9) := 4$. Both f and g are monotonic, and one can check that g is right adjoint to f (see Exercise 1.107). But g does not preserve joins because $1 \vee 2 = 4$ holds in Q, whereas $g(1) \vee g(2) = 1 \vee 2 = 3.9 \ne 4 = g(4)$ in P.

Exercise 1.107. To be sure that g really is right adjoint to f in Example 1.106, there are twelve tiny things to check; do so. That is, for every $p \in P$ and $q \in Q$, check that $f(p) \le q$ iff $p \le g(q)$. $\qquad\diamond$

Theorem 1.108 (Adjoint functor theorem for preorders). Suppose Q is a preorder that has all meets, and let P be any preorder. A monotone map $g: Q \to P$ preserves meets if and only if it is a right adjoint.

Similarly, if P has all joins and Q is any preorder, a monotone map $f: P \to Q$ preserves joins if and only if it is a left adjoint.

Proof. We will prove only the claim about meets; the claim about joins follows similarly.

We proved one direction in Proposition 1.104, namely that right adjoints preserve meets. For the other, suppose that g is a monotone map that preserves meets; we will construct a left adjoint f. We define our candidate $f: P \to Q$ on any $p \in P$ by

$$f(p) := \bigwedge\{q \in Q \mid p \le g(q)\}; \tag{1.8}$$

this meet is well defined because Q has all meets, but for f to really be a candidate, we need to show it is monotone. So suppose that $p \le p'$. Then $\{q' \in Q \mid p' \le g(q')\} \subseteq \{q \in Q \mid p \le g(q)\}$. By Proposition 1.86, this implies $f(p) \le f(p')$. Thus f is monotone.

By Proposition 1.104, it suffices to show that $p_0 \le g(f(p_0))$ and that $f(g(q_0)) \le q_0$ for all $p_0 \in P$ and $q_0 \in Q$. For the first, we have

$$p_0 \le \bigwedge \{g(q) \in P \mid p_0 \le g(q)\} \cong g\left(\bigwedge \{q \in Q \mid p_0 \le g(q)\}\right) = g(f(p_0)),$$

where the first inequality follows from the fact that if p_0 is below every element of a set, then it is below their meet, and the isomorphism is by definition of g preserving meets. For the second, we have

$$f(g(q_0)) = \bigwedge \{q \in Q \mid g(q_0) \le g(q)\} \le \bigwedge \{q_0\} = q_0,$$

where the first inequality follows from Proposition 1.86 since $\{q_0\} \subseteq \{q \in Q \mid g(q_0) \le g(q)\}$, and the fact that $\bigwedge \{q_0\} = q_0$. $\qquad \square$

Example 1.109. Let $f \colon A \to B$ be a function between sets. We can imagine A as a set of apples, B as a set of buckets, and f as putting each apple in a bucket.

Then we have the monotone map $f^* \colon \mathsf{P}(B) \to \mathsf{P}(A)$ that category theorists call "pullback along f." This map takes a subset $B' \subseteq B$ to its preimage $f^{-1}(B') \subseteq A$: that is, it takes a collection B' of buckets, and tells you all the apples that they contain in total. This operation is monotonic (more buckets means more apples) and it has both a left and a right adjoint. (There are two different adjunctions here involving f^*.)

The left adjoint $f_!(A)$ is given by the direct image: it maps a subset $A' \subseteq A$ to

$$f_!(A') := \{b \in B \mid \text{there exists } a \in A' \text{ such that } f(a) = b\}.$$

This map takes a set A' of apples, and tells you all the buckets that contain at least one of those apples.

The right adjoint f_* maps a subset $A' \subseteq A$ to

$$f_*(A') := \{b \in B \mid \text{for all } a \text{ such that } f(a) = b, \text{ we have } a \in A'\}.$$

This map takes a set A' of apples, and tells you all the buckets b that are all-A': all the apples in b are from the chosen subset A'. Note that if a bucket doesn't contain any apples at all, then vacuously all its apples are from A', so empty buckets count as far as f_* is concerned.

Notice that all three of these operations turn out to be interesting: start with a set B' of buckets and return all the apples in them, or start with a set A' of apples and find either the buckets that contain at least one apple from A', or the buckets whose only apples are from A'. But we did not invent these mappings f^*, $f_!$, and f_*: they were *induced* by the function f. They were automatic. It is one of the pleasures of category theory that adjoints so often turn out to have interesting semantic interpretations.

Exercise 1.110. Choose sets X and Y with between two and four elements each, and choose a function $f : X \to Y$.

1. Choose two different subsets $B_1, B_2 \subseteq Y$ and find $f^*(B_1)$ and $f^*(B_2)$.
2. Choose two different subsets $A_1, A_2 \subseteq X$ and find $f_!(A_1)$ and $f_!(A_2)$.
3. With the same $A_1, A_2 \subseteq X$, find $f_*(A_1)$ and $f_*(A_2)$. $\diamond$

1.4.4 Closure Operators

Given a Galois connection with $f : P \to Q$ left adjoint to $g : Q \to P$, we may compose f and g to arrive at a monotone map $f \mathbin{\mathring{\,}} g : P \to P$ from preorder P to itself. This monotone map has the property that $p \le (f \mathbin{\mathring{\,}} g)(p)$, and that $(f \mathbin{\mathring{\,}} g \mathbin{\mathring{\,}} f \mathbin{\mathring{\,}} g)(p) \cong (f \mathbin{\mathring{\,}} g)(p)$ for any $p \in P$. This is an example of a *closure operator*.[10]

Exercise 1.111. Suppose that f is left adjoint to g. Use Proposition 1.101 to show the following.

1. $p \le (f \mathbin{\mathring{\,}} g)(p)$.
2. $(f \mathbin{\mathring{\,}} g \mathbin{\mathring{\,}} f \mathbin{\mathring{\,}} g)(p) \cong (f \mathbin{\mathring{\,}} g)(p)$. To prove this, show inequalities in both directions, $\le$ and $\ge$. $\diamond$

Definition 1.112. A *closure operator* $j : P \to P$ on a preorder P is a monotone map such that for all $p \in P$ we have

(a) $p \le j(p)$,
(b) $j(j(p)) \cong j(p)$.

Example 1.113. Here is an example of closure operators from computation, very roughly presented. Imagine computation as a process of rewriting input expressions to output expressions. For example, a computer can rewrite the expression $7 + 2 + 3$ as the expression 12. The set of arithmetic expressions has a partial order according to whether one expression can be rewritten as another.

We might think of a computer program, then, as a method of taking an expression and reducing it to another expression. So it is a map $j : \mathtt{exp} \to \mathtt{exp}$. It furthermore is desirable to require that this computer program is a closure operator. Monotonicity means that if an expression x can be rewritten into expression y, then the reduction $j(x)$ can be rewritten into $j(y)$. Moreover, the requirement $x \le j(x)$ implies that j can only turn one expression into another if doing so is a permissible rewrite. The requirement $j(j(x)) = j(x)$ implies that if you try to reduce an expression that has already been reduced, the computer program leaves it as is. These properties provide useful structure in the analysis of program semantics.

---

[10] The other composite $g \mathbin{\mathring{\,}} f$ satisfies the dual properties: $(g \mathbin{\mathring{\,}} f)(q) \le q$ and $(g \mathbin{\mathring{\,}} f \mathbin{\mathring{\,}} g \mathbin{\mathring{\,}} f)(q) \cong (g \mathbin{\mathring{\,}} f)(q)$ for all $q \in Q$. This is called an *interior operator*, though we will not discuss this concept further.

Example 1.114 (Adjunctions from closure operators). Just as every adjunction gives rise to a closure operator, from every closure operator we may construct an adjunction.

Let P be a preorder and let $j \colon P \to P$ be a closure operator. We can define a preorder fix_j to have elements the fixed points of j; that is,

$$\mathrm{fix}_j := \{p \in P \mid j(p) \cong p\}.$$

This is a subset of P and inherits an order as a result; hence fix_j is a sub-preorder of P. Note that $j(p)$ is a fixed point for all $p \in P$, since $j(j(p)) \cong j(p)$.

We define an adjunction with left adjoint $j \colon P \to \mathrm{fix}_j$ sending p to $j(p)$, and right adjoint $g \colon \mathrm{fix}_j \to P$ simply the inclusion of the sub-preorder. To see it's really an adjunction, we need to see that for any $p \in P$ and $q \in \mathrm{fix}_j$, we have $j(p) \leq q$ if and only if $p \leq q$. Let's check it. Since $p \leq j(p)$, we have that $j(p) \leq q$ implies $p \leq q$ by transitivity. Conversely, since q is a fixed point, $p \leq q$ implies $j(p) \leq j(q) \cong q$.

Example 1.115. Another example of closure operators comes from logic. This will be discussed in the final chapter of the book, in particular Section 7.4.5, but we will give a quick overview here. In essence, logic is the study of when one formal statement – or proposition – implies another. For example, if n is prime then n is not a multiple of 6, or if it is raining then the ground is getting wetter. Here "n is prime," "n is not a multiple of 6," "it is raining," and "the ground is getting wetter" are propositions, and we gave two implications.

Take the set of all propositions and order them by $p \leq q$ iff p implies q, denoted $p \Rightarrow q$. Since $p \Rightarrow p$ and since whenever $p \Rightarrow q$ and $q \Rightarrow r$, we also have $p \Rightarrow r$, this is indeed a preorder.

A closure operator on it is often called a *modal operator*. It is a function j from propositions to propositions, for which $p \Rightarrow j(p)$ and $j(j(p)) = j(p)$. An example of a j is "assuming Bob is in San Diego … ." Think of this as a proposition B; so "assuming Bob is in San Diego, p" means $B \Rightarrow p$. Let's see why $B \Rightarrow -$ is a closure operator.

If p is true then "assuming Bob is in San Diego, p" is still true. Suppose that "assuming Bob is in San Diego it is the case that, 'assuming Bob is in San Diego, p' is true." It follows that "assuming Bob is in San Diego, p" is true. So we have seen, at least informally, that "assuming Bob is in San Diego …" is a closure operator.

1.4.5 Level Shifting

The last thing we want to discuss in this chapter is a phenomenon that happens often in category theory, something we might informally call "level shifting." It is easier to give an example of this than to explain it directly.

Given any set S, there is a set $\mathbf{Rel}(S)$ of binary relations on S. An element $R \in \mathbf{Rel}(S)$ is formally a subset $R \subseteq S \times S$. The set $\mathbf{Rel}(S)$ can be given an order via the subset relation, $R \subseteq R'$, i.e. if whenever $R(s_1, s_2)$ holds then so does $R'(s_1, s_2)$.

For example, the Hasse diagram for $\mathbf{Rel}(\{1\})$ is:

$$\overset{\varnothing}{\bullet} \longrightarrow \overset{\{(1,1)\}}{\bullet}$$

Exercise 1.116. Draw the Hasse diagram for the preorder $\mathbf{Rel}(\{1, 2\})$ of all binary relations on the set $\{1, 2\}$. ⬦

For any set S, there is also a set $\mathbf{Pos}(S)$, consisting of all the preorder relations on S. In fact there is a preorder structure $\sqsubseteq$ on $\mathbf{Pos}(S)$, again given by inclusion: $\leq$ is below $\leq'$ (we'll write $\leq \sqsubseteq \leq'$) if $a \leq b$ implies $a \leq' b$ for every $a, b \in S$. A preorder of preorder structures? That's what we mean by a level shift.

Every preorder relation is – in particular – a relation, so we have an inclusion $\mathbf{Pos}(S) \to \mathbf{Rel}(S)$. This is the right adjoint of a Galois connection. Its left adjoint is a monotone map $\mathrm{Cl}: \mathbf{Rel}(S) \to \mathbf{Pos}(S)$ given by taking any relation R, writing it in infix notation using $\leq$, and taking the reflexive and transitive closure, i.e. adding $s \leq s$ for every s and adding $s \leq u$ whenever $s \leq t$ and $t \leq u$.

Exercise 1.117. Let $S = \{1, 2, 3\}$. Let's try to understand the adjunction discussed above.

1. Come up with any preorder relation $\leq$ on S, and define $U(\leq)$ to be the subset $U(\leq) := \{(s_1, s_2) \mid s_1 \leq s_2\} \subseteq S \times S$, i.e. $U(\leq)$ is the image of $\leq$ under the inclusion $\mathbf{Pos}(S) \to \mathbf{Rel}(S)$, the relation "underlying" the preorder.
2. Come up with any two binary relations $Q \subseteq S \times S$ and $Q' \subseteq S \times S$ such that $Q \subseteq U(\leq)$ but $Q' \nsubseteq U(\leq)$. Note that your choice of Q, Q' do not have to come from preorders.

We now want to check that, in this case, the closure operation Cl is really left adjoint to the "underlying relation" map U.

3. Concretely (without using the assertion that there is some sort of adjunction), show that $\mathrm{Cl}(Q) \sqsubseteq \leq$, where $\sqsubseteq$ is the order on $\mathbf{Pos}(S)$, defined immediately above this exercise.
4. Concretely show that $\mathrm{Cl}(Q') \not\sqsubseteq \leq$. ⬦

1.5   Summary and Further Reading

In this first chapter, we set the stage for category theory by introducing one of the simplest interesting sorts of example: preorders. From this seemingly simple structure, a bunch of further structure emerges: monotone maps, meets, joins, and more. In terms of modeling real-world phenomena, we thought of preorders as the states of a system, and monotone maps as describing a way to use one system to observe another. From this point of view, generative effects occur when observations of the whole cannot be deduced by combining observations of the parts.

In the final section we introduced Galois connections. A Galois connection, or adjunction, is a pair of maps that are like inverses, but allowed to be more "relaxed" by getting the orders involved. Perhaps surprisingly, it turns out adjunctions are closely related to joins and meets: if a preorder P has all joins, then a monotone map out of P is a left adjoint if and only if it preserves joins; similarly for meets and right adjoints.

The next two chapters build significantly on this material, but in two different directions. Chapter 2 adds a new operation on the underlying set: it introduces the idea of a monoidal structure on preorders. This allows us to construct an element $a \otimes b$ of a preorder P from any elements $a, b \in P$, in a way that respects the order. On the other hand, Chapter 3 adds new structure on the order itself: it introduces the idea of a morphism, which describes not only whether $a \leq b$, but gives a name f for how a relates to b. This structure is known as a category. These generalizations are both fundamental to the story of compositionality, and in Chapter 4 we'll see them meet in the concept of a monoidal category. The lessons we have learned in this chapter will illuminate the more highly structured generalizations in the chapters to come. Indeed, it is a useful principle in studying category theory to try to understand concepts first in the setting of preorders – where often much of the complexity is stripped away and one can develop some intuition – before considering the general case.

But perhaps you might be interested in exploring some ideas in this chapter in other directions. While we won't return to them in this book, we learned about generative effects from Elie Adam's thesis [Ada17], and a much richer treatment of generative effect can be found there. In particular, he discusses abelian categories and cohomology, providing a way to detect generative effects in quite a general setting.

Another important application of preorders, monotone maps, and Galois connections is found in the analysis of programming languages. In this setting, preorders describe the possible states of a computer, and monotone maps describe the action of programs, or relationships between different ways of modeling computation states. Galois connections are useful for showing how different models may be closely related, and for transporting program analysis from one framework to another. For more detail on this, see Chapter 4 of the textbook [NNH99].

2 Resource Theories: Monoidal Preorders and Enrichment

2.1 Getting from a to b

You can't make an omelette without breaking an egg. To obtain the things we want requires resources, and the process of transforming what we have into what we want is often an intricate one. In this chapter, we will discuss how monoidal preorders can help us think about this matter.

Consider the following three questions you might ask yourself:

- Given what I have, is it *possible* to get what I want?
- Given what I have, what is the *minimum cost* to get what I want?
- Given what I have, what is the *set of ways* to get what I want?

These questions are about resources – those you have and those you want – but, perhaps more importantly, they are about moving from have to want: possibility of, cost of, and ways to.

Such questions come up not only in our lives, but also in science and industry. In chemistry, one asks whether a certain set of compounds can be transformed into another set, how much energy such a reaction will require, or what methods exist for making it happen. In manufacturing, one asks similar questions.

From an external point of view, both a chemist and an industrial firm might be regarded as store-houses of information on the above subjects. The chemist knows which compounds she can make given other ones, and how to do so; the firm has stored knowledge of the same sort. The research work of the chemist and the firm is to use what they know in order to derive – or discover – new knowledge.

This is roughly the first goal of this chapter: to discuss a formalism for expressing recipes – methods for transforming one set of resources into another – and for deriving new recipes from old. The idea here is not complicated, neither in life nor in our mathematical formalism. The value added then is to simply see how it works, so we can build on it within the book, and so others can build on it in their own work.

We briefly discuss the categorical approach to this idea – namely that of *monoidal preorders* – for building new recipes from old. The following *wiring diagram* shows, assuming one knows how to implement each of the interior boxes, how to implement the preparation of a lemon meringue pie:

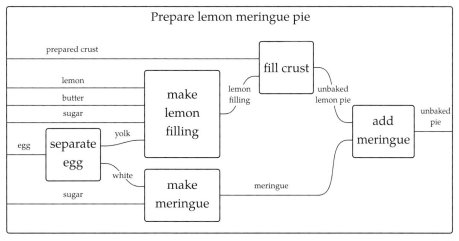

(2.1)

The wires show resources: we start with prepared crust, lemon, butter, sugar, and egg resources, and we end up with an unbaked pie resource. We could take this whole method and combine it with others, e.g. baking the pie:

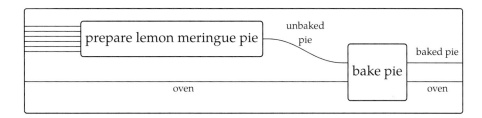

In the above example we see that resources are not always consumed when they are used. For example, we use an oven to convert – or catalyze the transformation of – an unbaked pie into a baked pie, and we get the oven back after we are done. It's a nice feature of ovens! To use economic terms, the oven is a "means of production" for pies.

String diagrams are important mathematical objects that will come up repeatedly in this book. They were invented in the mathematical context – more specifically in the context of monoidal categories – by Joyal and Street [JS93], but they have been used less formally by engineers and scientists in various contexts for a long time.

As we said above, our first goal in this chapter is to use monoidal preorders, and the corresponding wiring diagrams, as a formal language for building new recipes from old. Our second goal is to discuss something called $\mathcal{V}$-categories for various monoidal preorders $\mathcal{V}$.

A $\mathcal{V}$-category is a set of objects, which one may think of as points on a map, where $\mathcal{V}$ somehow "structures the question" of getting from point a to point b. The examples of monoidal preorders $\mathcal{V}$ that we will be most interested in are called **Bool** and **Cost**. Roughly speaking, a **Bool**-category is a set of points where the question of getting from point a to point b has a true / false answer. A **Cost**-category is a set of points where the question of getting from a to b has an answer $d \in [0, \infty]$, a cost.

This story works in more generality than monoidal preorders. Indeed, in Chapter 4 we will discuss something called a monoidal category, a notion which generalizes monoidal preorders, and we will generalize the definition of $\mathcal{V}$-category accordingly. In this more general setting, $\mathcal{V}$-categories can also address our third question above, describing *methods* of getting between points. For example a **Set**-category is a set of points where the question of getting from point a to point b has a set of answers (elements of which might be called methods).

We will begin in Section 2.2 by defining symmetric monoidal preorders, giving a few preliminary examples and discussing wiring diagrams. We then give many more examples of symmetric monoidal preorders, including some real-world examples, in the form of resource theories, and some mathematical examples that will come up again throughout the book. In Section 2.3 we discuss enrichment and $\mathcal{V}$-categories – how a monoidal preorder $\mathcal{V}$ can "structure the question" of getting from a to b – and then give some important constructions on $\mathcal{V}$-categories (Section 2.4), and analyze them using a sort of matrix multiplication technique (Section 2.5).

2.2   Symmetric Monoidal Preorders

In Section 1.2.2 we introduced preorders. The notation for a preorder, namely $(X, \leq)$, refers to two pieces of structure: a set called X and a relation called $\leq$ that is reflexive and transitive.

We want to add to the concept of preorders a way of combining elements in X, an operation taking two elements and adding or multiplying them together. However, the operation does not have to literally be addition or multiplication; it only needs to satisfy some of the properties one expects from them.

2.2.1   Definition and First Examples

We begin with a formal definition of symmetric monoidal preorders.

Definition 2.1.   A (strict) *symmetric monoidal structure* on a preorder $(X, \leq)$ consists of two constituents:

(i) an element $I \in X$, called the *monoidal unit*,
(ii) a function $\otimes \colon X \times X \to X$, called the *monoidal product*.

These constituents must satisfy the following properties, where we write $\otimes(x_1, x_2) = x_1 \otimes x_2$:

(a) for all $x_1, x_2, y_1, y_2 \in X$, if $x_1 \leq y_1$ and $x_2 \leq y_2$, then $x_1 \otimes x_2 \leq y_1 \otimes y_2$,
(b) for all $x \in X$, the equations $I \otimes x = x$ and $x \otimes I = x$ hold,
(c) for all $x, y, z \in X$, the equation $(x \otimes y) \otimes z = x \otimes (y \otimes z)$ holds,
(d) for all $x, y \in X$, the equation $x \otimes y = y \otimes x$ holds.

We call these conditions *monotonicity, unitality, associativity*, and *symmetry* respectively. A preorder equipped with a symmetric monoidal structure, $(X, \leq, I, \otimes)$, is called a *symmetric monoidal preorder*.

Anyone can propose a set X, an order $\leq$ on X, an element I in X, and a binary operation $\otimes$ on X and ask whether $(X, \leq, I, \otimes)$ is a symmetric monoidal preorder. And it will indeed be one, as long as it satisfies rules (a), (b), (c), and (d) of Definition 2.1.

Remark 2.2. It is often useful to replace $=$ with $\cong$ throughout Definition 2.1. The result is a perfectly good notion, called a *weak monoidal structure*. The reason we chose equality is that it makes equations look simpler, which we hope aids first-time readers.

The notation for the monoidal unit and the monoidal product may vary: monoidal units we have seen include I (as in the definition), 0, 1, `true`, `false`, $\{*\}$, and more. Monoidal products we have seen include $\otimes$ (as in the definition), $+$, $*$, $\wedge$, $\vee$, and $\times$. The *preferred notation* in a given setting is whatever best helps our brains remember what we're trying to do; the names I and $\otimes$ are just defaults.

Example 2.3. There is a well-known preorder structure, denoted $\leq$, on the set $\mathbb{R}$ of real numbers; e.g. $-5 \leq \sqrt{2}$. We propose 0 as a monoidal unit and $+\colon \mathbb{R} \times \mathbb{R} \to \mathbb{R}$ as a monoidal product. Does $(\mathbb{R}, \leq, 0, +)$ satisfy the conditions of Definition 2.1?

If $x_1 \leq y_1$ and $x_2 \leq y_2$, it is true that $x_1 + x_2 \leq y_1 + y_2$. It is also true that $0 + x = x$ and $x + 0 = x$, that $(x + y) + z = x + (y + z)$, and that $x + y = y + x$. Thus $(\mathbb{R}, \leq, 0, +)$ satisfies the conditions of being a symmetric monoidal preorder.

Exercise 2.4. Consider again the preorder $(\mathbb{R}, \leq)$ from Example 2.3. Someone proposes 1 as a monoidal unit and $*$ (usual multiplication) as a monoidal product. But an expert walks by and says "that won't work." Figure out why, or prove the expert wrong! $\diamond$

Example 2.5. A *monoid* consists of a set M, a function $*\colon M \times M \to M$ called the *monoid multiplication*, and an element $e \in M$ called the *monoid unit*, such that, when you write $*(m, n)$ as $m * n$, i.e. using infix notation, the equations

$$m * e = m, \qquad e * m = m, \qquad (m * n) * p = m * (n * p) \tag{2.2}$$

hold for all $m, n, p \in M$. It is called *commutative* if also $m * n = n * m$.

Every set S determines a discrete preorder $\mathbf{Disc}_S$ (where $m \leq n$ iff $m = n$; see Example 1.27), and it is easy to check that if $(M, e, *)$ is a commutative monoid then $(\mathbf{Disc}_M, =, e, *)$ is a symmetric monoidal preorder.

Exercise 2.6. We said it was easy to check that if $(M, *, e)$ is a commutative monoid then $(\mathbf{Disc}_M, =, *, e)$ is a symmetric monoidal preorder. Were we telling the truth? $\diamond$

Example 2.7. Here is a non-example for people who know the game "standard poker." Let H be the set of all poker hands, where a hand means a choice of five cards from the standard 52-card deck. As an order, put $h \leq h'$ if h' beats or equals h in poker.

One could propose a monoidal product $\otimes \colon H \times H \to H$ by assigning $h_1 \otimes h_2$ to be "the best hand one can form out of the ten cards in h_1 and h_2." If some cards are in both h_1 and h_2, just throw the duplicates away. So for example $\{2\heartsuit,\ 3\heartsuit,\ 4\heartsuit,\ 6\spadesuit,\ 7\spadesuit\} \otimes \{2\heartsuit,\ 5\heartsuit,\ 6\heartsuit,\ 6\spadesuit,\ 7\spadesuit\} = \{2\heartsuit,\ 3\heartsuit,\ 4\heartsuit,\ 5\heartsuit,\ 6\heartsuit\}$, because the latter is the best hand you can make with the former two.

This proposal for a monoidal structure will fail the condition (a) of Definition 2.1: it could be the case that $h_1 \leq i_1$ and $h_2 \leq i_2$, and yet *not* be the case that $h_1 \otimes h_2 \leq i_1 \otimes i_2$. For example, consider this case:

$$h_1 := \{2\heartsuit,\ 3\heartsuit,\ 10\spadesuit,\ J\spadesuit,\ Q\spadesuit\}, \qquad i_1 := \{4\clubsuit,\ 4\spadesuit,\ 6\heartsuit,\ 6\diamondsuit,\ 10\diamondsuit\},$$
$$h_2 := \{2\diamondsuit,\ 3\diamondsuit,\ 4\diamondsuit,\ K\spadesuit,\ A\spadesuit\}, \qquad i_2 := \{5\spadesuit,\ 5\heartsuit,\ 7\heartsuit,\ J\diamondsuit,\ Q\diamondsuit\}.$$

Here, $h_1 \leq i_1$ and $h_2 \leq i_2$, but $h_1 \otimes h_2 = \{10\spadesuit,\ J\spadesuit,\ Q\spadesuit,\ K\spadesuit,\ A\spadesuit\}$ is the best possible hand and beats $i_1 \otimes i_2 = \{5\spadesuit,\ 5\heartsuit,\ 6\heartsuit,\ 6\diamondsuit,\ Q\diamondsuit\}$.

Subsections 2.2.3 and 2.2.4 are dedicated to examples of symmetric monoidal preorders. Some are aligned with the notion of resource theories, others come from pure math. When discussing the former, we will use wiring diagrams, so here is a quick primer.

2.2.2   Introducing Wiring Diagrams

Wiring diagrams are visual representations for building new relationships from old. In a preorder without a monoidal structure, the only sort of relationship between objects is $\leq$, and the only way you build a new $\leq$ relationship from old ones is by chaining them together. We denote the relationship $x \leq y$ by

$$\boxed{x \boxed{\leq} y} \tag{2.3}$$

We can chain some number of these $\leq$-relationships – say 0, 1, 2, or 3 of them – together in series as shown here

$$\boxed{x_0} \qquad \boxed{x_0 \boxed{\leq} x_1} \qquad \boxed{x_0 \boxed{\leq} x_1 \boxed{\leq} x_2} \qquad \boxed{x_0 \boxed{\leq} x_1 \boxed{\leq} x_2 \boxed{\leq} x_3} \quad \cdots \tag{2.4}$$

If we add a symmetric monoidal structure, we can combine relationships not only in series but also in parallel. Here is an example:

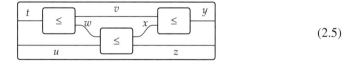

$$\tag{2.5}$$

Different styles of wiring diagrams

In fact, we shall see later that there are many styles of wiring diagrams. When we are dealing with preorders, the sort of wiring diagram we can draw is that with single-input, single-output boxes connected in series. When we are dealing with symmetric monoidal preorders, we can have more complex boxes and more complex wiring diagrams, including parallel composition. Later we shall see that for other sorts of categorical structures there are other styles of wiring diagrams:

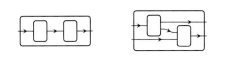

$$(2.6)$$

Wiring diagrams for symmetric monoidal preorders

The style of wiring diagram that makes sense in any symmetric monoidal preorder is that shown in Eq. (2.5): boxes can have multiple inputs and outputs, and they may be arranged in series and parallel. Symmetric monoidal preorders and their wiring diagrams are tightly coupled with each other. How so?

The answer is that a monoidal preorder $(X, \leq, I, \otimes)$ has some notion of element $(x \in X)$, relationship $(\leq)$, and combination (using transitivity and $\otimes$), and so do wiring diagrams: the wires represent elements, the boxes represent relationships, and the wiring diagrams themselves show how relationships can be combined. We call boxes and wires *icons*; we will encounter several more icons in this chapter, and throughout the book.

To get a bit more rigorous about the connection, let's start with a monoidal preorder $(X, \leq, I, \otimes)$ as in Definition 2.1. Wiring diagrams have wires on the left and the right. Each element $x \in X$ can be made the label of a wire. Note that, given two objects x, y, we can either draw two wires in parallel – one labeled x and one labeled y – or we can draw one wire labeled $x \otimes y$.

$$\frac{x}{y} \qquad \overline{x \otimes y}$$

We consider wires in parallel to represent the monoidal product of their labels, so we consider both cases above to represent the element $x \otimes y$. Note also that a wire labeled I or an absence of wires:

$$\overline{}_I \qquad\qquad nothing$$

both represent the monoidal unit I; another way of thinking of this is that the unit is the empty monoidal product.

A wiring diagram runs between a set of parallel wires on the left and a set of parallel wires on the right. We say that a wiring diagram is *valid* if the monoidal product of the elements on the left is less than the monoidal product of those on the right. For example, if we have the inequality $x \leq y$, the the diagram that is a box with a wire labeled x on the left and a wire labeled y on the right is valid; see the first box below:

The validity of the second box corresponds to the inequality $x_1 \otimes x_2 \leq y_1 \otimes y_2 \otimes y_3$. Before going on to the properties from Definition 2.1, let us pause for an example of what we've discussed so far.

Example 2.8. Recall the symmetric monoidal preorder $(\mathbb{R}, \leq, 0, +)$ from Example 2.3. The wiring diagrams for it allow wires labeled by real numbers. Drawing wires in parallel corresponds to adding their labels, and the wire labeled 0 is equivalent to no wires at all.

$$\overline{3.14} \qquad \overline{-1} \qquad \left.\begin{array}{c}\overline{3.14}\\[4pt]\overline{-1}\end{array}\right\} = \overline{2.14} \qquad \overline{} \; 0 = \textit{nothing}$$

And here we express a couple of facts about $(\mathbb{R}, \leq, 0, +)$ in this language: $4 \leq 7$ and $2 + 5 \leq -1 + 5 + 3$.

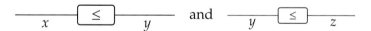

We now return to how the properties of symmetric monoidal preorders correspond to properties of this sort of wiring diagram. Let's first talk about the order structure: conditions (a) – reflexivity – and (b) – transitivity – from Definition 1.25. Reflexivity says that $x \leq x$, this means the diagram just consisting of a wire

$$\overline{\qquad x \qquad}$$

is always valid. Transitivity allows us to connect facts together: it says that if $x \leq y$ and $y \leq z$, then $x \leq z$. This means that if the diagrams

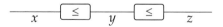 and

are valid, we can put them together and obtain the valid diagram

Next let's talk about the properties (a)–(d) from the definition of symmetric monoidal structure (Definition 2.1). Property (a) says that if $x_1 \leq y_1$ and $x_2 \leq y_2$ then $x_1 \otimes x_2 \leq y_1 \otimes y_2$. This corresponds to the idea that stacking any two valid boxes in parallel is still valid:

Condition (b), that $I \otimes x = x$ and $x \otimes I = x$, says we don't need to worry about I or blank space; in particular diagrams such as the following are valid:

$$\frac{\textit{nothing}}{x} \diagdown^{x}$$

Condition (c), that $(x \otimes y) \otimes z = x \otimes (y \otimes z)$, says that we don't have to worry about whether we build up diagrams from the top or from the bottom

$$\left.\frac{\dfrac{x}{y}}{z}\right\} = \left.\dfrac{x \otimes y}{\dfrac{}{y \otimes z}}\right\} = \left\{\dfrac{\dfrac{x}{}}{y \otimes z} = \left\{\dfrac{\dfrac{x}{y}}{z}\right.\right.$$

But this looks much harder than it is: the associative property should be thought of as saying that there is no distinction between the stuff on the very left above and the stuff on the very right, i.e.

$$\frac{\dfrac{x}{y}}{z} \qquad = \qquad \dfrac{x}{\dfrac{y}{z}}$$

and indeed a diagram that moves from one to the other is valid.

Finally, the symmetry condition (d), that $x \otimes y = y \otimes x$, says that a diagram is valid even if its wires cross:

$$x \diagdown\!\!\!\!\diagup y \qquad\qquad y \diagdown\!\!\!\!\diagup x$$
$$y \qquad\qquad x \qquad\qquad x \qquad\qquad y$$

One may regard the pair of crossing wires as another icon in our iconography, in addition to the boxes and wires we already have.

Wiring diagrams as graphical proofs

Given a monoidal preorder $\mathcal{X} = (X, \leq, I, \otimes)$, a wiring diagram is a graphical proof of something about $\mathcal{X}$. Each box in the diagram has a left side and a right side, say x and y, and represents the assertion that $x \leq y$.

A wiring diagram is a bunch of interior boxes connected together inside an exterior box. It represents a graphical proof that says: if all of the interior assertions are valid, then so is the exterior assertion.

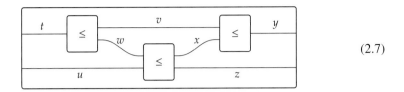

$$(2.7)$$

The inner boxes in diagram (2.7) translate into the assertions:

$$t \le v + w, \qquad w + u \le x + z, \qquad v + x \le y, \tag{2.8}$$

and the outer box translates into the assertion:

$$t + u \le y + z. \tag{2.9}$$

The whole wiring diagram (2.7) says "if you know that the assertions in (2.8) are true, then I am a proof that the assertion in (2.9) is also true." What exactly is the proof that diagram (2.7) represents? It is the proof

$$t + u \ \le \ v + w + u \ \le \ v + x + z \ \le \ y + z. \tag{2.10}$$

Indeed, each inequality here is a vertical slice of the diagram (2.7), and the transitivity of these inequalities is expressed by connecting these vertical slices together.

Example 2.9. Recall the lemon meringue pie wiring diagram from Eq. (2.1). It has five interior boxes, such as "separate egg" and "fill crust," and it has one exterior box called "Prepare lemon meringue pie." Each box is the assertion that, given the collection of resources on the left, say an egg, you can transform it into the collection of resources on the right, say an egg white and an egg yolk. The whole string diagram is a proof that if each of the interior assertions is true – i.e. you really do know how to separate eggs, make lemon filling, make meringue, fill crust, and add meringue – then the exterior assertion is true: you can prepare a lemon meringue pie.

Exercise 2.10. The string of inequalities in Eq. (2.10) is not quite a proof, because technically there is no such thing as $v + w + u$, for example. Instead, there is $(v + w) + u$ and $v + (w + u)$, and so on.

1. Formally prove, using only the rules of symmetric monoidal preorders (Definition 2.1) that, given the assertions in Eq. (2.8), the conclusion in Eq. (2.9) follows.
2. Reflexivity and transitivity should show up in your proof. Make sure you are explicit about where they do.
3. How can you look at the wiring diagram (2.5) and know that the symmetry axiom (Definition 2.1(d)) does not need to be invoked?   ◇

We next discuss some examples of symmetric monoidal preorders. We begin in Section 2.2.3 with some more concrete examples, from science, commerce, and informatics.

Then in Section 2.2.4 we discuss some examples arising from pure math, some of which will get a good deal of use later on, e.g. in Chapter 4.

2.2.3 Applied Examples

Resource theories are studies of how resources are exchanged in a given arena. For example, in social resource theory one studies a marketplace where combinations of goods can be traded for – as well as converted into – other combinations of goods.

Whereas marketplaces are very dynamic, and an apple might be tradable for an orange on Sunday but not on Monday, what we mean by resource theory in this chapter is a static notion: deciding "what buys what," once and for all.[1] This sort of static notion of conversion might occur in chemistry: the chemical reactions that are possible one day will quite likely be possible on a different day as well. Manufacturing may be somewhere in between: the set of production techniques – whereby a company can convert one set of resources into another – do not change much from day to day.

We learned about resource theories from [CFS16; Fri17], who go much further than we will; see Section 2.6 for more information. In this section we will focus only on the main idea. While there are many beautiful mathematical examples of symmetric monoidal preorders, as we shall see in Section 2.2.4, there are also ad hoc examples coming from life experience. In the next chapter, on databases, we shall see the same theme: while there are some beautiful mathematical categories out there, database schemas are ad hoc organizational patterns of information. Describing something as "ad hoc" is often considered derogatory, but it just means "formed, arranged, or done for a particular purpose only." There is nothing wrong with doing things for a particular purpose; it's common outside of pure math and pure art. Let's get to it.

Chemistry

In high school chemistry, we work with chemical equations, where material collections such as

$$H_2O, \quad NaCl, \quad 2NaOH, \quad CH_4 + 3O_2$$

are put together in the form of reaction equations, such as

$$2H_2O + 2Na \rightarrow 2NaOH + H_2.$$

The collection on the left, $2H_2O + 2Na$, is called the *reactant*, and the collection on the right, $2NaOH + H_2$, is called the *product*.

We can consider reaction equations such as the one above as taking place inside a single symmetric monoidal preorder (Mat, $\rightarrow$, 0, +). Here Mat is the set of all collections of atoms and molecules, sometimes called *materials*. So we have $NaCl \in$ Mat and $4H_2O + 6Ne \in$ Mat.

The set Mat has a preorder structure denoted by the $\rightarrow$ symbol, which is the preferred symbol in the setting of chemistry. To be clear, $\rightarrow$ is taking the place of the order relation

---

[1] Using some sort of temporal theory, e.g. the one presented in Chapter 7, one could take the notion here and have it change in time.

$\le$ from Definition 2.1. The $+$ symbol is the preferred notation for the monoidal product in the chemistry setting, taking the place of $\otimes$. While it does not come up in practice, we use 0 to denote the monoidal unit.

Exercise 2.11. Here is an exercise for people familiar with reaction equations: check that conditions (a), (b), (c), and (d) of Definition 2.1 hold. $\diamond$

An important notion in chemistry is that of catalysis: one compound *catalyzes* a certain reaction. For example, one might have the following set of reactions:

$$y + k \to y' + k', \qquad x + y' \to z', \qquad z' + k' \to z + k. \tag{2.11}$$

Using the laws of monoidal preorders, we obtain the composed reaction

$$x + y + k \to x + y' + k' \to z' + k' \to z + k. \tag{2.12}$$

Here k is the catalyst because it is found in both the reactant and the product of the reaction. It is said to catalyze the reaction $x + y \to z$. The idea is that the reaction $x + y \to z$ cannot take place given the reactions in Eq. (2.11). But if k is present, meaning if we add k to both sides, the resulting reaction can take place.

The wiring diagram for the reaction in Eq. (2.12) is shown in Eq. (2.13). The three interior boxes correspond to the three reactions given in Eq. (2.11), and the exterior box corresponds to the composite reaction $x + y + k \to z + k$.

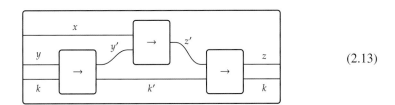

$$\tag{2.13}$$

Manufacturing

Whether we are talking about baking pies, building smart phones, or following pharmaceutical recipes, manufacturing firms need to store basic recipes, and build new recipes by combining simpler recipes in schemes like the one shown in Eq. (2.1) or Eq. (2.13).

The basic idea in manufacturing is exactly the same as that for chemistry, except there is an important assumption we can make in manufacturing that does not hold for chemical reactions:

You can trash anything you want, and it disappears from view.

This simple assumption has caused the world some significant problems, but it is still in effect. In our meringue pie example, we can ask: "what happened to the egg shell, or the paper wrapping the stick of butter"? The answer is they were trashed, i.e. thrown in the garbage bin. It would certainly clutter our diagram and our thinking if we had to carry these resources through the diagram:

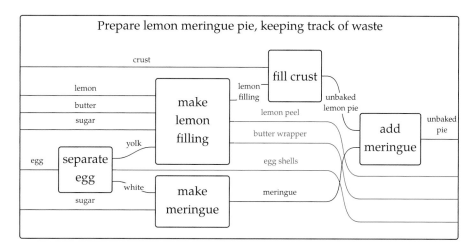

Instead, in our daily lives and in manufacturing, we do not have to hold on to something if we don't need it; we can just discard it. In terms of wiring diagrams, this can be shown using a new icon —●, as follows:

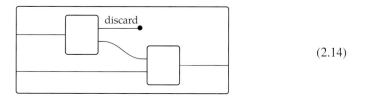

(2.14)

To model this concept of waste using monoidal categories, one just adds an additional axiom to (a), (b), (c), and (d) from Definition 2.1:

(e) $x \leq I$ for all $x \in X$.   (discard axiom)

It says that every x can be converted into the monoidal unit I. In the notation of the chemistry section, we would write instead $x \to 0$: any x yields nothing. But this is certainly not accepted in the chemistry setting. For example,

$$H_2O + NaCl \to^? H_2O$$

is certainly not a legal chemical equation. It is easy to throw things away in manufacturing, because we assume that we have access to – the ability to grab onto and directly manipulate – each item produced. In chemistry, when you have 10^{23} of substance A dissolved in something else, you cannot just simply discard A. So axiom (e) is valid in manufacturing but not in chemistry.

Recall that in Section 2.2.2 we said that there were many different styles of wiring diagrams. Now we're saying that adding the discard axiom changes the wiring diagram style, in that it adds this new discard icon that allows wires to terminate, as shown in Eq. (2.14). In informatics, we will change the wiring diagram style yet again.

Informatics

A major difference between information and a physical object is that information can
be copied. Whereas one cup of butter never becomes two, it is easy for a single email to
be sent to two different people. It is much easier to copy a music file than it is to copy a
CD. Here we do not mean "copy the information from one compact disc onto another"
– of course that's easy – instead, we mean that it's quite difficult to copy the physical
disc, thereby forming a second physical disc! In diagrams, the distinction is between the
relation

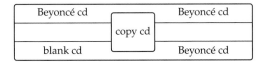

and the relation

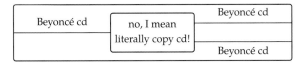

The former is possible, the latter is magic.

Of course material objects can sometimes be copied; cell mitosis is a case in point.
But this is a remarkable biological process, certainly not something that is expected for
ordinary material objects. In the physical world, we would make mitosis a box transforming one cell into two. But in (classical, not quantum) information, everything can
be copied, so we add a new icon to our repertoire.

Namely, in wiring diagram notation, copying information appears as a new icon, ⊸,
allowing us to split wires:

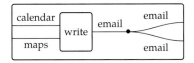

Now with two copies of the email, we can send one to Alice and one to Bob.

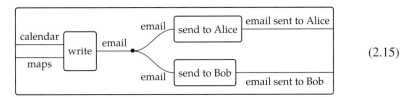

(2.15)

Information can also be discarded, at least in the conventional way of thinking, so in
addition to axioms (a) to (d) from Definition 2.1, we can keep axiom (e) from page 49
and add a new *copy axiom*:

(f) $x \leq x + x$ for all $x \in X$ (copy axiom)

allowing us to make mathematical sense of diagrams like Eq. (2.15).

Now that we have examples of monoidal preorders under our belts, let's discuss some nice mathematical examples.

2.2.4 Abstract Examples

In this section we discuss several mathematical examples of symmetric monoidal structures on preorders.

The booleans

The simplest nontrivial preorder is the booleans: $\mathbb{B} = \{\text{true}, \text{false}\}$ with $\text{false} \leq \text{true}$. There are two different symmetric monoidal structures on it.

Example 2.12 (Booleans with AND). We can define a monoidal structure on $\mathbb{B}$ by letting the monoidal unit be true and the monoidal product be $\wedge$ (AND). If one thinks of $\text{false} = 0$ and $\text{true} = 1$, then $\wedge$ corresponds to the usual multiplication operation $*$. That is, with this correspondence, the two tables below match up:

$\wedge$	false	true		$*$	0	1
false	false	false		0	0	0
true	false	true		1	0	1

$$(2.16)$$

One can check that all the properties in Definition 2.1 hold, so we have a monoidal preorder which we denote **Bool** $:= (\mathbb{B}, \leq, \text{true}, \wedge)$.

Bool will be important when we get to the notion of enrichment. Enriching in a monoidal preorder $\mathcal{V} = (V, \leq, I, \otimes)$ means "letting $\mathcal{V}$ structure the question of getting from a to b." All of the structures of a monoidal preorder – i.e. the set V, the ordering relation $\leq$, the monoidal unit I, and the monoidal product $\otimes$ – play a role in how enrichment works.

For example, let's look at the case of **Bool** $= (\mathbb{B}, \leq, \text{true}, \wedge)$. The fact that its underlying set is $\mathbb{B} = \{\text{false}, \text{true}\}$ will translate into saying that "getting from a to b is a true/false question." The fact that true is the monoidal unit will translate into saying "you can always get from a to a." The fact that $\wedge$ is the monoidal product will translate into saying "if you can get from a to b AND you can get from b to c then you can get from a to c." Finally, the "if–then" form of the previous sentence is coming from the order relation $\leq$. We will make this more precise in Section 2.3.

We will be able to play the same game with other monoidal preorders, as we shall see after we define a monoidal preorder called **Cost** in Example 2.21.

Some other monoidal preorders

It is a bit imprecise to call **Bool** "the" boolean monoidal preorder, because there is another monoidal structure on $(\mathbb{B}, \leq)$, which we describe in Exercise 2.13. The first structure, however, seems to be more useful in practice than the second.

Exercise 2.13. Let $(\mathbb{B}, \leq)$ be as above, but now consider the monoidal product to be $\vee$ (OR).

$\vee$	false	true
false	false	true
true	true	true

max	0	1
0	0	1
1	1	1

What must the monoidal unit be in order to satisfy the conditions of Definition 2.1? Does it satisfy the rest of the conditions?   ◇

In Example 2.14 and Exercise 2.15 we give two different monoidal structures on the preorder $(\mathbb{N}, \leq)$ of natural numbers, where $\leq$ is the usual ordering ($0 \leq 1$ and $5 \leq 16$).

Example 2.14 (Natural numbers with addition). There is a monoidal structure on $(\mathbb{N}, \leq)$ where the monoidal unit is 0 and the monoidal product is $+$, i.e. $6 + 4 = 10$. It is easy to check that $x_1 \leq y_1$ and $x_2 \leq y_2$ implies $x_1 + x_2 \leq y_1 + y_2$, as well as all the other conditions of Definition 2.1.

Exercise 2.15. Show there is a monoidal structure on $(\mathbb{N}, \leq)$ where the monoidal product is $*$, i.e. $6 * 4 = 24$. What should the monoidal unit be?   ◇

Example 2.16 (Divisibility and multiplication). Recall from Example 1.40 that there is a "divisibility" order on $\mathbb{N}$: we write $m|n$ to mean that m divides into n without remainder. So $1|m$ for all m and $4|12$.

There is a monoidal structure on $(\mathbb{N}, |)$, where the monoidal unit is 1 and the monoidal product is $*$, i.e. $6 * 4 = 24$. Then if $x_1|y_1$ and $x_2|y_2$, then $(x_1 * x_2)|(y_1 * y_2)$. Indeed, if there is some $p_1, p_2 \in \mathbb{N}$ such that $x_1 * p_1 = y_1$ and $x_2 * p_2 = y_2$, then $(p_1 * p_2) * (x_1 * x_2) = y_1 * y_2$.

Exercise 2.17. Again taking the divisibility order $(\mathbb{N}, |)$, someone proposes 0 as the monoidal unit and $+$ as the monoidal product. Does that proposal satisfy the conditions of Definition 2.1? Why or why not?   ◇

Exercise 2.18. Consider the preorder $(P, \leq)$ with Hasse diagram $\boxed{\text{no} \to \text{maybe} \to \text{yes}}$. We propose a monoidal structure with yes as the monoidal unit and "min" as the monoidal product.

1. Make sense of "min" by filling in the multiplication table with elements of P.

min	no	maybe	yes
no	?	?	?
maybe	?	?	?
yes	?	?	?

2. Check that the axioms of Definition 2.1 hold for **NMY** $:= (P, \leq, \text{yes}, \text{min})$, given your definition of min. If not, change your definition of min.   ◇

Exercise 2.19. Let S be a set and let $P(S)$ be its power set, the set of all subsets of S, including the empty subset, $\varnothing \subseteq S$, and the "everything" subset, $S \subseteq S$. We can give $P(S)$ an order: $A \leq B$ is given by the subset relation $A \subseteq B$, as discussed in Example 1.45. We propose a symmetric monoidal structure on $P(S)$ with monoidal unit S and monoidal product given by intersection $A \cap B$.

Does it satisfy the conditions of Definition 2.1?   ◇

Exercise 2.20. Let $\mathtt{Prop}^{\mathbb{N}}$ denote the set of all mathematical statements one can make about a natural number, where we consider two statements to be the same if one is true if and only if the other is true. For example "n is prime" is an element of $\mathtt{Prop}^{\mathbb{N}}$, and so are "$n = 2$" and "$n \geq 11$." The statements "$n + 2 = 5$" and "$n$ is the least odd prime" are considered the same. Given $P, Q \in \mathtt{Prop}^{\mathbb{N}}$, we say $P \leq Q$ if for all $n \in \mathbb{N}$, whenever $P(n)$ is true, so is $Q(n)$.

Define a monoidal unit and a monoidal product on $\mathtt{Prop}^{\mathbb{N}}$ that satisfy the conditions of Definition 2.1.   ◇

The Monoidal preorder cost

As we said above, when we enrich in monoidal preorders we see them as different ways to structure the question of "getting from here to there." We will explain this in more detail in Section 2.3. The following monoidal preorder will eventually structure a notion of distance or cost for getting from here to there.

Example 2.21 (Lawvere's monoidal preorder, **Cost**). Let $[0, \infty]$ denote the set of nonnegative real numbers – such as 0, 1, 15.33$\overline{3}$, and 2π – together with ∞. Consider the preorder $([0, \infty], \geq)$, with the usual notion of $\geq$, where of course $\infty \geq x$ for all $x \in [0, \infty]$.

There is a monoidal structure on this preorder, where the monoidal unit is 0 and the monoidal product is $+$. In particular, $x + \infty = \infty$ for any $x \in [0, \infty]$. Let's call this monoidal preorder

$$\mathbf{Cost} := ([0, \infty], \geq, 0, +),$$

because we can think of the elements of $[0, \infty]$ as costs. In terms of structuring "getting from here to there," **Cost** seems to say "getting from a to b is a question of cost." The monoidal unit being 0 will translate into saying that you can always get from a to a at no cost. The monoidal product being $+$ will translate into saying that the cost of getting from a to c is at most the cost of getting from a to b *plus* the cost of getting from b to c. Finally, the "at most" in the previous sentence comes from the $\geq$.

The opposite of a monoidal preorder

One can take the opposite of any preorder, just flip the order: $(X, \leq)^{\mathrm{op}} := (X, \geq)$; see Example 1.53. Proposition 2.22 says that if the preorder had a symmetric monoidal structure, so does its opposite.

Proposition 2.22. Suppose $\mathcal{X} = (X, \leq)$ is a preorder and $\mathcal{X}^{\mathrm{op}} = (X, \geq)$ is its opposite. If $(X, \leq, I, \otimes)$ is a symmetric monoidal preorder then so is its opposite, $(X, \geq, I, \otimes)$.

Proof. Let's first check monotonicity. Suppose $x_1 \geq y_1$ and $x_2 \geq y_2$ in $\mathcal{X}^{\mathrm{op}}$; we need to show that $x_1 \otimes x_2 \geq y_1 \otimes y_2$. But by definition of opposite order, we have $y_1 \leq x_1$ and $y_2 \leq x_2$ in $\mathcal{X}$, and thus $y_1 \otimes y_2 \leq x_1 \otimes x_2$ in $\mathcal{X}$. Thus indeed $x_1 \otimes x_2 \geq y_1 \otimes y_2$ in $\mathcal{X}^{\mathrm{op}}$. The other three conditions are even easier; see Exercise 2.23. □

Exercise 2.23. Complete the proof of Proposition 2.22 by proving that the three remaining conditions of Definition 2.1 are satisfied.   ◇

Exercise 2.24. Since **Cost** is a symmetric monoidal preorder, Proposition 2.22 says that **Cost**^{op} is too.

1. What is **Cost**^{op} as a preorder?
2. What is its monoidal unit?
3. What is its monoidal product?   ◇

2.2.5   Monoidal Monotone Maps

Recall from Example 1.44 that for any preorder $(X, \leq)$ there is an induced equivalence relation $\cong$ on X, where $x \cong x'$ iff both $x \leq x'$ and $x' \leq x$.

Definition 2.25. Let $\mathcal{P} = (P, \leq_P, I_P, \otimes_P)$ and $\mathcal{Q} = (Q, \leq_Q, I_Q, \otimes_Q)$ be monoidal preorders. A *monoidal monotone* from $\mathcal{P}$ to $\mathcal{Q}$ is a monotone map $f \colon (P, \leq_P) \to (Q, \leq_Q)$, satisfying two conditions:

(a) $I_Q \leq_Q f(I_P)$,
(b) $f(p_1) \otimes_Q f(p_2) \leq_Q f(p_1 \otimes_P p_2)$

for all $p_1, p_2 \in P$.
There are strengthenings of these conditions that are also important. If f satisfies the following conditions, it is called a *strong monoidal monotone*:

(a') $I_Q \cong f(I_P)$,
(b') $f(p_1) \otimes_Q f(p_2) \cong f(p_1 \otimes_P p_2)$;

and if it satisfies the following conditions it is called a *strict monoidal monotone*:

(a'') $I_Q = f(I_P)$,
(b'') $f(p_1) \otimes_Q f(p_2) = f(p_1 \otimes_P p_2)$.

Monoidal monotones are examples of *monoidal functors*, which we shall see various incarnations of throughout the book; see Definition 6.58. What we call monoidal monotones could also be called *lax monoidal monotones*, and there is a dual notion of *oplax monoidal monotones*, where the inequalities in (a) and (b) are reversed; we will not use oplaxity in this book.

Example 2.26. There is a monoidal monotone $i\colon (\mathbb{Z}, \leq, 0, +) \to (\mathbb{R}, \leq, 0, +)$, where $i(n) = n$ for all $n \in \mathbb{Z}$. It is clearly monotonic, $m \leq n$ implies $i(m) \leq i(n)$. It is even strict monoidal because $i(0) = 0$ and $i(m + n) = i(m) + i(n)$.

There is also a monoidal monotone $f\colon (\mathbb{R}, \leq, 0, +) \to (\mathbb{Z}, \leq, 0, +)$ going the other way. Here $f(x) := \lfloor x \rfloor$ is the floor function, e.g. $f(3.14) = 3$. It is monotonic because $x \leq y$ implies $f(x) \leq f(y)$. Also $f(0) = 0$ and $f(x) + f(y) \leq f(x + y)$, so it is a monoidal monotone. But it is not strict or even strong because $f(0.5) + f(0.5) \neq f(0.5 + 0.5)$.

Recall **Bool** $= (\mathbb{B}, \leq, \texttt{true}, \wedge)$ from Example 2.12 and **Cost** $= ([0, \infty], \geq, 0, +)$ from Example 2.21. There is a monoidal monotone $g\colon$ **Bool** $\to$ **Cost**, given by $g(\texttt{false}) := \infty$ and $g(\texttt{true}) := 0$.

Exercise 2.27.

1. Check that the map $g\colon (\mathbb{B}, \leq, \texttt{true}, \wedge) \to ([0, \infty], \geq, 0, +)$ presented above indeed
   - is monotonic,
   - satisfies condition (a) of Definition 2.25,
   - satisfies condition (b) of Definition 2.25.
2. Is g strict?   ◇

Exercise 2.28. Let **Bool** and **Cost** be as above, and consider the following quasi-inverse functions $d, u\colon [0, \infty] \to \mathbb{B}$ defined as follows:

$$d(x) := \begin{cases} \texttt{false} & \text{if } x > 0, \\ \texttt{true} & \text{if } x = 0, \end{cases} \qquad u(x) := \begin{cases} \texttt{false} & \text{if } x = \infty, \\ \texttt{true} & \text{if } x < \infty. \end{cases}$$

1. Is d monotonic?
2. Does d satisfy conditions (a) and (b) of Definition 2.25?
3. Is d strict?
4. Is u monotonic?
5. Does u satisfy conditions (a) and (b) of Definition 2.25?
6. Is u strict?   ◇

Exercise 2.29.

1. Is $(\mathbb{N}, \leq, 1, *)$ a monoidal preorder, where $*$ is the usual multiplication of natural numbers?
2. If not, why not? If so, does there exist a monoidal monotone $(\mathbb{N}, \leq, 0, +) \to (\mathbb{N}, \leq, 1, *)$? If not, why not? If so, find it.
3. Is $(\mathbb{Z}, \leq, 1, *)$ a monoidal preorder?   ◇

2.3   Enrichment

In this section we will introduce $\mathcal{V}$-categories, where $\mathcal{V}$ is a symmetric monoidal pre-order. We shall see that **Bool**-categories are preorders, and that **Cost**-categories are a nice generalization of the notion of metric space.

2.3.1 $\mathcal{V}$-Categories

While $\mathcal{V}$-categories can be defined even when $\mathcal{V}$ is not symmetric, i.e. just obeys conditions (a)–(c) of Definition 2.1, certain things don't work quite right. For example, we shall see later in Exercise 2.54 that the symmetry condition is necessary in order for products of $\mathcal{V}$-categories to exist. Anyway, here's the definition.

Definition 2.30.   Let $\mathcal{V} = (V, \leq, I, \otimes)$ be a symmetric monoidal preorder. A $\mathcal{V}$-*category* $\mathcal{X}$ consists of two constituents, satisfying two properties. To specify $\mathcal{X}$,

(i) one specifies a set $\mathrm{Ob}(\mathcal{X})$, elements of which are called *objects*;
(ii) for every two objects x, y, one specifies an element $\mathcal{X}(x, y) \in V$, called the *hom-object*.[2]

The above constituents are required to satisfy two properties:

(a) for every object $x \in \mathrm{Ob}(\mathcal{X})$ we have $I \leq \mathcal{X}(x, x)$,
(b) for every three objects $x, y, z \in \mathrm{Ob}(\mathcal{X})$, we have $\mathcal{X}(x, y) \otimes \mathcal{X}(y, z) \leq \mathcal{X}(x, z)$.

We call $\mathcal{V}$ the *base of the enrichment* for $\mathcal{X}$ or say that $\mathcal{X}$ is *enriched* in $\mathcal{V}$.

Example 2.31. As we shall see in the next subsection, from every preorder we can construct a **Bool**-category, and vice versa. So, to get a feel for $\mathcal{V}$-categories, let us consider the preorder generated by the Hasse diagram

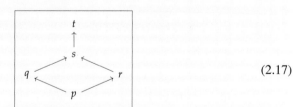

$$(2.17)$$

How does this correspond to a **Bool**-category $\mathcal{X}$? Well, the objects of $\mathcal{X}$ are simply the elements of the preorder, i.e. $\mathrm{Ob}(\mathcal{X}) = \{p, q, r, s, t\}$. Next, for every pair of objects

---

[2] The word "hom" is short for *homomorphism* and reflects the origins of this subject. A more descriptive name for $\mathcal{X}(x, y)$ might be *mapping object*, but we use "hom" mainly because it is an important jargon word to know in the field.

(x, y) we need an element of $\mathbb{B} = \{\texttt{false}, \texttt{true}\}$: simply take $\texttt{true}$ if $x \leq y$, and $\texttt{false}$ otherwise. So, for example, since $s \leq t$ and $t \not\leq s$, we have $\mathcal{X}(s, t) = \texttt{true}$ and $\mathcal{X}(t, s) = \texttt{false}$. Recalling from Example 2.12 that the monoidal unit I of **Bool** is $\texttt{true}$, it's straightforward to check that this obeys both (a) and (b), so we have a **Bool**-category.

In general, it's sometimes convenient to represent a $\mathcal{V}$-category $\mathcal{X}$ with a square matrix. The rows and columns of the matrix correspond to the objects of $\mathcal{X}$, and the (x, y)th entry is simply the hom-object $\mathcal{X}(x, y)$. So, for example, the above preorder in Eq. (2.17) can be represented by the matrix

$\cdot \leq \cdot$	p	q	r	s	t
p	true	true	true	true	true
q	false	true	false	true	true
r	false	false	true	true	true
s	false	false	false	true	true
t	false	false	false	false	true

2.3.2 Preorders as **Bool**-Categories

Our colleague Peter Gates has called category theory "a primordial ooze," because so much of it can be defined in terms of other parts of it. There is nowhere to rightly call the beginning, because that beginning can be defined in terms of something else. So be it; this is part of the fun.

Theorem 2.32. There is a one-to-one correspondence between preorders and **Bool**-categories.

Here we find ourselves in the ooze, because we are saying that preorders are the same as **Bool**-categories, whereas **Bool** is itself a preorder. "So then **Bool** is like ... enriched in itself?" Yes, every preorder, including **Bool**, is enriched in **Bool**, as we shall now see.

Proof of Theorem 2.32. Let's check that we can construct a preorder from any **Bool**-category. Since $\mathbb{B} = \{\texttt{false}, \texttt{true}\}$, Definition 2.30 says a **Bool**-category consists of two things:

(i) a set $\mathrm{Ob}(\mathcal{X})$,
(ii) for every $x, y \in \mathrm{Ob}(\mathcal{X})$ an element $\mathcal{X}(x, y) \in \mathbb{B}$, i.e. the element $\mathcal{X}(x, y)$ is either $\texttt{true}$ or $\texttt{false}$.

We will use these two things to begin forming a preorder whose elements are the objects of $\mathcal{X}$. So let's call the preorder $(X, \leq)$, and let $X := \mathrm{Ob}(\mathcal{X})$. For the $\leq$ relation, let's declare $x \leq y$ iff $\mathcal{X}(x, y) = \texttt{true}$. We have the makings of a preorder, but for it to work, the $\leq$ relation must be reflexive and transitive. Let's see if we get these from the properties guaranteed by Definition 2.30:

(a) for every element $x \in X$ we have `true` $\leq \mathfrak{X}(x, x)$,
(b) for every three elements $x, y, z \in X$ we have $\mathfrak{X}(x, y) \wedge \mathfrak{X}(y, z) \leq \mathfrak{X}(x, z)$.

For $b \in$ **Bool**, if `true` $\leq b$ then $b =$ `true`, so the first statement says $\mathfrak{X}(x, x) =$ `true`, which means $x \leq x$. For the second statement, one can consult Eq. (2.16). Since `false` $\leq b$ for all $b \in \mathbb{B}$, the only way statement (b) has any force is if $\mathfrak{X}(x, y) =$ `true` and $\mathfrak{X}(y, z) =$ `true`, in which case it forces $\mathfrak{X}(x, z) =$ `true`. This condition exactly translates as saying that $x \leq y$ and $y \leq z$ implies $x \leq z$. Thus we have obtained reflexivity and transitivity from the two axioms of **Bool**-categories.

In Example 2.31, we constructed a **Bool**-category from a preorder. We leave it to the reader to generalize this example and show that the two constructions are inverses; see Exercise 2.33.   □

Exercise 2.33.

1. Start with a preorder $(P, \leq)$, and use it to define a **Bool**-category as we did in Example 2.31. In the proof of Theorem 2.32 we showed how to turn that **Bool**-category back into a preorder. Show that, doing so, you get the preorder you started with.
2. Similarly, show that if you turn a **Bool**-category into a preorder using the above proof, and then turn the preorder back into a **Bool**-category using your method, you get the **Bool**-category you started with.   ◇

We now discuss a beautiful application of the notion of enriched categories: metric spaces.

2.3.3   Lawvere Metric Spaces

Metric spaces offer a precise way to describe spaces of points, each pair of which is separated by some distance. Here is the usual definition:

Definition 2.34.   A *metric space* (X, d) consists of:

(i) a set X, elements of which are called *points*,
(ii) a function $d \colon X \times X \to \mathbb{R}_{\geq 0}$, where $d(x, y)$ is called the *distance between x and y*.

These constituents must satisfy four properties:

(a) for every $x \in X$, we have $d(x, x) = 0$,
(b) for every $x, y \in X$, if $d(x, y) = 0$ then $x = y$,
(c) for every $x, y \in X$, we have $d(x, y) = d(y, x)$,
(d) for every $x, y, z \in X$, we have $d(x, y) + d(y, z) \geq d(x, z)$.

The fourth property is called the *triangle inequality*.
    If we ask instead in (ii) for a function $d \colon X \times X \to [0, \infty] = \mathbb{R}_{\geq 0} \cup \{\infty\}$, we call (X, d) an *extended* metric space.

The triangle inequality says that when plotting a route from x to z, the distance is always at most what you get by choosing an intermediate point y and going $x \to y \to z$.

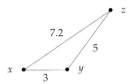

It can be invoked in three different ways in the above picture: $3 + 5 \geq 7.2$, but also $5 + 7.2 \geq 3$, and $3 + 7.2 \geq 5$. Oh yeah, and $5 + 3 \geq 7.2$, $7.2 + 5 \geq 3$, and $7.2 + 3 \geq 5$.

The triangle inequality wonderfully captures something about distance, as does the fact that $d(x, x) = 0$ for any x. However, the other two conditions are not quite as general as we would like. Indeed, there are many examples of things that "should" be metric spaces, but which do not satisfy conditions (b) or (c) of Definition 2.34.

For example, what if we take X to be places in your neighborhood, but instead of measuring distance, you want $d(x, y)$ to measure *effort* to get from x to y. Then if there are any hills, the symmetry axiom, $d(x, y) \stackrel{?}{=} d(y, x)$, fails: it's easier to get from x downhill to y then to go from y uphill to x.

Another way to find a model that breaks the symmetry axiom is to imagine that the elements of X are not points, but whole regions such as the US, Spain, and Boston. Say that the distance from region A to region B is understood using the setup "I will put you in an arbitrary part of A and you just have to get anywhere in B; what is the distance in the worst-case scenario?" So $d(\text{US}, \text{Spain})$ is the distance from somewhere in the western US to the western tip of Spain: you just have to get into Spain, but you start in the worst possible part of the US for doing so.

Exercise 2.35. Which distance is bigger under the above description, $d(\text{Spain}, \text{US})$ or $d(\text{US}, \text{Spain})$? ◇

This notion of distance, which is strongly related to something called *Hausdorff distance*,[3] will again satisfy the triangle inequality, but it violates the symmetry condition. It also violates another condition, because $d(\text{Boston}, \text{US}) = 0$. No matter where you are in Boston, the distance to the nearest point of the US is 0. On the other hand, $d(\text{US}, \text{Boston}) \neq 0$.

Finally, one can imagine a use for distances that are not finite. In terms of my effort, the distance from here to Pluto is ∞, and it would not be any better if Pluto was still a

---

[3] The Hausdorff distance gives a metric on the set of all subsets $U \subseteq X$ of a given metric space (X, d). One first defines

$$d_L(U, V) := \sup_{u \in U} \inf_{v \in V} d(u, v),$$

and this is exactly the formula we intend above; the result will be a Lawvere metric space. However, if one wants the Hausdorff distance to define a (symmetric) metric, as in Definition 2.34, one must take the above formula and symmetrize it: $d(U, V) := \max(d_L(U, V), d_L(V, U))$. We happen to see the unsymmetrized notion as more interesting.

planet. Similarly, in terms of Hausdorff distance, discussed above, the distance between two regions is often infinite, e.g. the distance between $\{r \in \mathbb{R} \mid r < 0\}$ and $\{0\}$ as subsets of $(\mathbb{R}, d)$ is infinite.

When we drop conditions (b) and (c) and allow for infinite distances, we get the following relaxed notion of metric space, first proposed by Lawvere. Recall the symmetric monoidal preorder $\mathbf{Cost} = ([0, \infty], \geq, 0, +)$ from Example 2.21.

Definition 2.36. A *Lawvere metric space* is a $\mathbf{Cost}$-category.

This is a very compact definition, but it packs a punch. Let's work out what it means, by relating it to the usual definition of metric space. By Definition 2.30, a $\mathbf{Cost}$-category $\mathcal{X}$ consists of:

(i) a set $\mathrm{Ob}(\mathcal{X})$,
(ii) for every $x, y \in \mathrm{Ob}(\mathcal{X})$ an element $\mathcal{X}(x, y) \in [0, \infty]$.

Here the set $\mathrm{Ob}(\mathcal{X})$ is playing the role of the set of points, and $\mathcal{X}(x, y) \in [0, \infty]$ is playing the role of distance, so let's write a little translator:

$$X := \mathrm{Ob}(\mathcal{X}), \qquad d(x, y) := \mathcal{X}(x, y).$$

The properties of a category enriched in $\mathbf{Cost}$ are:

(a) $0 \geq d(x, x)$ for all $x \in X$,
(b) $d(x, y) + d(y, z) \geq d(x, z)$ for all $x, y, z \in X$.

Since $d(x, x) \in [0, \infty]$, if $0 \geq d(x, x)$ then $d(x, x) = 0$. So the first condition is equivalent to the first condition from Definition 2.34, namely $d(x, x) = 0$. The second condition is the triangle inequality.

Example 2.37. The set $\mathbb{R}$ of real numbers can be given a metric space structure, and hence a Lawvere metric space structure. Namely $d(x, y) := |y - x|$, the absolute value of the difference. So $d(3, 7) = 4$.

Exercise 2.38. Consider the symmetric monoidal preorder $(\mathbb{R}_{\geq 0}, \geq, 0, +)$, which is almost the same as $\mathbf{Cost}$, except it does not include ∞. How would you characterize the difference between a Lawvere metric space and a $(\mathbb{R}_{\geq 0}, \geq, 0, +)$-category in the sense of Definition 2.30? ◇

Presenting metric spaces with weighted graphs

Just as one can convert a Hasse diagram into a preorder, one can convert any weighted graph – a graph whose edges are labeled with numbers $w \geq 0$ – into a Lawvere metric space. In fact, we will consider these as graphs labeled with elements of $[0, \infty]$, and more precisely call them **Cost**-weighted graphs.[4]

One might think of a **Cost**-weighted graph as describing a city with some one-way roads (a two-way road is modeled as two one-way roads), each having some effort-to-traverse, which for simplicity we just call distance. For example, consider the following weighted graphs:

$$X := \quad \text{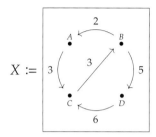} \qquad\qquad \text{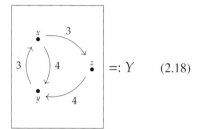} =: Y \qquad (2.18)$$

Given a weighted graph, one forms a metric d_X on its set X of vertices by setting $d(p, q)$ to be the length of the shortest path from p to q. For example, here is the the table of distances for Y

$$
\begin{array}{c|ccc}
d(\nearrow) & x & y & z \\
\hline
x & 0 & 4 & 3 \\
y & 3 & 0 & 6 \\
z & 7 & 4 & 0
\end{array}
\qquad (2.19)
$$

Exercise 2.39. Copy and complete the following table of distances in the weighted graph X from Eq. (2.18).

$$
\begin{array}{c|cccc}
d(\nearrow) & A & B & C & D \\
\hline
A & 0 & ? & ? & ? \\
B & 2 & ? & 5 & ? \\
C & ? & ? & ? & ? \\
D & ? & ? & ? & ?
\end{array}
$$

$\diamond$

Above we converted a weighted graph G, e.g. as shown in Eq. (2.18), into a table of distances, but this takes a bit of thinking. There is a more direct construction for taking G and getting a square matrix M_G, whose rows and columns are indexed by the vertices of G. To do so, set M_G to be 0 along the diagonal, to be ∞ wherever an edge is missing, and to be the edge weight if there is an edge.

---

[4] This generalizes Hasse diagrams, which we could call **Bool**-weighted graphs – the edges of a Hasse diagram are thought of as weighted with `true`; we simply ignore any edges that are weighted with `false`, and neglect to even draw them!

For example, the matrix associated to Y in Eq. (2.18) would be

$$
M_Y := \begin{array}{c|ccc}
\nearrow & x & y & z \\
\hline
x & 0 & 4 & 3 \\
y & 3 & 0 & \infty \\
z & \infty & 4 & 0
\end{array}
\tag{2.20}
$$

As soon as you see how we did this, you'll understand that it takes no thinking to turn a weighted graph G into a matrix M_G in this way. We shall see in Section 2.5.3 that the more difficult "distance matrices" d_Y, such as (2.19), can be obtained from the easy graph matrices M_Y, such as (2.20), by repeating a certain sort of "matrix multiplication."

Exercise 2.40.   Copy and complete the matrix M_X associated to the graph X in Eq. (2.18):

$$
M_X = \begin{array}{c|cccc}
\nearrow & A & B & C & D \\
\hline
A & 0 & ? & ? & ? \\
B & 2 & 0 & \infty & ? \\
C & ? & ? & ? & ? \\
D & ? & ? & ? & ?
\end{array}
$$

$\diamond$

2.3.4 $\mathcal{V}$-Variations on Preorders and Metric Spaces

We have told the story of **Bool** and **Cost**. But in Section 2.2.4 we gave examples of many other monoidal preorders, and each one serves as the base of enrichment for a kind of enriched category. Which of them are useful? Something only becomes useful when someone finds a use for it. We will find uses for some and not others, though we encourage readers to think about what it would mean to enrich in the various monoidal categories discussed above; maybe they can find a use we have not explored.

Exercise 2.41.   Recall the monoidal preorder **NMY** $:= (P, \leq, \mathrm{yes}, \min)$ from Exercise 2.18. Interpret what a **NMY**-category is. $\diamond$

In the next two exercises, we use $\mathcal{V}$-weighted graphs to construct $\mathcal{V}$-categories. This is possible because we will use preorders that, like **Bool** and **Cost**, have joins.

Exercise 2.42.   Let M be a set and let $\mathcal{M} := (P(M), \subseteq, M, \cap)$ be the monoidal preorder whose elements are subsets of M.

Someone gives the following interpretation, "for any set M, imagine it as the set of modes of transportation (e.g. car, boat, foot). Then an $\mathcal{M}$-category X tells you all the modes that will get you from a all the way to b, for any two points $a, b \in \mathrm{Ob}(X)$."

1. Draw a graph with four vertices and four or five edges, each labeled with a subset of $M = \{\text{car, boat, foot}\}$.
2. From this graph is it possible to construct an $\mathcal{M}$-category, where the hom-object from x to y is computed as follows: for each path p from x to y, take the intersection of the sets labeling the edges in p. Then, take the union of the these sets over all paths p from x to y. Write out the corresponding 4×4-matrix of hom-objects, and convince yourself that this is indeed an $\mathcal{M}$-category.
3. Does the person's interpretation look right, or is it subtly mistaken somehow?   ◇

Exercise 2.43.   Consider the monoidal preorder $\mathcal{W} := (\mathbb{N} \cup \{\infty\}, \leq, \infty, \min)$.

1. Draw a small graph labeled by elements of $\mathbb{N} \cup \{\infty\}$.
2. Write out the matrix whose rows and columns are indexed by the nodes in the graph, and whose (x, y)th entry is given by the *maximum* over all paths p from x to y of the *minimum* edge label in p.
3. Prove that this matrix is the matrix of hom-objects for a $\mathcal{W}$-category. This will give you a feel for how $\mathcal{W}$ works.
4. Make up an interpretation, like that in Exercise 2.42, for how to imagine enrichment in $\mathcal{W}$.   ◇

2.4   Constructions on $\mathcal{V}$-Categories

Now that we have a good intuition for what $\mathcal{V}$-categories are, we give three examples of what can be done with $\mathcal{V}$-categories. The first (Section 2.4.1) is known as change of base. This allows us to use a monoidal monotone $f : \mathcal{V} \to \mathcal{W}$ to construct $\mathcal{W}$-categories from $\mathcal{V}$-categories. The second construction (Section 2.4.2), that of $\mathcal{V}$-functors, allows us to complete the analogy: a preorder is to a **Bool**-category as a monotone map is to what? The third construction (Section 2.4.2) is known as a $\mathcal{V}$-product and gives us a way of combining two $\mathcal{V}$-categories.

2.4.1   Changing the Base of Enrichment

Any monoidal monotone $\mathcal{V} \to \mathcal{W}$ between symmetric monoidal preorders lets us convert $\mathcal{V}$-categories into $\mathcal{W}$-categories.

Construction 2.44.   Let $f : \mathcal{V} \to \mathcal{W}$ be a monoidal monotone. Given a $\mathcal{V}$-category $\mathcal{C}$, one forms the associated $\mathcal{W}$-category, say $\mathcal{C}_f$, as follows.

(i) We take the same objects: $\mathrm{Ob}(\mathcal{C}_f) := \mathrm{Ob}(\mathcal{C})$.
(ii) For any $c, d \in \mathrm{Ob}(\mathcal{C})$, put $\mathcal{C}_f(c, d) := f(\mathcal{C}(c, d))$.

This construction $\mathcal{C}_f$ does indeed obey the definition of a $\mathcal{W}$-category, as can be seen by applying Definition 2.25 (of monoidal monotone) and Definition 2.30 (of $\mathcal{V}$-category):

(a) for every $c \in \mathcal{C}$, we have

$$
\begin{aligned}
I_W &\leq f(I_V) && (f \text{ is monoidal monotone}) \\
&\leq f(\mathcal{C}(c, c)) && (\mathcal{C} \text{ is } \mathcal{V}\text{-category}) \\
&= \mathcal{C}_f(c, c) && (\text{definition of } \mathcal{C}_f);
\end{aligned}
$$

(b) for every $c, d, e \in \mathrm{Ob}(\mathcal{C})$, we have

$$
\begin{aligned}
\mathcal{C}_f(c, d) \otimes_W \mathcal{C}_f(d, e) &= f(\mathcal{C}(c, d)) \otimes_W f(\mathcal{C}(d, e)) && (\text{definition of } \mathcal{C}_f) \\
&\leq f\big(\mathcal{C}(c, d) \otimes_V \mathcal{C}(d, e)\big) && (f \text{ is monoidal monotone}) \\
&\leq f(\mathcal{C}(c, e)) && (\mathcal{C} \text{ is } \mathcal{V}\text{-category}) \\
&= \mathcal{C}_f(c, e) && (\text{definition of } \mathcal{C}_f).
\end{aligned}
$$

Example 2.45. As an example, consider the function $f \colon [0, \infty] \to \{\texttt{true}, \texttt{false}\}$ given by

$$
f(x) := \begin{cases} \texttt{true} & \text{if } x = 0, \\ \texttt{false} & \text{if } x > 0. \end{cases} \tag{2.21}
$$

It is easy to check that f is monotonic and that f preserves the monoidal product and monoidal unit; that is, it's easy to show that f is a monoidal monotone. (Recall Exercise 2.28.)

Thus f lets us convert Lawvere metric spaces into preorders.

Exercise 2.46.   Recall the "regions of the world" Lawvere metric space from Exercise 2.35 and the text above it. We just learned that, using the monoidal monotone f in Eq. (2.21), we can convert it to a preorder. Draw the Hasse diagram for the preorder corresponding to the regions US, Spain, and Boston. How could you interpret this preorder relation?   ◇

Exercise 2.47.

1. Find another monoidal monotone $g \colon \mathbf{Cost} \to \mathbf{Bool}$ different from the one defined in Eq. (2.21).
2. Using Construction 2.44, both your monoidal monotone g and the monoidal monotone f in Eq. (2.21) can be used to convert a Lawvere metric space into a preorder. Find a Lawvere metric space $\mathcal{X}$ on which they give different answers, $\mathcal{X}_f \neq \mathcal{X}_g$.   ◇

2.4.2   Enriched Functors

The notion of functor provides the most important type of relationship between categories.

Definition 2.48. Let $\mathcal{V}$ be a symmetric monoidal preorder and let $\mathcal{X}$ and $\mathcal{Y}$ be $\mathcal{V}$-categories. A $\mathcal{V}$-*functor from* $\mathcal{X}$ *to* $\mathcal{Y}$, denoted $F: \mathcal{X} \to \mathcal{Y}$, consists of one constituent

(i) a function $F: \mathrm{Ob}(\mathcal{X}) \to \mathrm{Ob}(\mathcal{Y})$

subject to one constraint

(a) for all $x_1, x_2 \in \mathrm{Ob}(\mathcal{X})$, one has $\mathcal{X}(x_1, x_2) \leq \mathcal{Y}(F(x_1), F(x_2))$.

Example 2.49. For example, we have said several times – e.g. in Theorem 2.32 – that preorders are **Bool**-categories, where $\mathcal{X}(x_1, x_2) = \texttt{true}$ is denoted $x_1 \leq x_2$. One would hope that monotone maps between preorders would correspond exactly to **Bool**-functors, and that's true. A monotone map $(X, \leq_X) \to (Y, \leq_Y)$ is a function $F: X \to Y$ such that, for every $x_1, x_2 \in X$, if $x_1 \leq_X x_2$ then $F(x_1) \leq_Y F(x_2)$. In other words, we have

$$\mathcal{X}(x_1, x_2) \leq \mathcal{Y}(F(x_1), F(x_2)),$$

where the above $\leq$ takes place in the enriching category $\mathcal{V} = \textbf{Bool}$; this is exactly the condition from Definition 2.48.

Remark 2.50. In fact, we have what is called an *equivalence* of categories between the category of preorders and the category of **Bool**-categories. In the next chapter we will develop the ideas necessary to state what this means precisely (Remark 3.50).

Example 2.51. Lawvere metric spaces are **Cost**-categories. The definition of **Cost**-functor should hopefully return a nice notion – a "friend" – from the theory of metric spaces, and it does: it recovers the notion of Lipschitz function. A Lipschitz (or more precisely, 1-Lipschitz) function is one under which the distance between any pair of points does not increase. That is, given Lawvere metric spaces (X, d_X) and (Y, d_Y), a **Cost**-functor between them is a function $F: X \to Y$ such that for every $x_1, x_2 \in X$ we have $d_X(x_1, x_2) \geq d_Y(F(x_1), F(x_2))$.

Exercise 2.52. The concepts of opposite, dagger, and skeleton (see Examples 1.53 and 1.67 and Remark 1.30) extend from preorders to $\mathcal{V}$-categories. The *opposite* of a $\mathcal{V}$-category $\mathcal{X}$ is denoted $\mathcal{X}^{\mathrm{op}}$ and is defined by

(i) $\mathrm{Ob}(\mathcal{X}^{\mathrm{op}}) := \mathrm{Ob}(\mathcal{X})$, and
(ii) for all $x, y \in \mathcal{X}$, we have $\mathcal{X}^{\mathrm{op}}(x, y) := \mathcal{X}(y, x)$.

A $\mathcal{V}$-category $\mathcal{X}$ is a *dagger* $\mathcal{V}$-category if the identity function is a $\mathcal{V}$-functor $\dagger: \mathcal{X} \to \mathcal{X}^{\mathrm{op}}$. And a *skeletal* $\mathcal{V}$-category is one in which if $I \leq \mathcal{X}(x, y)$ and $I \leq \mathcal{X}(y, x)$, then $x = y$.

Recall that an extended metric space (X, d) is a Lawvere metric space with two extra properties; see properties (b) and (c) in Definition 2.34.

1. Show that a skeletal dagger **Cost**-category is an extended metric space.
2. Use Exercise 1.68 to make sense of the following analogy: "preorders are to sets as Lawvere metric spaces are to extended metric spaces."   ◇

2.4.3   Product $\mathcal{V}$-Categories

If $\mathcal{V} = (V, \leq, I, \otimes)$ is a symmetric monoidal preorder and $\mathcal{X}$ and $\mathcal{Y}$ are $\mathcal{V}$-categories, then we can define their $\mathcal{V}$-product, which is a new $\mathcal{V}$-category.

Definition 2.53.   Let $\mathcal{X}$ and $\mathcal{Y}$ be $\mathcal{V}$-categories. Define their $\mathcal{V}$-*product*, or simply *product*, to be the $\mathcal{V}$-category $\mathcal{X} \times \mathcal{Y}$ with

(i) $Ob(\mathcal{X} \times \mathcal{Y}) := Ob(\mathcal{X}) \times Ob(\mathcal{Y})$,
(ii) $(\mathcal{X} \times \mathcal{Y})\big((x, y), (x', y')\big) := \mathcal{X}(x, x') \otimes \mathcal{Y}(y, y')$,

for two objects (x, y) and (x', y') in $Ob(\mathcal{X} \times \mathcal{Y})$.

Product $\mathcal{V}$-categories are indeed $\mathcal{V}$-categories (Definition 2.30); see Exercise 2.54.

Exercise 2.54.   Let $\mathcal{X} \times \mathcal{Y}$ be the $\mathcal{V}$-product of $\mathcal{V}$-categories as in Definition 2.53.

1. Check that for every object $(x, y) \in Ob(\mathcal{X} \times \mathcal{Y})$ we have $I \leq (\mathcal{X} \times \mathcal{Y})\big((x, y), (x, y)\big)$.
2. Check that for every three objects (x_1, y_1), (x_2, y_2), and (x_3, y_3), we have

$$(\mathcal{X} \times \mathcal{Y})\big((x_1, y_1), (x_2, y_2)\big) \otimes (\mathcal{X} \times \mathcal{Y})\big((x_2, y_2), (x_3, y_3)\big) \leq (\mathcal{X} \times \mathcal{Y})\big((x_1, y_1), (x_3, y_3)\big).$$

3. We said at the start of Section 2.3.1 that the symmetry of $\mathcal{V}$ (condition (d) of Definition 2.1) would be required here. Point out exactly where that condition is used.   ◇

When taking the product of two preorders $(P, \leq_P) \times (Q, \leq_Q)$, as first described in Example 1.51, we say that $(p_1, q_1) \leq (p_2, q_2)$ iff both $p_1 \leq p_2$ AND $q_1 \leq q_2$; the AND is the monoidal product $\otimes$ from **Bool**. Thus the product of preorders is an example of a **Bool**-product.

Example 2.55. Let $\mathcal{X}$ and $\mathcal{Y}$ be the Lawvere metric spaces (i.e. **Cost**-categories) defined by the following weighted graphs:

$$\mathcal{X} := \boxed{\begin{array}{ccccc} A & 2 & B & 3 & C \\ \bullet & \longrightarrow & \bullet & \longrightarrow & \bullet \end{array}} \qquad 5 \left(\begin{array}{c} \overset{p}{\bullet} \\ \circlearrowleft \\ \underset{q}{\bullet} \end{array}\right) 8 \ =: \mathcal{Y} \qquad (2.22)$$

Their product is defined by taking the product of their sets of objects, so there are six objects in $\mathcal{X} \times \mathcal{Y}$. And the distance $d_{X \times Y}((x, y), (x', y'))$ between any two points is given by the sum $d_X(x, x') + d_Y(y, y')$.

Examine the following graph, and make sure you understand how easy it is to derive from the weighted graphs for $\mathcal{X}$ and $\mathcal{Y}$ in Eq. (2.22):

$$
\mathcal{X} \times \mathcal{Y} =
\begin{array}{|c|}
\hline
\begin{array}{ccccc}
(A,p) & \xrightarrow{2} & (B,p) & \xrightarrow{3} & (C,p) \\
\bullet & & \bullet & & \bullet \\
5 \Big(\Big) 8 & 5 & \Big(\Big) 8 & 5 & \Big(\Big) 8 \\
\bullet & \xrightarrow{2} & \bullet & \xrightarrow{3} & \bullet \\
(A,q) & & (B,q) & & (C,q)
\end{array} \\
\hline
\end{array}
$$

Exercise 2.56. Consider $\mathbb{R}$ as a Lawvere metric space, i.e. as a **Cost**-category (see Example 2.37). Form the **Cost**-product $\mathbb{R} \times \mathbb{R}$. What is the distance from $(5, 6)$ to $(-1, 4)$? Hint: apply Definition 2.53; the answer is not $\sqrt{40}$. $\diamond$

In terms of matrices, $\mathcal{V}$-products are also quite straightforward. They generalize what is known as the Kronecker product of matrices. The matrices for $\mathcal{X}$ and $\mathcal{Y}$ in Eq. (2.22) are shown below

$\mathcal{X}$	A	B	C
A	0	2	5
B	∞	0	3
C	∞	∞	0

$\mathcal{Y}$	p	q
p	0	5
q	8	0

and their product is:

$\mathcal{X} \times \mathcal{Y}$	(A, p)	(B, p)	(C, p)	(A, q)	(B, q)	(C, q)
(A, p)	0	2	5	5	7	10
(B, p)	∞	0	3	∞	5	8
(C, p)	∞	∞	0	∞	∞	5
(A, q)	8	10	13	0	2	5
(B, q)	∞	8	11	∞	0	3
(C, q)	∞	∞	8	∞	∞	0

We have drawn the product matrix as a block matrix, where there is one block – shaped like $\mathcal{X}$ – for every entry of $\mathcal{Y}$. Make sure you can see each block as the $\mathcal{X}$-matrix shifted by an entry in $\mathcal{Y}$. This comes directly from the formula from Definition 2.53 and the fact that the monoidal product in **Cost** is $+$.

2.5   Computing Presented $\mathcal{V}$-Categories with Matrix Multiplication

In Section 2.3.3 we promised a straightforward way to construct the matrix representation of a **Cost**-category from a **Cost**-weighted graph. To do this, we use a generalized matrix multiplication. We shall show that this works, not just for **Cost**, but also for **Bool**, and many other monoidal preorders. The property required of the preorder is that of being a unital, commutative quantale. These are preorders with all joins, plus one additional ingredient, being *monoidal closed*, which we define next, in Section 2.5.1. The definition of a quantale will be given in Section 2.5.2.

2.5.1   Monoidal Closed Preorders

The definition of $\mathcal{V}$-category makes sense for any symmetric monoidal preorder $\mathcal{V}$. But that does not mean that any base of enrichment $\mathcal{V}$ is as useful as any other. In this section we define closed monoidal categories, which in particular enrich themselves! "Before you can really enrich others, you should really enrich yourself."

> **Definition 2.57.**   A symmetric monoidal preorder $\mathcal{V} = (V, \leq, I, \otimes)$ is called *symmetric monoidal closed* (or just *closed*) if, for every two elements $v, w \in V$, there is an element $v \multimap w$ in $\mathcal{V}$, called the *hom-element*, with the property
>
> $$(a \otimes v) \leq w \quad \text{iff} \quad a \leq (v \multimap w) \tag{2.23}$$
>
> for all $a, v, w \in V$.

Remark 2.58.   The term 'closed' refers to the fact that a hom-element can be constructed for any two elements, so the preorder can be seen as closed under the operation of "taking homs." In later chapters we'll meet the closely related concepts of compact closed categories (Definition 4.44) and cartesian closed categories (Section 7.2.1) that make this idea more precise. See especially Exercise 7.8.

One can consider the hom-element $v \multimap w$ as a kind of "single-use v-to-w converter." So Eq. (2.23) says that a and v are enough to get w if and only if a is enough to get a single-use v-to-w converter.

Exercise 2.59.   Condition (2.23) says precisely that there is a Galois connection in the sense of Definition 1.90. Let's prove this fact. In particular, we'll prove that a monoidal preorder is monoidal closed iff, given any $v \in V$, the map $(- \otimes v) \colon V \to V$ given by multiplying with v has a right adjoint. We write this right adjoint $(v \multimap -) \colon V \to V$.

1. Using Definition 2.1, show that $(- \otimes v)$ is monotone.
2. Supposing that $\mathcal{V}$ is closed, show that for all $v, w \in V$ we have $\big((v \multimap w) \otimes v\big) \leq w$.
3. Using part 2, show that $(v \multimap -)$ is monotone.
4. Conclude that a symmetric monoidal preorder is closed if and only if the monotone map $(- \otimes v)$ has a right adjoint.   ◇

Example 2.60. The monoidal preorder **Cost** $= ([0, \infty], \geq, 0, +)$ is monoidal closed. Indeed, for any $x, y \in [0, \infty]$, define $x \multimap y := \max(0, y - x)$. Then, for any $a, x, y \in [0, \infty]$, we have

$$a + x \geq y \quad \text{iff} \quad a \geq y - x \quad \text{iff} \quad \max(0, a) \geq \max(0, y - x) \quad \text{iff} \quad a \geq (x \multimap y)$$

so $\multimap$ satisfies the condition of Eq. (2.23).

Note that we have not considered subtraction in **Cost** before; we can in fact use monoidal closure to *define* subtraction in terms of the order and monoidal structure!

Exercise 2.61. Show that **Bool** $= (\mathbb{B}, \leq, \texttt{true}, \wedge)$ is monoidal closed. ◇

Example 2.62. A non-example is $(\mathbb{B}, \leq, \texttt{false}, \vee)$. Indeed, suppose we had a $\multimap$ operator as in Definition 2.57. Note that $\texttt{false} \leq p \multimap q$, for any p, q no matter what $\multimap$ is, because $\texttt{false}$ is less than everything. But using $a = \texttt{false}$, $p = \texttt{true}$, and $q = \texttt{false}$, we then get a contradiction: $(a \vee p) \not\leq q$ and yet $a \leq (p \multimap q)$.

Example 2.63. We started this chapter by talking about resource theories. What does the closed structure look like from that perspective? For example, in chemistry it would say that for every two material collections c, d one can form a material collection $c \multimap d$ with the property that, for any a, one has

$$a + c \to d \quad \text{iff} \quad a \to (c \multimap d).$$

Or, more down to earth, since we have the reaction $2H_2O + 2Na \to 2NaOH + H_2$, we must also have

$$2H_2O \to (2Na \multimap (2NaOH + H_2)).$$

So, from just two molecules of water, you can form a certain substance, and not many substances fit the bill – our preorder Mat of chemical materials is not closed.

But it is not so far-fetched: this hypothetical new substance $(2Na \multimap (2NaOH + H_2))$ is not really a substance, but a potential reaction: namely that of converting sodium to sodium-hydroxide-plus-hydrogen. Two molecules of water unlock that potential.

Proposition 2.64. Suppose $\mathcal{V} = (V, \leq, I, \otimes, \multimap)$ is a symmetric monoidal preorder that is closed. Then the following hold.

(a) For every $v \in V$, the monotone map $- \otimes v : (V, \leq) \to (V, \leq)$ is left adjoint to $v \multimap - : (V, \leq) \to (V, \leq)$.

(b) For any element $v \in V$ and set of elements $A \subseteq V$, if the join $\bigvee_{a \in A} a$ exists then so does $\bigvee_{a \in A} v \otimes a$ and we have

$$\left(v \otimes \bigvee_{a \in A} a \right) \cong \bigvee_{a \in A} (v \otimes a). \tag{2.24}$$

(c) For any $v, w \in V$, we have $v \otimes (v \multimap w) \leq w$.

(d) For any $v \in V$, we have $v \cong (I \multimap v)$.

(e) For any $u, v, w \in V$, we have $(u \multimap v) \otimes (v \multimap w) \leq (u \multimap w)$.

Proof. We go through the claims in order.

(a) The definition of $(- \otimes v)$ being left adjoint to $(v \multimap -)$ is exactly the condition Eq. (2.23); see Definition 1.90 and Exercise 2.59.

(b) This follows from (a), using the fact that left adjoints preserve joins (Proposition 1.104).

(c) This follows from (a), using the equivalent characterization of Galois connection in Proposition 1.101. More concretely, from reflexivity $(v \multimap w) \leq (v \multimap w)$, we obtain $(v \multimap w) \otimes v \leq w$ Eq. (2.23), and we are done by symmetry, which says $v \otimes (v \multimap w) = (v \multimap w) \otimes v$.

(d) Since $v \otimes I = v \leq v$, Eq. (2.23) says $v \leq (I \multimap v)$. For the other direction, we have $(I \multimap v) = I \otimes (I \multimap v) \leq v$ by (c).

(e) To obtain this inequality, we just need $u \otimes (u \multimap v) \otimes (v \multimap w) \leq w$. But this follows by two applications of (c). $\square$

One might read (c) as saying "if I have a v and a single-use v-to-w converter, I can have a w." One might read (d) as saying "having a v is the same as having a single-use nothing-to-v converter." And one might read (e) as saying "if I have a single-use u-to-v converter and a single-use v-to-w converter, I can get a single-use u-to-w converter."

Remark 2.65. We can consider $\mathcal{V}$ to be enriched in itself. That is, for every $v, w \in \mathrm{Ob}(\mathcal{V})$, we can define $\mathcal{V}(v, w) := (v \multimap w) \in V$. For this to really be an enrichment, we just need to check the two conditions of Definition 2.30. The first condition $I \leq \mathcal{X}(x, x) = (x \multimap x)$ is satisfied because $I \otimes x \leq x$. The second condition is satisfied by Proposition 2.64(e).

2.5.2   Quantales

To perform matrix multiplication over a monoidal preorder, we need one more thing: joins. These were first defined in Definition 1.76.

Definition 2.66. A *unital commutative quantale* is a symmetric monoidal closed preorder $\mathcal{V} = (V, \leq, I, \otimes, \multimap)$ that has all joins: $\bigvee A$ exists for every $A \subseteq V$. In particular, we often denote the empty join by $0 := \bigvee \varnothing$.

Whenever we speak of quantales in this book, we mean unital commutative quantales. We will try to remind the reader of that. There are also very interesting applications of noncommutative quantales; see Section 2.6.

Example 2.67. In Example 2.60, we saw that **Cost** is monoidal closed. To check whether **Cost** is a quantale, we take an arbitrary set of elements $A \subseteq [0, \infty]$ and ask if it has a join $\bigvee A$. To be a join, it needs to satisfy two properties:

a. $a \geq \bigvee A$ for all $a \in A$, and
b. if $b \in [0, \infty]$ is any element such that $a \geq b$ for all $a \in A$, then $\bigvee A \geq b$.

In fact we can define such a join: it is typically called the *infimum*, or greatest lower bound, of A.[5] For example, if $A = \{2, 3\}$ then $\bigvee A = 2$. We have joins for infinite sets too: if $B = \{2.5, 2.05, 2.005, \ldots\}$, its infimum is 2. Finally, in order to say that $([0, \infty], \geq)$ has all joins, we need a join to exist for the empty set $A = \varnothing$ too. The first condition becomes vacuous – there are no as in A – but the second condition says that for any $b \in [0, \infty]$ we have $\bigvee \varnothing \geq b$; this means $\bigvee \varnothing = \infty$.

Thus indeed $([0, \infty], \geq)$ has all joins, so **Cost** is a quantale.

Exercise 2.68.

1. What is $\bigvee \varnothing$, which we generally denote 0, in the case
   a. $\mathcal{V} = \textbf{Bool} = (\mathbb{B}, \leq, \texttt{true}, \wedge)$?
   b. $\mathcal{V} = \textbf{Cost} = ([0, \infty], \geq, 0, +)$?
2. What is the join $x \vee y$ in the case
   a. $\mathcal{V} = \textbf{Bool}$, and $x, y \in \mathbb{B}$ are booleans?
   b. $\mathcal{V} = \textbf{Cost}$, and $x, y \in [0, \infty]$ are distances? ◇

Exercise 2.69. Show that $\textbf{Bool} = (\mathbb{B}, \leq, \texttt{true}, \wedge)$ is a quantale. ◇

Exercise 2.70. Let S be a set and recall the power set monoidal preorder $(P(S), \subseteq, S, \cap)$ from Exercise 2.19. Is it a quantale? ◇

Remark 2.71. One can personify the notion of unital, commutative quantale as a kind of navigator. A navigator is someone who understands "getting from one place to another." Different navigators may care about or understand different aspects – whether one can get from A to B, how much time it will take, what modes of travel will work, etc. – but they certainly have some commonalities. Most importantly, a navigator needs to be able to read a map: given routes A to B and B to C, they understand how to get a route A to C. And they know how to search over the space of way-points to get from A to C. These will correspond to the monoidal product and the join operations, respectively.

---

[5] Here, by the infimum of a subset $A \subseteq [0, \infty]$, we mean infimum in the usual order on $[0, \infty]$: the largest number that is $\leq$ everything in A. For example, the infimum of $\{3.1, 3.01, 3.001, \ldots\}$ is 3. But note that this is the *supremum* in the reversed, $\geq$, order of **Cost**.

Proposition 2.72. Let $\mathcal{P} = (P, \leq)$ be a preorder. It has all joins iff it has all meets.

Proof. The joins (resp. meets) in $\mathcal{P}$ are the meets (resp. joins) in $\mathcal{P}^{\text{op}}$, so the two claims are dual: it suffices to show that if $\mathcal{P}$ has all joins then it has all meets.

Suppose $\mathcal{P}$ has all joins and suppose that $A \subseteq \mathcal{P}$ is a subset for which we want the meet. Consider the set $M_A := \{p \in P \mid p \leq a \text{ for all } a \in A\}$ of elements below everything in A. Let $m_A := \bigvee_{p \in M_A} p$ be their join. We claim that m_A is a meet for A.

We first need to know that for any $a \in A$ we have $m_A \leq a$, but this is by definition of join: since all $p \in M_A$ satisfy $p \leq a$, so does their join $m_A \leq a$. We second need to know that for any $m' \in P$ with $m' \leq a$ for all $a \in A$, we have $m' \leq m$. But every such m' is actually an element of M_A and m is their join, so $m' \leq m$. This completes the proof. $\qquad\square$

In particular, a quantale has all meets and all joins, even though we only define it to have all joins.

Remark 2.73. The notion of Hausdorff distance can be generalized, allowing the role of **Cost** to be taken by any quantale $\mathcal{V}$. If $\mathcal{X}$ is a $\mathcal{V}$-category with objects X, and $U \subseteq X$ and $V \subseteq X$, we can generalize the usual Hausdorff distance, on the left below, to the formula on the right:

$$d(U, V) := \sup_{u \in U} \inf_{v \in V} d(u, v), \qquad \mathcal{X}(U, V) := \bigwedge_{u \in U} \bigvee_{v \in V} \mathcal{X}(u, v).$$

For example, if $\mathcal{V} = \textbf{Bool}$, the Hausdorff distance between sub-preorders U and V answers the question "can I get into V from every $u \in U$," i.e. $\forall_{u \in U}.\, \exists_{v \in V}.\, u \leq v$.[6] Or for another example, use $\mathcal{V} = \mathsf{P}(M)$ with its interpretation as modes of transportation, as in Exercise 2.42. Then the Hausdorff distance $d(U, V) \in \mathsf{P}(M)$ tells us those modes of transportation that will get us into V from every point in U.

Proposition 2.74. Suppose $\mathcal{V} = (V, \leq, I, \otimes)$ is any symmetric monoidal preorder that has all joins. Then $\mathcal{V}$ is closed – i.e. it has a $\multimap$ operation and hence is a quantale – if and only if $\otimes$ distributes over joins, i.e. if Eq. (2.24) holds for all $v \in V$ and $A \subseteq V$.

Proof. We showed one direction in Proposition 2.64(b): if $\mathcal{V}$ is monoidal closed then Eq. (2.24) holds. We need to show that if Eq. (2.24) holds then $- \otimes v \colon V \to V$ has a right adjoint $v \multimap -$. This is just the adjoint functor theorem, Theorem 1.108. It says we can define $v \multimap w$ to be

$$v \multimap w := \bigvee_{\{a \in V \mid a \otimes v \leq w\}} a. \qquad\qquad\square$$

[6] Here $\forall$ is the universal quantifier, "for all," and $\exists$ is the existential quantifier, "there exists."

2.5.3 Matrix Multiplication in a Quantale

A quantale $\mathcal{V} = (V, \leq, I, \otimes, \multimap)$, as defined in Definition 2.57, provides what is necessary to perform matrix multiplication.[7] The usual formula for matrix multiplication is:

$$(M * N)(i, k) = \sum_j M(i, j) * N(j, k). \tag{2.25}$$

We will get a formula where joins stand in for the sum operation $\sum$, and $\otimes$ stands in for the product operation $*$. Recall our convention of writing $0 := \bigvee \varnothing$.

Definition 2.75. Let $\mathcal{V} = (V, \leq, \otimes, I)$ be a quantale. Given sets X and Y, a *matrix with entries in $\mathcal{V}$*, or simply a *$\mathcal{V}$-matrix*, is a function $M: X \times Y \to V$. For any $x \in X$ and $y \in Y$, we call $M(x, y)$ the *(x, y)-entry*.

Here is how you multiply $\mathcal{V}$-matrices $M: X \times Y \to V$ and $N: Y \times Z \to V$. Their product is defined to be the matrix $(M * N): X \times Z \to V$, whose entries are given by the formula

$$(M * N)(x, z) := \bigvee_{y \in Y} M(x, y) \otimes N(y, z). \tag{2.26}$$

Note how similar this is to Eq. (2.25).

Example 2.76. Let $\mathcal{V} = \textbf{Bool}$. Here is an example of matrix multiplication $M * N$. Here $X = \{1, 2, 3\}$, $Y = \{1, 2\}$, and $Z = \{1, 2, 3\}$, matrices $M: X \times Y \to \mathbb{B}$ and $N: Y \times Z \to \mathbb{B}$ are shown to the left below, and their product is shown to the right:

$$\begin{pmatrix} \text{false} & \text{false} \\ \text{false} & \text{true} \\ \text{true} & \text{true} \end{pmatrix} * \begin{pmatrix} \text{true} & \text{true} & \text{false} \\ \text{true} & \text{false} & \text{true} \end{pmatrix} = \begin{pmatrix} \text{false} & \text{false} & \text{false} \\ \text{true} & \text{false} & \text{true} \\ \text{true} & \text{true} & \text{true} \end{pmatrix}$$

The identity V-matrix on a set X is $I_X: X \times X \to V$ given by

$$I_X(x, y) := \begin{cases} I & \text{if } x = y, \\ 0 & \text{if } x \neq y. \end{cases}$$

Exercise 2.77. Write down the (2×2)-identity matrix for each of the quantales $(\mathbb{N}, \leq, 1, *)$, $\textbf{Bool} = (\mathbb{B}, \leq, \texttt{true}, \wedge)$, and $\textbf{Cost} = ([0, \infty], \geq, 0, +)$. $\diamond$

Exercise 2.78. Let $\mathcal{V} = (V, \leq, I, \otimes, \multimap)$ be a quantale. Use Eq. (2.26) and Proposition 2.64 to do the following.

---

[7] This works for noncommutative quantales as well.

1. Prove the *identity law*: for any sets X and Y and V-matrix $M : X \times Y \to V$, one has $I_X * M = M$.
2. Prove the *associative law*: for any matrices $M : W \times X \to V$, $N : X \times Y \to V$, and $P : Y \times Z \to V$, one has $(M * N) * P = M * (N * P)$. ◇

Recall the weighted graph Y from Eq. (2.18). One can read off the associated matrix M_Y, and one can calculate the associated metric d_Y:

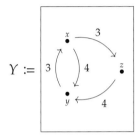

$$Y :=$$

M_Y	x	y	z
x	0	4	3
y	3	0	∞
z	∞	4	0

d_Y	x	y	z
x	0	4	3
y	3	0	6
z	7	4	0

Here we fully explain how to compute d_Y using only M_Y.

The matrix M_Y can be thought of as recording the length of paths that traverse either 0 or 1 edges: the diagonals being 0 mean we can get from x to x without traversing any edges. When we can get from x to y in one edge we record its length in M_Y, otherwise we use ∞.

When we multiply M_Y by itself using the formula Eq. (2.26), the result M_Y^2 tells us the length of the shortest path traversing two edges or fewer. Similarly M_Y^3 tells us about the shortest path traversing three edges or fewer:

$$M_Y^2 = $$

$\nearrow$	x	y	z
x	0	4	3
y	3	0	6
z	7	4	0

$$M_Y^3 = $$

$\nearrow$	x	y	z
x	0	4	3
y	3	0	6
z	7	4	0

One sees that the powers stabilize: $M_Y^2 = M_Y^3$; as soon as that happens one has the matrix of distances, d_Y. Indeed M_Y^n records the lengths of the shortest path traverse n edges or fewer, and the powers will always stabilize if the set of vertices is finite, since the shortest path from one vertex to another will never visit a given vertex more than once.[8]

Exercise 2.79. Recall from Exercise 2.40 the matrix M_X, for X as in Eq. (2.18). Calculate M_X^2, M_X^3, and M_X^4. Check that M_X^4 is what you got for the distance matrix in Exercise 2.39. ◇

This procedure gives an algorithm for computing the $\mathcal{V}$-category presented by any $\mathcal{V}$-weighted graph using matrix multiplication.

---

[8] The method works even in the infinite case: one takes the infimum of all powers M_Y^n. The result always defines a Lawvere metric space.

2.6   Summary and Further Reading

In this chapter we thought of elements of preorders as describing resources, with the order detailing whether one resource could be obtained from another. This naturally led to the question of how to describe what could be built from a pair of resources, which led us to consider monoid structures on preorders. More abstractly, these monoidal preorders were seen to be examples of enriched categories, or $\mathcal{V}$-categories, over the symmetric monoidal preorder **Bool**. Changing **Bool** to the symmetric monoidal preorder **Cost**, we arrived upon Lawvere metric spaces, a slight generalization of the usual notion of metric space. In terms of resources, **Cost**-categories tell us the cost of obtaining one resource from another.

At this point, we sought to get a better feel for $\mathcal{V}$-categories in two ways. First, we introduced various important constructions: base change, functors, products. Second, we looked at how to present $\mathcal{V}$-categories using labeled graphs; here, perhaps surprisingly, we saw that matrix multiplication gives an algorithm to compute the hom-objects from a labeled graph.

Resource theories are discussed in much more detail in [CFS16; Fri17]. The authors provide many more examples of resource theories in science, including in thermodynamics, Shannon's theory of communication channels, and quantum entanglement. They also discuss more of the numerical theory than we did, including calculating the asymptotic rate of conversion from one resource into another.

Enrichment is a fundamental notion in category theory, and we will return to it in Chapter 4, generalizing the definition so that categories, rather than mere preorders, can serve as bases of enrichment. In this more general setting we can still perform the constructions we introduced in Section 2.4 – base change, functors, products – and many others; the authoritative, but by no means easy, reference on this is the book by Kelly [Kel05].

While preorders were familiar before category theory came along, Lawvere metric spaces are a beautiful generalization of the previous notion of (symmetric) metric space, that is due to, well, Lawvere. A deeper exploration than the taste we gave here can be found in his classic paper [Law73], where he also discusses ideas like Cauchy completeness in category-theoretic terms, and which hence generalize to other categorical settings.

We observed that, while any symmetric monoidal preorder can serve as a base for enrichment, certain preorders – quantales – are better than others. Quantales are well known for links to other parts of mathematics too. The word quantale is in fact a portmanteau of 'quantum locale', where quantum refers to quantum physics, and locale is a fundamental structure in topology. For a book-length introduction to quantales and their applications, one might check [Ros90]. The notion of cartesian closed categories, later generalized to monoidal closed categories, is due to Ronnie Brown [Bro61].

Note that while we have considered only commutative quantales, the noncommutative variety also arise naturally. For example, the power set of any monoid forms a quantale that is commutative iff the monoid is. Another example is the set of all binary relations on a set X, where multiplication is relational composition; this is noncommutative. Such noncommutative quantales have application to concurrency theory, and in particular process semantics and automata; see [AV93] for details.

3 Databases: Categories, Functors, and Universal Constructions

3.1 What is a Database?

Integrating data from disparate sources is a major problem in industry today. A study in 2008 [BH08] showed that data integration accounts for 40% of IT (information technology) budgets, and that the market for data integration software was $2.5 billion in 2007 and increasing at a rate of more than 8% per year. In other words, it is a major problem; but what is it?

A database is a system of interlocking tables

Data becomes information when it is stored *in* a given *formation*. That is, the numbers and letters don't mean anything until they are organized, often into a system of interlocking tables. An organized system of interlocking tables is called a database. Here is a favorite example:

Employee	FName	WorksIn	Mngr		Department	DName	Secr
1	Alan	101	2		101	Sales	1
2	Ruth	101	2		102	IT	3
3	Kris	102	3				

$$(3.1)$$

These two tables interlock by use of a special left-hand column, demarcated by a vertical line; it is called the ID column. The ID column of the first table is called "Employee," and the ID column of the second table is called "Department." The entries in the ID column – e.g. 1, 2, 3 or 101, 102 – are like row labels; they indicate a whole row of the table they're in. Thus each row label must be unique (no two rows in a table can have the same label), so that it can unambiguously specify its row.

Each table's ID column, and the set of unique identifiers found therein, is what allows for the interlocking mentioned above. Indeed, other entries in various tables can reference rows in a given table by use of its ID column. For example, each entry in the WorksIn column references a department for each employee; each entry in the Mngr (manager) column references an employee for each employee, and each entry in the Secr (secretary) column references an employee for each department. Managing all this cross-referencing is the purpose of databases.

Looking back at tables (3.1), one might notice that every non-ID column, found in either table, is a reference to a label of some sort. Some of these, namely WorksIn,

Mngr, and Secr, are *internal references*, often called *foreign keys*; they refer to rows (keys) in the ID column of another (foreign) table. Others, namely FName and DName, are *external references*; they refer to strings or integers, which can also be thought of as labels, whose meaning is known more broadly. Internal reference labels can be changed as long as the change is consistent – 1 could be replaced by 1001 every-where without changing the meaning – whereas external reference labels certainly cannot! Changing Ruth to Bruce everywhere would change how people understood the data.

The reference structure for a given database – i.e. how tables interlock via foreign keys – tells us something about what information was intended to be stored in it. One may visualize the reference structure for Eq. (3.1) graphically as follows:

easySchema :=
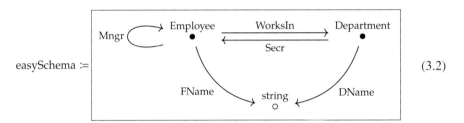
(3.2)

This is a kind of "Hasse diagram for a database," much like the Hasse diagrams for preorders in Remark 1.34. How should you read it?

The two tables from Eq. (3.1) are represented in the graph (3.2) by the two black nodes, which are given the same name as the ID columns: Employee and Department. There is another node – drawn white rather than black – which represents the external reference type of strings, like "Alan," "Alpha," and "Sales." The arrows in the diagram represent non-ID columns of the tables; they point in the direction of reference: WorksIn refers an employee to a department.

Exercise 3.1. Count the number of non-ID columns in Eq. (3.1). Count the number of arrows (foreign keys) in Eq. (3.2). They should be the same number in this case; is this a coincidence?   ◇

A Hasse-style diagram like the one in Eq. (3.2) can be called a *database schema*; it represents how the information is being organized, the formation in which the data is kept. One may add rules, sometimes called "business rules" to the schema, in order to ensure the integrity of the data. If these rules are violated, one knows that data being entered does not conform to the way the database designers intended. For example, the designers may enforce rules saying

- every department's secretary must work in that department;
- every employee's manager must work in the same department as the employee.

Doing so changes the schema, say from 'easySchema' (3.2) to 'mySchema' below.

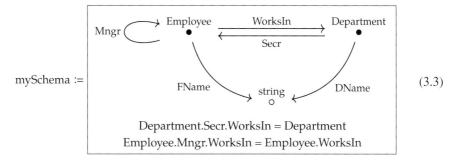

$$\text{mySchema} := \qquad\qquad\qquad\qquad\qquad\qquad\qquad\qquad\qquad\qquad\qquad (3.3)$$

In other words, the difference is that easySchema plus constraints equals mySchema.

We will soon see that database schemas are categories $\mathcal{C}$, that the data itself is given by a "set-valued" functor $\mathcal{C} \to \mathbf{Set}$, and that databases can be mapped to each other via functors $\mathcal{C} \to \mathcal{D}$. In other words, there is a relatively large overlap between database theory and category theory. This has been worked out in a number of papers; see Section 3.6. It has also been implemented in working software, called FQL, which stands for *functorial query language*. Here is example FQL code for the schema shown above:

```
schema mySchema = {
    nodes
        Employee, Department;
    attributes
        DName : Department -> string,
        FName : Employee   -> string;
    arrows
        Mngr    : Employee   -> Employee,
        WorksIn : Employee   -> Department,
        Secr    : Department -> Employee;
    equations
        Department.Secr.WorksIn = Department,
        Employee.Mngr.WorksIn   = Employee.WorksIn;
}
```

Communication between databases

We have said that databases are designed to store information about something. But different people or organizations might view the same sort of thing in different ways. For example, one bank stores its financial records according to European standards and another does so according to Japanese standards. If these two banks merge into one, they will need to be able to share their data despite differences in the shape of their database schemas.

Such problems are huge and intricate in general, because databases often comprise hundreds or thousands of interlocking tables. Moreover, these problems occur more frequently than just when companies want to merge. It is quite common that a given company moves data between databases on a daily basis. The reason is that different ways of organizing information are convenient for different purposes. Just like we pack our clothes in a suitcase when traveling but use a closet at home, there is generally not one best way to organize anything.

Category theory provides a mathematical approach for translating between these different organizational forms. That is, it formalizes a sort of automated reorganizing process called *data migration*, which takes data that fits snugly in one schema and moves it into another.

Here is a simple case. Imagine an airline company has two different databases, perhaps created at different times, that hold roughly the same data.

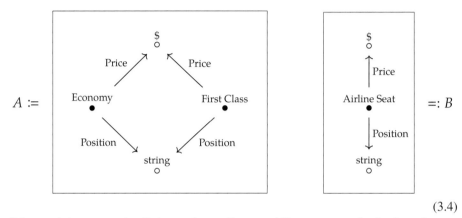

(3.4)

Schema A has more detail than schema B – an airline seat may be in first class or economy – but they are roughly the same. We shall see that they can be connected by a functor, and that data conforming to A can be migrated through this functor to schema B and vice versa.

The statistics at the beginning of this section show that this sort of problem – when occurring at enterprise scale – continues to prove difficult and expensive. If one attempts to move data from a source schema to a target schema, the migrated data could fail to fit into the target schema or fail to satisfy some of its constraints. This happens surprisingly often in the world of business: a night may be spent moving data, and the next morning it is found to have arrived broken and unsuitable for further use. In fact, it is believed that over half of database migration projects fail.

In this chapter, we will discuss a category-theoretic method for migrating data. Using categories and functors, one can prove up front that a given data migration will not fail, i.e. that the result is guaranteed to fit into the target schema and satisfy all its constraints.

The material in this chapter gets to the heart of category theory: in particular, we discuss categories, functors, natural transformations, adjunctions, limits, and colimits. In fact, many of these ideas have been present in the discussion above:

- The schema pictures, e.g. Eq. (3.3), depict categories $\mathcal{C}$.
- The instances, e.g. Eq. (3.1), are functors from $\mathcal{C}$ to a certain category called **Set**.
- The implicit mapping in Eq. (3.4), which takes economy and first class seats in A to airline seats in B, constitutes a functor $A \to B$.
- The notion of data migration for moving data between schemas is formalized by adjoint functors.

We begin in Section 3.2 with the definition of categories and a bunch of different sorts of examples. In Section 3.3 we bring back databases, in particular their instances and the maps between them, by discussing functors and natural transformations. In Section 3.4 we discuss data migration by way of adjunctions, which generalize the Galois connections we introduced in Section 1.4. Finally in Section 3.5 we give a bonus section on limits and colimits.[1]

3.2 Categories

A category $\mathcal{C}$ consists of four pieces of data – objects, morphisms, identities, and a composition rule – satisfying two properties, as described below.

Definition 3.2. To specify a *category* $\mathcal{C}$:

(i) one specifies a collection[2] $\mathrm{Ob}(\mathcal{C})$, elements of which are called *objects*;

(ii) for every two objects c, d, one specifies a set $\mathcal{C}(c,d)$,[3] elements of which are called *morphisms* from c to d;

(iii) for every object $c \in \mathrm{Ob}(\mathcal{C})$, one specifies a morphism $\mathrm{id}_c \in \mathcal{C}(c,c)$, called the *identity morphism* on c;

(iv) for every three objects $c, d, e \in \mathrm{Ob}(\mathcal{C})$ and morphisms $f \in \mathcal{C}(c,d)$ and $g \in \mathcal{C}(d,e)$, one specifies a morphism $f \,\mathring{,}\, g \in \mathcal{C}(c,e)$, called *the composite of f and g*.

We will sometimes write an object $c \in \mathcal{C}$, instead of $c \in \mathrm{Ob}(\mathcal{C})$. It will also be convenient to denote elements $f \in \mathcal{C}(c,d)$ as $f \colon c \to d$. Here, c is called the *domain* of f, and d is called the *codomain* of f.

These constituents are required to satisfy two conditions:

(a) *unitality*: for any morphism $f \colon c \to d$, composing with the identities at c or d does nothing: $\mathrm{id}_c \,\mathring{,}\, f = f$ and $f \,\mathring{,}\, \mathrm{id}_d = f$;

(b) *associativity*: for any three morphisms $f \colon c_0 \to c_1$, $g \colon c_1 \to c_2$, and $h \colon c_2 \to c_3$, the following are equal: $(f \,\mathring{,}\, g) \,\mathring{,}\, h = f \,\mathring{,}\, (g \,\mathring{,}\, h)$. We write this composite simply as $f \,\mathring{,}\, g \,\mathring{,}\, h$.

Our next goal is to give lots of examples of categories. Our first source of examples is that of free and finitely presented categories, which generalize the notion of Hasse diagram from Remark 1.34.

---

[1] By "bonus," we mean that although not strictly essential to the understanding of this particular chapter, limits and colimits will show up throughout the book and throughout one's interaction with category theory, and we think the reader will especially benefit from this material in the long run.

[2] Here, a *collection* can be thought of as a bunch of things, just like a set, but that may be too large to formally be a set. An example is the collection of all sets, which would run afoul of Russell's paradox if it were itself a set.

[3] This set $\mathcal{C}(c,d)$ is often denoted $\mathrm{Hom}_{\mathcal{C}}(c,d)$, and called the "hom-set from c to d." The word "hom" stands for homomorphism, of which the word "morphism" is a shortened version.

3.2.1   Free Categories

Recall from Definition 1.31 that a graph consists of two types of thing: vertices and arrows. From there one can define paths, which are just head-to-tail sequences of arrows. Every path p has a start vertex and an end vertex; if p goes from v to w, we write $p: v \to w$. To every vertex v, there is a trivial path, containing no arrows, starting and ending at v; we often denote it by id_v or simply by v. We may also concatenate paths: given $p: v \to w$ and $q: w \to x$, their concatenation is denoted $p \,\fatsemi\, q$, and it goes $v \to x$.

In Chapter 1, we used graphs to depict preorders $(V, \leq)$: the vertices form the elements of the preorder, and we say that $v \leq w$ if there is a path from v to w in G. We will now use graphs in a very similar way to depict certain categories, known as *free categories*. Then we will explain a strong relationship between preorders and categories in Section 3.2.3.

Definition 3.3.   For any graph $G = (V, A, s, t)$, we can define a category **Free**(G), called the *free category on* G, whose objects are the vertices V and whose morphisms from c to d are the paths from c to d. The identity morphism on an object c is simply the trivial path at c. Composition is given by concatenation of paths.

For example, we define **2** to be the free category generated by the graph shown below:

$$\mathbf{2} := \mathbf{Free}\left(\boxed{ \overset{v_1}{\bullet} \overset{f_1}{\longrightarrow} \overset{v_2}{\bullet} } \right) \tag{3.5}$$

It has two objects v_1 and v_2, and three morphisms: $\mathrm{id}_{v_1}: v_1 \to v_1$, $f_1: v_1 \to v_2$, and $\mathrm{id}_{v_2}: v_2 \to v_2$. Here id_{v_1} is the path of length 0 starting and ending at v_1, f_1 is the path of length 1 consisting of just the arrow f_1, and id_{v_2} is the length 0 path at v_2. As our notation suggests, id_{v_1} is the identity morphism for the object v_1, and similarly id_{v_2} for v_2. As composition is given by concatenation, we have, for example, $\mathrm{id}_{v_1} \,\fatsemi\, f_1 = f_1$, $\mathrm{id}_{v_2} \,\fatsemi\, \mathrm{id}_{v_2} = \mathrm{id}_{v_2}$, and so on.

From now on, we may elide the difference between a graph and the corresponding free category **Free**(G), at least when the one we mean is clear enough from context.

Exercise 3.4.   For **Free**(G) to really be a category, we must check that this data we specified obeys the unitality and associativity properties. Check that these are obeyed for any graph G.   ◇

Exercise 3.5.   The free category on the graph shown here:[4]

$$\mathbf{3} := \mathrm{Free}\left(\boxed{ \overset{v_1}{\bullet} \overset{f_1}{\longrightarrow} \overset{v_2}{\bullet} \overset{f_2}{\longrightarrow} \overset{v_3}{\bullet} } \right) \tag{3.6}$$

---

[4] As mentioned above, we elide the difference between the graph and the corresponding free category.

has three objects and six morphisms: the three vertices and six paths in the graph.

Create six names, one for each of the six morphisms in **3**. Write down a six-by-six table, label the rows and columns by the six names you chose.

1. Copy and complete the table by writing the name of the composite in each cell, when there is a composite.
2. Where are the identities?   ◇

Exercise 3.6. Let's make some definitions, based on the pattern above:

1. What is the category **1**? That is, what are its objects and morphisms?
2. What is the category **0**?
3. What is the formula for the number of morphisms in **n** for arbitrary $n \in \mathbb{N}$?   ◇

Example 3.7 (Natural numbers as a free category). Consider the following graph:

$$
\begin{array}{c}
s \\
\circlearrowleft \\
\bullet \\
z
\end{array}
\tag{3.7}
$$

It has only one vertex and one arrow, but it has infinitely many paths. Indeed, it has a unique path of length n for every natural number $n \in \mathbb{N}$. That is, Path $= \{z, s, (s \mathbin{\fatsemi} s), (s \mathbin{\fatsemi} s \mathbin{\fatsemi} s), \ldots\}$, where we write z for the length 0 path on z; it represents the morphism id_z. There is a one-to-one correspondence between Path and the natural numbers, $\mathbb{N} = \{0, 1, 2, 3, \ldots\}$.

This is an example of a category with one object. A category with one object is called a *monoid*, a notion we first discussed in Example 2.5. There we said that a monoid is a tuple $(M, *, e)$ where $*\colon M \times M \to M$ is a function and $e \in M$ is an element, and $m * 1 = m = 1 * m$ and $(m * n) * p = m * (n * p)$.

The two notions may superficially look different, but it is easy to describe the connection. Given a category $\mathcal{C}$ with one object, say $\bullet$, let $M := \mathcal{C}(\bullet, \bullet)$, let $e = \mathrm{id}_\bullet$, and let $*\colon \mathcal{C}(\bullet, \bullet) \times \mathcal{C}(\bullet, \bullet) \to \mathcal{C}(\bullet, \bullet)$ be the composition operation $* = \mathbin{\fatsemi}$. The associativity and unitality requirements for the monoid will be satisfied because $\mathcal{C}$ is a category.

Exercise 3.8. In Example 3.7 we identified the paths of the loop graph (3.7) with numbers $n \in \mathbb{N}$. Paths can be concatenated. Given numbers $m, n \in \mathbb{N}$, what number corresponds to the concatenation of their associated paths?   ◇

3.2.2   Presenting Categories via Path Equations

So for any graph G, there is a free category on G. But we don't have to stop there: we can add equations between paths in the graph, and still get a category. We are only allowed to equate two paths p and q when they are *parallel*, meaning they have the same source vertex and the same target vertex.

A finite graph with path equations is called a *finite presentation* for a category, and the category that results is known as a *finitely presented category*. Here are two examples:

$$\text{Free\_square} := \begin{array}{ccc} A & \xrightarrow{\ f\ } & B \\ {\scriptstyle g}\downarrow & & \downarrow{\scriptstyle h} \\ C & \xrightarrow{\ i\ } & D \end{array} \qquad \text{Comm\_square} := \begin{array}{ccc} A & \xrightarrow{\ f\ } & B \\ {\scriptstyle g}\downarrow & & \downarrow{\scriptstyle h} \\ C & \xrightarrow{\ i\ } & D \end{array}$$

no equations $f \fatsemi h = g \fatsemi i$

Both of these are presentations of categories: in the left-hand one, there are no equations so it presents a free category, as discussed in Section 3.2.1. The free square category has ten morphisms, because every path is a unique morphism.

Exercise 3.9.

1. Write down the ten paths in the free square category above.
2. Name two different paths that are parallel.
3. Name two different paths that are not parallel.   ◇

On the other hand, the category presented on the right has only nine morphisms, because $f \fatsemi h$ and $g \fatsemi i$ are made equal. This category is called the "commutative square." Its morphisms are

$$\{A, B, C, D, f, g, h, i, f \fatsemi h\}.$$

One might say "the missing one is $g \fatsemi i$," but that is not quite right: $g \fatsemi i$ is there too, because it is equal to $f \fatsemi h$. As usual, A denotes id_A, etc.

Exercise 3.10.   Write down all the morphisms in the category presented by the following diagram:

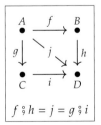

$f \fatsemi h = j = g \fatsemi i$ ◇

Example 3.11. We should also be aware that enforcing an equation between two morphisms often implies additional equations. Here are two more examples of presentations, in which this phenomenon occurs:

$$\mathcal{C} := \boxed{\begin{array}{c} s \\ \circlearrowleft \\ \bullet \\ z \\ \hline s \,\mathbin{\fatsemi}\, s = z \end{array}} \qquad\qquad \mathcal{D} := \boxed{\begin{array}{c} s \\ \circlearrowleft \\ \bullet \\ z \\ \hline s \,\mathbin{\fatsemi}\, s \,\mathbin{\fatsemi}\, s \,\mathbin{\fatsemi}\, s = s \,\mathbin{\fatsemi}\, s \end{array}}$$

In $\mathcal{C}$ we have the equation $s \,\mathbin{\fatsemi}\, s = z$. But this implies $s \,\mathbin{\fatsemi}\, s \,\mathbin{\fatsemi}\, s = z \,\mathbin{\fatsemi}\, s = s$! And similarly we have $s \,\mathbin{\fatsemi}\, s \,\mathbin{\fatsemi}\, s \,\mathbin{\fatsemi}\, s = z \,\mathbin{\fatsemi}\, z = z$. The set of morphisms in $\mathcal{C}$ is in fact merely $\{z, s\}$, with composition described by $s \,\mathbin{\fatsemi}\, s = z \,\mathbin{\fatsemi}\, z = z$, and $z \,\mathbin{\fatsemi}\, s = s \,\mathbin{\fatsemi}\, z = s$. In group theory, one would speak of a group called $\mathbb{Z}/2\mathbb{Z}$.

Exercise 3.12. Write down all the morphisms in the category $\mathcal{D}$ from Example 3.11. ◇

Remark 3.13. We can now see that the schemas in Section 3.1, e.g. Eqs. (3.2) and (3.3), are finite presentations of categories. We will come back to this idea in Section 3.3.

3.2.3 Preorders and Free Categories: Two Ends of a Spectrum

Now that we have used graphs to depict preorders in Chapter 1 and categories above, we may want to know the relationship between these two uses. The main idea we want to explain now is that

"A preorder is a category where every two parallel arrows are the same."

Thus any preorder can be regarded as a category, and any category can be somehow "crushed down" into a preorder. Let's discuss these ideas.

Preorders as categories

Suppose $(P, \leq)$ is a preorder. It specifies a category $\mathcal{P}$ as follows. The objects of $\mathcal{P}$ are precisely the elements of P; that is, $\mathrm{Ob}(\mathcal{P}) = P$. As for morphisms, $\mathcal{P}$ has exactly one morphism $p \to q$ if $p \leq q$ and no morphisms $p \to q$ if $p \not\leq q$. The fact that $\leq$ is reflexive ensures that every object has an identity, and the fact that $\leq$ is transitive ensures that morphisms can be composed. We call $\mathcal{P}$ the *category corresponding to the preorder* $(P, \leq)$.

In fact, a Hasse diagram for a preorder can be thought of a presentation of a category where, for all vertices p and q, every two paths from $p \to q$ are declared equal. For example, in Eq. (1.3) we saw a Hasse diagram that was like the graph on the left:

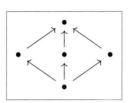

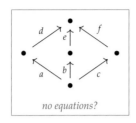

no equations?

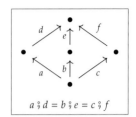

$a \,\r{9}\, d = b \,\r{9}\, e = c \,\r{9}\, f$

The Hasse diagram (left) might look the most like the free category presentation (middle) which has no equations, but that is not correct. The free category has three morphisms (paths) from bottom object to top object, whereas preorders are categories with *at most one* morphism between two given objects. Instead, the diagram on the right, with these paths from bottom to top made equal, is the correct presentation for the preorder on the left.

Exercise 3.14.   What equations would you need to add to the graphs below in order to present the associated preorders?

$$G_1 = \boxed{\bullet \;\underset{g}{\overset{f}{\rightrightarrows}}\; \bullet} \qquad G_2 = \boxed{\begin{array}{c} f \\ \circlearrowleft \\ \bullet \end{array}} \qquad G_3 = \boxed{\begin{array}{ccc} \bullet & \xrightarrow{\;f\;} & \bullet \\ {\scriptstyle g}\downarrow & & \downarrow{\scriptstyle h} \\ \bullet & \xrightarrow[i]{} & \bullet \end{array}} \qquad G_4 = \boxed{\begin{array}{ccc} \bullet & \xrightarrow{\;f\;} & \bullet \\ {\scriptstyle g}\downarrow & & \downarrow{\scriptstyle h} \\ \bullet & & \bullet \end{array}}$$

$\diamond$

The preorder reflection of a category

Given any category $\mathcal{C}$, one can obtain a preorder $(C, \le)$ from it by destroying the distinction between any two parallel morphisms. That is, let $C := \mathrm{Ob}(\mathcal{C})$, and put $c_1 \le c_2$ iff $\mathcal{C}(c_1, c_2) \ne \varnothing$. If there is one, or two, or fifty, or infinitely many morphisms $c_1 \to c_2$ in $\mathcal{C}$, the preorder reflection does not see the difference. But it does see the difference between some morphisms and no morphisms.

Exercise 3.15.   What is the preorder reflection of the category $\mathbb{N}$ from Example 3.7? $\diamond$

We have only discussed adjoint functors between preorders, but soon we will discuss adjoints in general. Here is a statement you might not understand exactly, but it's true; you can ask a category theory expert about it and they should be able to explain it to you:

Considering a preorder as a category is right adjoint to turning a category into a preorder by preorder reflection.

Remark 3.16 (Ends of a spectrum).   The main point of this subsection is that both preorders and free categories are specified by a graph without path equations, but they denote opposite ends of a spectrum. In both cases, the vertices of the graph become the objects of a category and the paths become morphisms. But in the case of free categories, there are no equations so each path becomes a different morphism. In the case of preorders, all parallel paths become the same morphism. Every category presentation, i.e. graph with some equations, lies somewhere in between the free category (no equations) and its preorder reflection (all possible equations).

3.2.4 Important Categories in Mathematics

We have been talking about category presentations, but there are categories that are best understood directly, not by way of presentations. Recall the definition of category from Definition 3.2. The most important category in mathematics is the category of sets.

Definition 3.17. The *category of sets*, denoted **Set**, is defined as follows.

(i) Ob(**Set**) is the collection of all sets.
(ii) If S and T are sets, then $\mathbf{Set}(S, T) = \{f : S \to T \mid f \text{ is a function}\}$.
(iii) For each set S, the identity morphism is the function $\mathrm{id}_S : S \to S$ given by $\mathrm{id}_S(s) := s$ for each $s \in S$.
(iv) Given $f : S \to T$ and $g : T \to U$, their composite is the function $f \, \mathring{,} \, g : S \to U$ given by $(f \, \mathring{,} \, g)(s) := g(f(s))$.

These definitions satisfy the unitality and associativity conditions, so **Set** is indeed a category.

Closely related is the category **FinSet**. This is the category whose objects are finite sets and whose morphisms are functions between them.

Exercise 3.18. Let $\underline{2} = \{1, 2\}$ and $\underline{3} = \{1, 2, 3\}$. These are objects in the category **Set** discussed in Definition 3.17. Write down all the elements of the set $\mathbf{Set}(\underline{2}, \underline{3})$; there should be nine. ◇

Remark 3.19. You may have wondered what categories have to do with $\mathcal{V}$-categories (Definition 2.30); perhaps you think the definitions hardly look alike. Despite the term "enriched category," $\mathcal{V}$-categories are not categories with extra structure. While some sorts of $\mathcal{V}$-categories, such as **Bool**-categories, i.e. preorders, can naturally be seen as categories, other sorts, such as **Cost**-categories, cannot.

The reason for the importance of **Set** is that, if we generalize the definition of enriched category (Definition 2.30), we find that categories in the sense of Definition 3.2 are exactly **Set**-categories – so categories are $\mathcal{V}$-categories for a very special choice of $\mathcal{V}$. We'll come back to this in Section 4.4.4. For now, we simply remark that just like a deep understanding of the category **Cost** – e.g. knowing that it is a quantale – yields insight into Lawvere metric spaces, so the study of **Set** yields insights into categories.

There are many other categories that mathematicians care about:

- **Top**: the category of topological spaces (neighborhood),
- **Grph**: the category of graphs (connection),
- **Meas**: the category of measure spaces (amount),
- **Mon**: the category of monoids (action),
- **Grp**: the category of groups (reversible action, symmetry),
- **Cat**: the category of categories (action in context, structure).

But in fact, this does not at all do justice to the diversity of categories mathematicians think about. They work with whatever category they find fits their purpose at the time, like "the category of connected Riemannian manifolds of dimension at most 4."

Here is one more source of examples: take any category you already have and reverse all its morphisms; the result is again a category.

Example 3.20. Let $\mathcal{C}$ be a category. Its *opposite*, denoted $\mathcal{C}^{\mathrm{op}}$, is the category with the same objects, $\mathrm{Ob}(\mathcal{C}^{\mathrm{op}}) := \mathrm{Ob}(\mathcal{C})$, and for any two objects $c, d \in \mathrm{Ob}(\mathcal{C})$, one has $\mathcal{C}^{\mathrm{op}}(c, d) := \mathcal{C}(d, c)$. Identities and composition are as in $\mathcal{C}$.

3.2.5   Isomorphisms in a Category

The previous sections have all been about examples of categories: free categories, presented categories, and important categories in math. In this section, we briefly switch gears and talk about an important concept in category theory, namely the concept of isomorphism.

In a category, there is often the idea that two objects are interchangeable. For example, in the category **Set**, one can exchange the set $\{\blacksquare, \square\}$ for the set $\{0, 1\}$ and everything will be the same, other than the names for the elements. Similarly, if one has a preorder with elements a, b, such that $a \leq b$ and $b \leq a$, i.e. $a \cong b$, then a and b are essentially the same. How so? Well, they act the same, in that for any other object c, we know that $c \leq a$ iff $c \leq b$, and $c \geq a$ iff $c \geq b$. The notion of isomorphism formalizes this notion of interchangeability.

Definition 3.21.   An *isomorphism* is a morphism $f : A \to B$ such that there exists a morphism $g : B \to A$ satisfying $f \,\fatsemi\, g = \mathrm{id}_A$ and $g \,\fatsemi\, f = \mathrm{id}_B$. In this case we call f and g *inverses*, and we often write $g = f^{-1}$, or equivalently $f = g^{-1}$. We also say that A and B are *isomorphic* objects.

Example 3.22. The set $A := \{a, b, c\}$ and the set $\underline{3} = \{1, 2, 3\}$ are isomorphic, i.e. there exists an isomorphism $f : A \to \underline{3}$ given by $f(a) = 2$, $f(b) = 1$, $f(c) = 3$. The isomorphisms in the category **Set** are bijections.

Recall that the cardinality of a finite set is the number of elements in it. This can be understood in terms of isomorphisms in **FinSet**. Namely, for any finite set $A \in$ **FinSet**, its cardinality is the number $n \in \mathbb{N}$ such that there exists an isomorphism $A \cong \underline{n}$. Georg Cantor defined the cardinality of any set X to be its isomorphism class, meaning the equivalence class consisting of all sets that are isomorphic to X.

Exercise 3.23.

1. What is the inverse $f^{-1} \colon \underline{3} \to A$ of the function f given in Example 3.22?
2. How many distinct isomorphisms are there $A \to \underline{3}$?   ◇

Exercise 3.24.   Show that in any given category $\mathcal{C}$, for any given object $c \in \mathcal{C}$, the identity id_c is an isomorphism.   ◇

Exercise 3.25.   Recall Examples 3.7 and 3.11. A monoid in which every morphism is an isomorphism is known as a *group*.

1. Is the monoid in Example 3.7 a group?
2. What about the monoid $\mathcal{C}$ in Example 3.11?   ◇

Exercise 3.26.   Let G be a graph, and let **Free**(G) be the corresponding free category. Somebody tells you that the only isomorphisms in **Free**(G) are the identity morphisms. Is that person correct? Why or why not?   ◇

Example 3.27.   In this example, we shall see that it is possible for g and f to be almost – but not quite – inverses, in a certain sense.
 Consider the functions $f \colon \underline{2} \to \underline{3}$ and $g \colon \underline{3} \to \underline{2}$ drawn below:

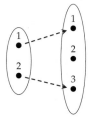 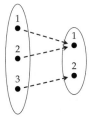

Then the reader should be able to check instantly that $f \,\mathring{,}\, g = \mathrm{id}_{\underline{2}}$ but $g \,\mathring{,}\, f \neq \mathrm{id}_{\underline{3}}$. Thus f and g are not inverses and hence not isomorphisms. We won't need this terminology, but category theorists would say that f and g form a *retraction*.

3.3   Functors, Natural Transformations, and Databases

In Section 3.1 we showed some database schemas: graphs with path equations. Then in Section 3.2.2 we said that graphs with path equations correspond to finitely presented categories. Now we want to explain what the data in a database is, as a way to introduce functors. To do so, we begin by noticing that sets and functions – the objects and morphisms in the category **Set** – can be captured by particularly simple databases.

3.3.1   Sets and Functions as Databases

The first observation is that any set can be understood as a table with only one column: the ID column.

Planet of Sol
Mercury
Venus
Earth
Mars
Jupiter
Saturn
Uranus
Neptune

Prime number
2
3
5
7
11
13
17
⋮

Flying pig

Rather than put the elements of the set between braces, e.g. $\{2, 3, 5, 7, 11, \ldots\}$, we write them down as rows in a table.

In databases, single-column tables are often called controlled vocabularies, or master data. Now to be honest, we can only write out every single entry in a table when its set of rows is finite. A database practitioner might find the idea of our prime number table a bit unrealistic. But we're mathematicians, so since the idea makes perfect sense abstractly, we will continue to think of sets as one-column tables.

The above databases have schemas consisting of just one vertex:

Planet of Sol	Prime number	Flying pig
●	●	●

Obviously, there's really not much difference between these schemas, other than the label of the unique vertex. So we could say "sets are databases whose schema consists of a single vertex." Let's move on to functions.

A function $f : A \to B$ can almost be depicted as a two-column table

Beatle	Played
George	lead guitar
John	rhythm guitar
Paul	bass guitar
Ringo	drums

except it is unclear whether the elements of the right-hand column exhaust all of B. What if there are rock-and-roll instruments out there that none of the Beatles played? So a function $f : A \to B$ requires two tables, one for A and its f column, and one for B:

Beatle	Played	Rock-and-roll instrument
George	lead guitar	bass guitar
John	rhythm guitar	drums
Paul	bass guitar	keyboard
Ringo	drums	lead guitar
		rhythm guitar

Thus the database schema for any function is just a labeled version of **2**:

The lesson is that an instance of a database takes a presentation of a category, and turns every vertex into a set, and every arrow into a function. As such, it describes a map from the presented category to the category **Set**. In Section 2.4.2 we saw that maps of $\mathcal{V}$-categories are known as $\mathcal{V}$-functors. Similarly, we call maps of plain old categories functors.

3.3.2 Functors

A functor is a mapping between categories. It sends objects to objects and morphisms to morphisms, while preserving identities and composition. Here is the formal definition.

Definition 3.28.   Let $\mathcal{C}$ and $\mathcal{D}$ be categories. To specify a *functor from $\mathcal{C}$ to $\mathcal{D}$*, denoted $F \colon \mathcal{C} \to \mathcal{D}$,

(i) for every object $c \in \mathrm{Ob}(\mathcal{C})$, one specifies an object $F(c) \in \mathrm{Ob}(\mathcal{D})$;
(ii) for every morphism $f \colon c_1 \to c_2$ in $\mathcal{C}$, one specifies a morphism $F(f) \colon F(c_1) \to F(c_2)$ in $\mathcal{D}$.

The above constituents must satisfy two properties:

(a) for every object $c \in \mathrm{Ob}(\mathcal{C})$, we have $F(\mathrm{id}_c) = \mathrm{id}_{F(c)}$;
(b) for every three objects $c_1, c_2, c_3 \in \mathrm{Ob}(\mathcal{C})$ and two morphisms $f \in \mathcal{C}(c_1, c_2)$, $g \in \mathcal{C}(c_2, c_3)$, the equation $F(f \,\mathring{\,}\, g) = F(f) \,\mathring{\,}\, F(g)$ holds in $\mathcal{D}$.

Example 3.29. For example, here we draw three functors $F \colon \mathbf{2} \to \mathbf{3}$:

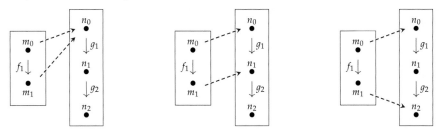

In each case, the dotted arrows show what the functor F does to the vertices in $\mathbf{2}$; once that information is specified, it turns out – in this special case – that what F does to the three paths in $\mathbf{2}$ is completely determined. In the left-hand diagram, F sends every path to the trivial path, i.e. the identity on n_0. In the middle diagram $F(m_0) = n_0$, $F(f_1) = g_1$, and $F(m_1) = n_1$. In the right-hand diagram, $F(m_0) = n_0$, $F(m_1) = n_2$, and $F(f_1) = g_1 \, \mathbin{\mathring{\,}} \, g_2$.

Exercise 3.30.   Above we wrote down three functors $\mathbf{2} \to \mathbf{3}$. Find and write down all the remaining functors $\mathbf{2} \to \mathbf{3}$.   ◇

Example 3.31.   Recall the categories presented by Free_square and Comm_square in Section 3.2.2. Here they are again, with $'$ added to the labels in Free_square to help distinguish them:

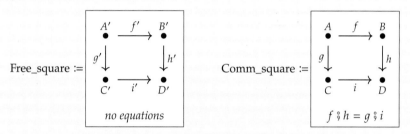

There are lots of functors from the free square category (let's call it $\mathcal{F}$) to the commutative square category (let's call it $\mathcal{C}$).

However, there is exactly one functor $F \colon \mathcal{F} \to \mathcal{C}$ that sends A' to A, B' to B, C' to C, and D' to D. That is, once we have made this decision about how F acts on objects, each of the ten paths in $\mathcal{F}$ is forced to go to a certain path in $\mathcal{C}$: the one with the right source and target.

Exercise 3.32.   Say where each of the ten morphisms in $\mathcal{F}$ is sent under the functor F from Example 3.31.   ◇

All of our example functors so far have been completely determined by what they do on objects, but this is usually not the case.

Exercise 3.33.   Consider the free categories $\mathcal{C} = \boxed{\bullet \to \bullet}$ and $\mathcal{D} = \boxed{\bullet \rightrightarrows \bullet}$. Give two functors $F, G \colon \mathcal{C} \to \mathcal{D}$ that act the same on objects but differently on morphisms.   ◇

Example 3.34.   There are also lots of functors from the commutative square category $\mathcal{C}$ to the free square category $\mathcal{F}$, but *none* that sends A to A', B to B', C to C', and D to D'. The reason is that if F were such a functor, then since $f \, \mathbin{\mathring{\,}} \, h = g \, \mathbin{\mathring{\,}} \, i$ in $\mathcal{C}$, we would have $F(f \,\mathring{,}\, h) = F(g \,\mathring{,}\, i)$, but then the rules of functors would let us reason as follows:

$$f' \,\mathring{,}\, h' = F(f) \,\mathring{,}\, F(h) = F(f \,\mathring{,}\, h) = F(g \,\mathring{,}\, i) = F(g) \,\mathring{,}\, F(i) = g' \,\mathring{,}\, i'.$$

The resulting equation, $f' \,\mathring{,}\, h' = g' \,\mathring{,}\, i'$ does not hold in $\mathcal{F}$ because it is a free category (there are "no equations"): every two paths are considered different morphisms. Thus our proposed F is not a functor.

Example 3.35 (Functors between preorders are monotone maps). Recall from Section 3.2.3 that preorders are categories with at most one morphism between any two objects. A functor between preorders is exactly a monotone map.

For example, consider the preorder $(\mathbb{N}, \leq)$ considered as a category $\mathcal{N}$ with objects $\mathrm{Ob}(\mathcal{N}) = \mathbb{N}$ and a unique morphism $m \to n$ iff $m \leq n$. A functor $F \colon \mathcal{N} \to \mathcal{N}$ sends each object $n \in \mathbb{N}$ to an object $F(n) \in \mathbb{N}$. It must send morphisms in $\mathcal{N}$ to morphisms in $\mathcal{N}$. This means if there is to be a morphism $m \to n$ then there must also be a morphism $F(m) \to F(n)$. In other words, if $m \leq n$, then we must also have $F(m) \leq F(n)$. But as long as $m \leq n$ implies $F(m) \leq F(n)$, we have a functor.

Thus a functor $F \colon \mathcal{N} \to \mathcal{N}$ and a monotone map $\mathbb{N} \to \mathbb{N}$ are the same thing.

Exercise 3.36 (The category of categories). Back in the primordial ooze, there is a category **Cat** in which *the objects are themselves categories*. Your task here is to construct this category.

1. Given any category $\mathcal{C}$, show that there exists a functor $\mathrm{id}_{\mathcal{C}} \colon \mathcal{C} \to \mathcal{C}$, known as the *identity functor on* $\mathcal{C}$, that maps each object to itself and each morphism to itself.

Note that a functor $\mathcal{C} \to \mathcal{D}$ consists of a function from $\mathrm{Ob}(\mathcal{C})$ to $\mathrm{Ob}(\mathcal{D})$ and for each pair of objects $c_1, c_2 \in \mathcal{C}$ a function from $\mathcal{C}(c_1, c_2)$ to $\mathcal{D}(F(c_1), F(c_2))$.

2. Show that given $F \colon \mathcal{C} \to \mathcal{D}$ and $G \colon \mathcal{D} \to \mathcal{E}$, we can define a new functor $(F \,\mathring{,}\, G) \colon \mathcal{C} \to \mathcal{E}$ just by composing functions.
3. Show that there is a category, call it **Cat**, where the objects are categories, morphisms are functors, and identities and composition are given as above. $\diamond$

3.3.3 Database Instances as **Set**-Valued Functors

Let $\mathcal{C}$ be a category, and recall the category **Set** from Definition 3.17. A functor $F \colon \mathcal{C} \to$ **Set** is known as a *set-valued functor* on $\mathcal{C}$. Much of database theory (not how to make them fast, but what they are and what you do with them) can be cast in this light.

Indeed, we already saw in Remark 3.13 that any database schema can be regarded as (presenting) a category $\mathcal{C}$. The next thing to notice is that the data itself – any instance of the database – is given by a set-valued functor $I \colon \mathcal{C} \to$ **Set**. The only additional detail is that for any white node, such as $c = \overset{\text{string}}{\circ}$, we want to force I to map to the set of strings. We suppress this detail in the following definition.

Definition 3.37. Let $\mathcal{C}$ be a schema, i.e. a finitely presented category. A $\mathcal{C}$-*instance* is a functor $I \colon \mathcal{C} \to \textbf{Set}$.[5]

Exercise 3.38. Let $\mathbf{1}$ denote the category with one object, called 1, one identity morphism id_1, and no other morphisms. For any functor $F \colon \mathbf{1} \to \textbf{Set}$ one can extract a set $F(1)$. Show that, for any set S, there is a functor $F_S \colon \mathbf{1} \to \textbf{Set}$ such that $F_S(1) = S$. ◇

The above exercise reaffirms that the set of planets, the set of prime numbers, and the set of flying pigs are all set-valued functors – instances – on the schema $\mathbf{1}$. Similarly, set-valued functors on the category $\mathbf{2}$ are functions. All our examples so far are for the situation where the schema is a free category (no equations). Let's try an example of a category that is not free.

Example 3.39. Consider the following category:

$$\mathcal{C} := \boxed{\begin{array}{c} s \\ \curvearrowleft \\ \bullet \\ z \\ \\ s \mathbin{\overset{\circ}{,}} s = s \end{array}} \tag{3.8}$$

What is a set-valued functor $F \colon \mathcal{C} \to \textbf{Set}$? It will consist of a set $Z := F(z)$ and a function $S := F(s) \colon Z \to Z$, subject to the requirement that $S \mathbin{\overset{\circ}{,}} S = S$. Here are some examples.

- Z is the set of US citizens, and S sends each citizen to her or his president. The president's president is her- or himself.
- $Z = \mathbb{N}$ is the set of natural numbers and S sends each number to 0. In particular, 0 goes to itself.
- Z is the set of all well-formed arithmetic expressions, such as $13 + (2 * 4)$ or -5, that one can write using integers and the symbols $+, -, *, (,)$. The function S evaluates the expression to return an integer, which is itself a well-formed expression. The evaluation of an integer is itself.
- $Z = \mathbb{N}_{\geq 2}$, and S sends n to its smallest prime factor. The smallest prime factor of a prime is itself.

---

[5] Warning: a $\mathcal{C}$-instance is a state of the database "at an instant in time." The term "instance" should not be confused with its usage in object-oriented programming, which would correspond more to what we call a row $r \in I(c)$.

$\mathbb{N}_{\geq 2}$	Smallest prime factor
2	2
3	3
4	2
$\vdots$	$\vdots$
49	7
50	2
51	3
$\vdots$	$\vdots$

Exercise 3.40. Above, we thought of the sort of data that would make sense for the schema (3.8). Give an example of the sort of data that would make sense for the following schemas:

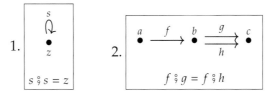

1.   2.

$\diamond$

The main idea is this: a database schema is a category, and an instance on that schema – the data itself – is a set-valued functor. All the constraints, or business rules, are ensured by the rules of functors, namely that functors preserve composition.[6]

3.3.4   Natural Transformations

If $\mathcal{C}$ is a schema – i.e. a finitely presented category – then there are many database instances on it, which we can organize into a category. But this is part of a larger story, namely that of natural transformations. An abstract picture to have in mind is this:

[6] One can put more complex constraints, called *embedded dependencies*, on a database; these correspond category-theoretically to what are called "lifting problems" in category theory. See [Spi14b] for more on this.

Definition 3.41.   Let $\mathcal{C}$ and $\mathcal{D}$ be categories, and let $F, G \colon \mathcal{C} \to \mathcal{D}$ be functors. To specify a *natural transformation* $\alpha \colon F \Rightarrow G$,

(i) for each object $c \in \mathcal{C}$, one specifies a morphism $\alpha_c \colon F(c) \to G(c)$ in $\mathcal{D}$, called the *c-component of* α.

These components must satisfy the following, called the *naturality condition*:

(a) for every morphism $f \colon c \to d$ in $\mathcal{C}$, the following equation must hold:

$$F(f) \, \mathring{\,}\, \alpha_d = \alpha_c \, \mathring{\,}\, G(f).$$

A natural transformation $\alpha \colon F \to G$ is called a *natural isomorphism* if each component α_c is an isomorphism in $\mathcal{D}$.

The naturality condition can also be written as a so-called *commutative diagram*. A diagram in a category is drawn as a graph whose vertices and arrows are labeled by objects and morphisms in the category. For example, here is a diagram that's relevant to the naturality condition in Definition 3.41:

$$
\begin{array}{ccc}
F(c) & \xrightarrow{\ \alpha_c\ } & G(c) \\
{\scriptstyle F(f)}\downarrow & & \downarrow{\scriptstyle G(f)} \\
F(d) & \xrightarrow[\ \alpha_d\ ]{} & G(d)
\end{array}
\tag{3.9}
$$

Definition 3.42.   A *diagram D* in $\mathcal{C}$ is a functor $D \colon \mathcal{J} \to \mathcal{C}$ from any category $\mathcal{J}$, called the *indexing category* of the diagram D. We say that D *commutes* if $D(f) = D(f')$ holds for every parallel pair of morphisms $f, f' \colon a \to b$ in $\mathcal{J}$.[7]

In terms of Eq. (3.9), the only case of two parallel morphisms is that of $F(c) \rightrightarrows G(d)$, so to say that the diagram commutes is to say that $F(f) \, \mathring{\,}\, \alpha_d = \alpha_c \, \mathring{\,}\, G(f)$. This is exactly the naturality condition from Definition 3.41.

Example 3.43. A representative picture is as follows:

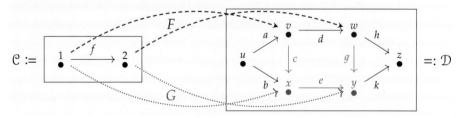

---

[7] We could package this formally by saying that D commutes iff it factors through the preorder reflection of $\mathcal{J}$.

We have depicted, in blue and red respectively, two functors $F, G: \mathcal{C} \to \mathcal{D}$. A natural transformation $\alpha: F \Rightarrow G$ is given by choosing components $\alpha_1: v \to x$ and $\alpha_2: w \to y$. We have highlighted the only choice for each in green: namely, $\alpha_1 = c$ and $\alpha_2 = g$.

The key point is that the functors F and G are ways of viewing the category $\mathcal{C}$ as lying inside the category $\mathcal{D}$. The natural transformation α, then, is a way of relating these two views using the morphisms in $\mathcal{D}$. Does this help you to see and appreciate the notation $\mathcal{C} \overset{F}{\underset{G}{\Longrightarrow}} \alpha \Downarrow \mathcal{D}$?

Example 3.44. We said in Exercise 3.38 that a functor $\mathbf{1} \to \mathbf{Set}$ can be identified with a set. So suppose A and B are sets considered as functors $A, B: \mathbf{1} \to \mathbf{Set}$. A natural transformation between these functors is just a function between the sets.

Definition 3.45. Let $\mathcal{C}$ and $\mathcal{D}$ be categories. We denote by $\mathcal{D}^{\mathcal{C}}$ the category whose objects are functors $F: \mathcal{C} \to \mathcal{D}$ and whose morphisms $\mathcal{D}^{\mathcal{C}}(F, G)$ are the natural transformations $\alpha: F \to G$. This category $\mathcal{D}^{\mathcal{C}}$ is called the *functor category*, or the *category of functors from $\mathcal{C}$ to $\mathcal{D}$*.

Exercise 3.46. Let's look more deeply at how $\mathcal{D}^{\mathcal{C}}$ is a category.

1. Figure out how to compose natural transformations. (Hint: an expert tells you "for each object $c \in \mathcal{C}$, compose the c-components.")
2. Propose an identity natural transformation on any object $F \in \mathcal{D}^{\mathcal{C}}$, and check that it is unital (i.e. that it obeys condition (a) of Definition 3.2). ◇

Example 3.47. In our new language, Example 3.44 says that $\mathbf{Set}^{\mathbf{1}}$ is isomorphic to $\mathbf{Set}$.

Example 3.48. Let $\mathcal{N}$ denote the category associated to the preorder $(\mathbb{N}, \leq)$, and recall from Example 3.35 that we can identify a functor $F: \mathcal{N} \to \mathcal{N}$ with a nondecreasing sequence $(F_0, F_1, F_2, \ldots)$ of natural numbers, i.e. $F_0 \leq F_1 \leq F_2 \leq \cdots$. If G is another functor, considered as a nondecreasing sequence, then what is a natural transformation $\alpha: F \to G$?

Since there is at most one morphism between two objects in a preorder, each component $\alpha_n: F_n \to G_n$ has no data, it just tells us a fact: that $F_n \leq G_n$. And the naturality condition is vacuous: every square in a preorder commutes. So a natural transformation between F and G exists iff $F_n \le G_n$ for each n, and any two natural transformations $F \Rightarrow G$ are the same. In other words, the category $\mathcal{N}^{\mathcal{N}}$ is itself a preorder, namely the preorder of monotone maps $\mathbb{N} \to \mathbb{N}$.

Exercise 3.49. Let $\mathcal{C}$ be an arbitrary category and let $\mathcal{P}$ be a preorder, thought of as a category. Consider the following statements:

1. For any two functors $F, G \colon \mathcal{C} \to \mathcal{P}$, there is at most one natural transformation $F \to G$.
2. For any two functors $F, G \colon \mathcal{P} \to \mathcal{C}$, there is at most one natural transformation $F \to G$.

For each, if it is true, say why; if it is false, give a counterexample.   ◇

Remark 3.50.   Recall that in Remark 2.50 we said the category of preorders is equivalent to the category of **Bool**-categories. We can now state the precise meaning of this sentence. First, there exists a category **PrO** in which the objects are preorders and the morphisms are monotone maps. Second, there exists a category **Bool-Cat** in which the objects are **Bool**-categories and the morphisms are **Bool**-functors. We call these two categories equivalent because there exist functors $F \colon \mathbf{PrO} \to \mathbf{Bool\text{-}Cat}$ and $G \colon \mathbf{Bool\text{-}Cat} \to \mathbf{PrO}$ such that there exist natural isomorphisms $F \,\mathring{,}\, G \cong \mathrm{id}_{\mathbf{PrO}}$ and $G \,\mathring{,}\, F \cong \mathrm{id}_{\mathbf{Bool\text{-}Cat}}$ in the sense of Definition 3.41.

3.3.5   The Category of Instances on a Schema

Definition 3.51.   Suppose that $\mathcal{C}$ is a database schema and $I, J \colon \mathcal{C} \to \mathbf{Set}$ are database instances. An *instance homomorphism* between them is a natural transformation $\alpha \colon I \to J$. Write $\mathcal{C}\text{-}\mathbf{Inst} := \mathbf{Set}^{\mathcal{C}}$ to denote the functor category as defined in Definition 3.45.

We saw in Example 3.44 that **1-Inst** is isomorphic to the category **Set**. In this subsection, we will show that there is a schema whose instances are graphs and whose instance homomorphisms are graph homomorphisms.

Extended example: the category of graphs as a functor category

You may find yourself back in the primordial ooze (first discussed in Section 2.3.2), because while previously we have been using graphs to present categories, now we obtain graphs themselves as database instances on a specific schema (which is itself a graph):

$$\mathbf{Gr} := \boxed{\begin{array}{c} \text{Arrow} \xrightarrow[\text{target}]{\text{source}} \text{Vertex} \\[2mm] \textit{no equations} \end{array}}$$

Here is an example of a **Gr**-instance, i.e. a set-valued functor $I : \mathbf{Gr} \to \mathbf{Set}$, in table form:

Arrow	source	target
a	1	2
b	1	3
c	1	3
d	2	2
e	2	3

Vertex
1
2
3
4

(3.10)

Here $I(\text{Arrow}) = \{a, b, c, d, e\}$, and $I(\text{Vertex}) = \{1, 2, 3, 4\}$. One can draw the instance I as a graph:

$$I = \boxed{\begin{array}{c} \text{graph} \end{array}}$$

Every row in the Vertex table is drawn as a vertex, and every row in the Arrow table is drawn as an arrow, connecting its specified source and target. Every possible graph can be written as a database instance on the schema **Gr**, and every possible **Gr**-instance can be represented as a graph.

Exercise 3.52. In Eq. (3.2), a graph is shown (forget the distinction between white and black nodes). Write down the corresponding **Gr**-instance, as in Eq. (3.10). (Do not be concerned that you are in the primordial ooze.) ◇

Thus the objects in the category **Gr-Inst** are graphs. The morphisms in **Gr-Inst** are called *graph homomorphisms*. Let's unwind this. Suppose that $G, H : \mathbf{Gr} \to \mathbf{Set}$ are functors (i.e. **Gr**-instances); that is, they are objects $G, H \in \mathbf{Gr\text{-}Inst}$. A morphism $G \to H$ is a natural transformation $\alpha : G \to H$ between them; what does that entail?

By Definition 3.41, since **Gr** has two objects, α consists of two components,

$$\alpha_{\text{Vertex}} : G(\text{Vertex}) \to H(\text{Vertex}) \qquad \text{and} \qquad \alpha_{\text{Arrow}} : G(\text{Arrow}) \to H(\text{Arrow}),$$

both of which are morphisms in **Set**. In other words, α consists of a function from vertices of G to vertices of H and a function from arrows of G to arrows of H. For these functions to constitute a graph homomorphism, they must "respect source and target" in the precise sense that the naturality condition, Eq. (3.9), holds. That is, for every morphism in **Gr**, namely source and target, the following diagrams must commute:

$$
\begin{array}{ccc}
G(\text{Arrow}) & \xrightarrow{\alpha_{\text{Arrow}}} & H(\text{Arrow}) \\
{\scriptstyle G(\text{source})}\downarrow & & \downarrow{\scriptstyle H(\text{source})} \\
G(\text{Vertex}) & \xrightarrow[\alpha_{\text{Vertex}}]{} & H(\text{Vertex})
\end{array}
\qquad
\begin{array}{ccc}
G(\text{Arrow}) & \xrightarrow{\alpha_{\text{Arrow}}} & H(\text{Arrow}) \\
{\scriptstyle G(\text{target})}\downarrow & & \downarrow{\scriptstyle H(\text{target})} \\
G(\text{Vertex}) & \xrightarrow[\alpha_{\text{Vertex}}]{} & H(\text{Vertex})
\end{array}
$$

These may look complicated, but they say exactly what we want. We want the functions α_{Vertex} and α_{Arrow} to respect source and targets in G and H. The left diagram says "start with an arrow in G. You can either apply α to the arrow and then take its source in H, or you can take its source in G and then apply α to that vertex; either way you get the same answer." The right-hand diagram says the same thing about targets.

Example 3.53. Consider the graphs G and H shown below

$$
G := \boxed{\overset{1}{\bullet} \xrightarrow{\ a\ } \overset{2}{\bullet} \xrightarrow{\ b\ } \overset{3}{\bullet}}
\qquad
H := \boxed{\overset{4}{\bullet} \underset{c}{\overset{d}{\rightrightarrows}} \overset{5}{\bullet} \circlearrowright e}
$$

Here they are, written as database instances – i.e. set-valued functors – on **Gr**:

$G :=$

Arrow	source	target
a	1	2
b	2	3

Vertex
1
2
3

$H :=$

Arrow	source	target
c	4	5
d	4	5
e	5	5

Vertex
4
5

The top row is G and the bottom row is H. They are offset so you can more easily complete the following exercise.

Exercise 3.54. We claim that – with G, H as in Example 3.53 – there is exactly one graph homomorphism $\alpha : G \to H$ such that $\alpha_{\text{Arrow}}(a) = d$.

1. What is the other value of α_{Arrow}, and what are the three values of α_{Vertex}?
2. In your own copy of the tables of Example 3.53, draw α_{Arrow} as two lines connecting the cells in the ID column of $G(\text{Arrow})$ to those in the ID column of $H(\text{Arrow})$. Similarly, draw α_{Vertex} as connecting lines.
3. Check the source column and target column and make sure that the matches are natural, i.e. that "alpha-then-source equals source-then-alpha" and similarly for "target." $\diamond$

3.4 Adjunctions and Data Migration

We have talked about how set-valued functors on a schema can be understood as filling that schema with data. But there are also functors between schemas. When the two sorts of functors are composed, data is migrated. This is the simplest form of data migration; more complex ways to migrate data come from using adjoints. All of the above is the subject of this section.

3.4.1 Pulling Back Data along a Functor

To begin, we will migrate data between the graph-indexing schema **Gr** and the loop schema, which we call **DDS**, shown below

We begin by writing down a sample instance $I : \mathbf{DDS} \to \mathbf{Set}$ on this schema:

State	next
1	4
2	4
3	5
4	5
5	5
6	7
7	6

(3.11)

We call the schema **DDS**, standing for discrete dynamical system. Indeed, we may think of the data in the **DDS**-instance of table (3.11) as listing the states and movements of a deterministic machine: at every point in time the machine is in one of the listed states, and given the machine in one of the states, in the next instant it moves to a uniquely determined next state.

Our goal is to migrate the data in (3.11) to data on **Gr**; this will give us the data of a graph and so allow us to visualize our machine.

We will use a functor connecting these schemas in order to move data between them. The reader can create any functor she likes, but we will use a specific functor $F : \mathbf{Gr} \to \mathbf{DDS}$ to migrate data in a way that makes sense to us, the authors. Here we draw F, using colors to aid understanding:

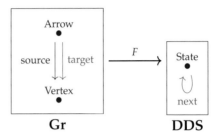

Gr   **DDS**

The functor F sends both objects of **Gr** to the "State" object of **DDS** (as it must). On morphisms, it sends the "source" morphism to the identity morphism on "State," and the "target" morphism to the morphism "next."

A sample database instance on **DDS** was given in Eq. (3.11); recall that this is a functor $I : \textbf{DDS} \to \textbf{Set}$. So now we have two functors as follows:

$$\textbf{Gr} \xrightarrow{\ F\ } \textbf{DDS} \xrightarrow{\ I\ } \textbf{Set}. \tag{3.12}$$

Objects in **Gr** are sent by F to objects in **DDS**, which are sent by I to objects in **Set**, which are sets. Morphisms in **Gr** are sent by F to morphisms in **DDS**, which are sent by I to morphisms in **Set**, which are functions. This defines a composite functor $F \mathbin{\unicode{x2A1F}} I : \textbf{Gr} \to \textbf{Set}$. Both F and I respect identities and composition, so $F \mathbin{\unicode{x2A1F}} I$ does too. Thus we have obtained an instance on **Gr**, i.e. we have converted our discrete dynamical system from Eq. (3.11) into a graph! What graph is it?

For an instance on **Gr**, we need to fill an Arrow table and a Vertex table. Both of these are sent by F to State, so let's fill both with the rows of State in Eq. (3.11). Similarly, since F sends "source" to the identity and sends "target" to "next," we obtain the following tables:

Arrow	source	target		Vertex
1	1	4		1
2	2	4		2
3	3	5		3
4	4	5		4
5	5	5		5
6	6	7		6
7	7	6		7

Now that we have a graph, we can draw it.

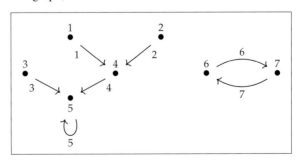

Each arrow is labeled by its source vertex, as if to say, "What I do next is determined by what I am now."

Exercise 3.55. Consider the functor $G \colon \mathbf{Gr} \to \mathbf{DDS}$ given by sending "source" to "next" and sending "target" to the identity on "State." Migrate the same data, called I in Eq. (3.11), using the functor G. Write down the tables and draw the corresponding graph. ◇

We refer to the above procedure – basically just composing functors as in Eq. (3.12) – as "pulling back data along a functor." We just now pulled back data I along functor F.

Definition 3.56. Let $\mathcal{C}$ and $\mathcal{D}$ be categories and let $F \colon \mathcal{C} \to \mathcal{D}$ be a functor. For any set-valued functor $I \colon \mathcal{D} \to \mathbf{Set}$, we refer to the composite functor $F \mathbin{\mathring{,}} I \colon \mathcal{C} \to \mathbf{Set}$ as the *pullback of I along F*. Given a natural transformation $\alpha \colon I \Rightarrow J$, there is a natural transformation $\alpha_F \colon F \mathbin{\mathring{,}} I \Rightarrow F \mathbin{\mathring{,}} J$, whose component $(F \mathbin{\mathring{,}} I)(c) \to (F \mathbin{\mathring{,}} J)(c)$ for any $c \in \mathrm{Ob}(\mathcal{C})$ is given by $(\alpha_F)_c := \alpha_{Fc}$.

$$\mathcal{C} \xrightarrow{\;F\;} \mathcal{D} \underset{J}{\overset{I}{\rightrightarrows}} \Downarrow\alpha \; \mathbf{Set} \qquad \rightsquigarrow \qquad \mathcal{C} \underset{F \mathbin{\mathring{,}} J}{\overset{F \mathbin{\mathring{,}} I}{\rightrightarrows}} \Downarrow\alpha_F \; \mathbf{Set}$$

This uses the data of F to define a functor $\Delta_F \colon \mathcal{D}\text{-}\mathbf{Inst} \to \mathcal{C}\text{-}\mathbf{Inst}$.

Note that the term pullback is also used for a certain sort of limit; for more details see Remark 3.85.

3.4.2 Adjunctions

In Section 1.4 we discussed Galois connections, which are adjunctions between preorders. Now that we have defined categories and functors, we can discuss adjunctions in general. The relevance to databases is that the data migration functor Δ from Definition 3.56 always has two adjoints of its own: a left adjoint which we denote Σ and a right adjoint which we denote Π.

Recall that an adjunction between preorders P and Q is a pair of monotone maps $f \colon P \to Q$ and $g \colon Q \to P$ that are *almost* inverses: we have

$$f(p) \le q \text{ iff } p \le g(q). \tag{3.13}$$

Recall from Section 3.2.3 that in a preorder P, a hom-set $P(a, b)$ has one element when $a \le b$, and no elements otherwise. We can thus rephrase Eq. (3.13) as an isomorphism of sets $Q(f(p), q) \cong P(p, g(q))$: either both are one-element sets or both are 0-element sets. This suggests how to define adjunctions in the general case.

Definition 3.57. Let $\mathcal{C}$ and $\mathcal{D}$ be categories, and $L\colon \mathcal{C} \to \mathcal{D}$ and $R\colon \mathcal{D} \to \mathcal{C}$ be functors. We say that L *is left adjoint to* R (and that R *is right adjoint to* L) if, for any $c \in \mathcal{C}$ and $d \in \mathcal{D}$, there is an isomorphism of hom-sets

$$\alpha_{c,d}\colon \mathcal{C}(c, R(d)) \overset{\cong}{\to} \mathcal{D}(L(c), d)$$

that is natural in c and d.[8]

Given a morphism $f\colon c \to R(d)$ in $\mathcal{C}$, its image $g := \alpha_{c,d}(f)$ is called its *mate*. Similarly, the mate of $g\colon L(c) \to d$ is f.

To denote an adjunction we write $L \dashv R$, or in diagrams

$$\mathcal{C} \underset{R}{\overset{L}{\rightleftarrows}} \mathcal{D}$$

with the $\Rightarrow$ in the direction of the left adjoint.

Example 3.58. Recall that every preorder $\mathcal{P}$ can be regarded as a category. Galois connections between preorders and adjunctions between the corresponding categories are exactly the same thing.

Example 3.59. Let $B \in \mathrm{Ob}(\mathbf{Set})$ be any set. There is an adjunction called "currying B," after the logician Haskell Curry:

$$\mathbf{Set} \underset{(-)^B}{\overset{-\times B}{\rightleftarrows}} \mathbf{Set} \qquad\qquad \mathbf{Set}(A \times B, C) \cong \mathbf{Set}(A, C^B).$$

Abstractly we write it as on the left, but what this means is that, for any sets A, C, there is a natural isomorphism as on the right.

To explain this, we need to talk about exponential objects in **Set**. Suppose that B and C are sets. Then the set of functions $B \to C$ is also a set; let's denote it C^B. It's written this way because if C has 10 elements and B has 3 elements then C^B has 10^3 elements, and more generally for any two finite sets $|C^B| = |C|^{|B|}$.

The idea of currying is that, given sets A, B, and C, there is a one-to-one correspondence between functions $(A \times B) \to C$ and functions $A \to C^B$. Intuitively, if I have a function f of two variables a, b, I can "put off" entering the second variable: if you give

---

[8] This naturality is between functors $\mathcal{C}^{\mathrm{op}} \times \mathcal{D} \to \mathbf{Set}$. It says that for any morphisms $f\colon c' \to c$ in $\mathcal{C}$ and $g\colon d \to d'$ in $\mathcal{D}$, the following diagram commutes:

$$
\begin{array}{ccc}
\mathcal{C}(c, Rd) & \xrightarrow{\;\alpha_{c,d}\;} & \mathcal{D}(Lc, d) \\
{\scriptstyle \mathcal{C}(f, Rg)}\downarrow & & \downarrow{\scriptstyle \mathcal{D}(Lf, g)} \\
\mathcal{C}(c', Rd') & \xrightarrow[\;\alpha_{c',d'}\;]{} & \mathcal{D}(Lc', d')
\end{array}
$$

me just a, I'll return a function $B \to C$ that's waiting for the B input. This is the curried version of f. As one might guess, there is a formal connection between exponential objects and what we called hom-elements $b \multimap c$ in Definition 2.57.

Exercise 3.60. In Example 3.59, we discussed an adjunction between functors $- \times B$ and $(-)^B$. But we only said how these functors worked on objects: for an arbitrary set X, they return sets $X \times B$ and X^B respectively.

1. Given a morphism $f \colon X \to Y$, what morphism should $- \times B \colon X \times B \to Y \times B$ return?
2. Given a morphism $f \colon X \to Y$, what morphism should $(-)^B \colon X^B \to Y^B$ return?
3. Consider the function $+ \colon \mathbb{N} \times \mathbb{N} \to \mathbb{N}$, which sends $(a, b) \mapsto a + b$. Currying $+$, we get a certain function $p \colon \mathbb{N} \to \mathbb{N}^{\mathbb{N}}$. What is $p(3)$?   ◇

Example 3.61. If you know some abstract algebra or topology, here are some other examples of adjunctions.

1. Free constructions: given any set you get a free group, free monoid, free ring, free vector space, etc.; each of these is a left adjoint. The corresponding right adjoint takes a group, a monoid, a ring, a vector space, etc. and forgets the algebraic structure to return the underlying set.
2. Similarly, given a graph you get a free preorder or a free category, as we discussed in Section 3.2.3; each is a left adjoint. The corresponding right adjoint is the underlying graph of a preorder or of a category.
3. Discrete things: given any set you get a discrete preorder, discrete graph, discrete metric space, discrete category, discrete topological space; each of these is a left adjoint. The corresponding right adjoint is again underlying set.
4. Codiscrete things: given any set you get a codiscrete preorder, complete graph, codiscrete category, codiscrete topological space; each of these is a right adjoint. The corresponding left adjoint is the underlying set.
5. Given a group, you can quotient by its commutator subgroup to get an abelian group; this is a left adjoint. The right adjoint is the inclusion of abelian groups into groups.

3.4.3 Left and Right Pushforward Functors, Σ and Π

Given $F \colon \mathcal{C} \to \mathcal{D}$, the data migration functor Δ_F turns $\mathcal{D}$-instances into $\mathcal{C}$-instances. This functor has both a left and a right adjoint:

$$\mathcal{C}\text{-}\mathbf{Inst} \xleftarrow{\quad \overset{\Sigma_F}{\underset{\Delta_F}{\rightleftarrows}} \quad} \mathcal{D}\text{-}\mathbf{Inst}$$

Using the names Σ and Π in this context is fairly standard in category theory. In the case of databases, they have the following helpful mnemonic:

Migration functor	pronounced	reminiscent of	database idea
Δ	delta	duplicate or destroy	duplicate or destroy tables or columns
Σ	sigma	sum	union (sum up) data
Π	pi	product	pair[9] and query data

Just like we used Δ_F to pull back any discrete dynamical system along $F \colon \mathbf{Gr} \to \mathbf{DDS}$ and get a graph, the migration functors Σ_F and Π_F can be used to turn any graph into a discrete dynamical system. That is, given an instance $J \colon \mathbf{Gr} \to \mathbf{Set}$, we can get instances $\Sigma_F(J)$ and $\Pi_F(J)$ on $\mathbf{DDS}$. This, however, is quite technical, and we leave it to the adventurous reader to compute an example, with help perhaps from [Spi14a], which explores the definitions of Σ and Π in detail. A less technical shortcut is simply to code up the computation in the open-source FQL software.

To get the basic idea across without getting mired in technical details, here we will instead discuss a very simple example. Recall the schemas from Eq. (3.4). We can set up a functor between them, the one sending black dots to black dots and white dots to white dots:

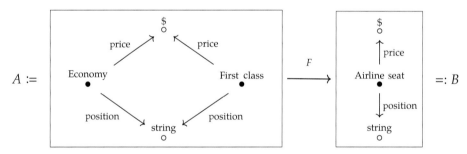

With this functor F in hand, we can transform any B-instance into an A-instance using Δ_F. Whereas Δ was interesting in the case of turning discrete dynamical systems into graphs in Section 3.4.1, it is not very interesting in this case. Indeed, it will just copy – Δ for duplicate – the rows in Airline Seat into both Economy and First Class.

Δ_F has two adjoints, Σ_F and Π_F, both of which transform any A-instance I into a B-instance. The functor Σ_F does what one would most expect from reading the names on each object: it will put into Airline Seat the union of Economy and First Class:

$$\Sigma_F(I)(\text{Airline Seat}) = I(\text{Economy}) \sqcup I(\text{First Class}).$$

The functor Π_F puts into Airline Seat the set of those pairs (e, f) where e is an Economy seat, f is a First Class seat, and e and f have the same price and position. In this particular example, one imagines that there should be no such seats in a valid instance

---

[9] This is more commonly called "join" by database programmers.

I, in which case $\Pi_F(I)(\text{Airline Seat})$ would be empty. But in other uses of these same schemas, Π_F can be a useful operation. For example, in the schema A replace the label 'Economy' by 'Rewards Program', and in B replace 'Airline Seat' by 'First Class Seats'. Then the operation Π_F finds those first class seats that are also rewards program seats. This operation is a kind of database query; querying is the operation that databases are built for.

The moral is that complex data migrations can be specified by constructing functors F between schemas and using the "induced" functors Δ_F, Σ_F, and Π_F. Indeed, in practice essentially all useful migrations can be built up from these. Hence the language of categories provides a framework for specifying and reasoning about data migrations.

3.4.4 Single Set Summaries of Databases

To give a stronger idea of the flavor of Σ and Π, we consider another special case, namely where the target category $\mathcal{D}$ is equal to $\mathbf{1}$; see Exercise 3.6. In this case, there is exactly one functor $\mathcal{C} \to \mathbf{1}$ for any $\mathcal{C}$; let's denote it

$$!\colon \mathcal{C} \to \mathbf{1}. \tag{3.14}$$

Exercise 3.62. Describe the functor $!\colon \mathcal{C} \to \mathbf{1}$ from Eq. (3.14). Where does it send each object? What about each morphism? ◇

We want to consider the data migration functors $\Sigma_!\colon \mathcal{C}\text{-}\mathbf{Inst} \to \mathbf{1}\text{-}\mathbf{Inst}$ and $\Pi_!\colon \mathcal{C}\text{-}\mathbf{Inst} \to \mathbf{1}\text{-}\mathbf{Inst}$. In Example 3.44, we saw that an instance on $\mathbf{1}$ is the same thing as a set. So let's identify $\mathbf{1}\text{-}\mathbf{Inst}$ with $\mathbf{Set}$, and hence discuss

$$\Sigma_!\colon \mathcal{C}\text{-}\mathbf{Inst} \to \mathbf{Set} \qquad \text{and} \qquad \Pi_!\colon \mathcal{C}\text{-}\mathbf{Inst} \to \mathbf{Set}.$$

Given any schema $\mathcal{C}$ and instance $I\colon \mathcal{C} \to \mathbf{Set}$, we will get sets $\Sigma_!(I)$ and $\Pi_!(I)$. Thinking of these sets as database instances, each corresponds to a single one-column table – a controlled vocabulary – summarizing an entire database instance on the schema $\mathcal{C}$.

Consider the following schema

$$\mathcal{G} := \boxed{\begin{array}{ccc} \text{Email} & \xrightarrow[\text{received\_by}]{\text{sent\_by}} & \text{Address} \\ \bullet & & \bullet \\ & \textit{no equations} & \end{array}} \tag{3.15}$$

Here's a sample instance $I\colon \mathcal{G} \to \mathbf{Set}$:

Email	sent_by	received_by		Address
Em_1	Bob	Grace		Bob
Em_2	Grace	Pat		Doug
Em_3	Bob	Emmy		Emmy
Em_4	Sue	Doug		Grace
Em_5	Doug	Sue		Pat
Em_6	Bob	Bob		Sue

Exercise 3.63.   Note that $\mathcal{G}$ from Eq. (3.15) is isomorphic to the schema **Gr**. In Section 3.3.5 we saw that instances on **Gr** are graphs. Draw the above instance I as a graph. $\diamond$

Now we have a unique functor $!: \mathcal{G} \to \mathbf{1}$, and we want to say what $\Sigma_!(I)$ and $\Pi_!(I)$ give us as single-set summaries. First, $\Sigma_!(I)$ tells us all the emailing groups – the "connected components" – in I:

$$
\begin{array}{|l}
\hline
\mathbf{1} \\
\hline
\text{Bob-Grace-Pat-Emmy} \\
\text{Sue-Doug} \\
\hline
\end{array}
$$

This form of summary, involving identifying entries into common groups, or quotients, is typical of Σ-operations.

The functor $\Pi_!(I)$ lists the emails from I which were sent from a person to her- or him-self.

$$
\begin{array}{|l}
\hline
\mathbf{1} \\
\hline
\text{Em\_6} \\
\hline
\end{array}
$$

This is again a sort of query, selecting the entries that fit the criterion of self-to-self emails. Again, this is typical of Π-operations.

Where do these facts – that $\Pi_!$ and $\Sigma_!$ act the way we said – come from? Everything follows from the definition of adjoint functors (3.57): indeed we hope this, together with the examples given in Example 3.61, give the reader some idea of how general and useful adjunctions are, both in mathematics and in database theory.

One more point: while we will not spell out the details, we note that these operations are also examples of constructions known as colimits and limits in **Set**. We end this chapter with bonus material, exploring these key category theoretic constructions. The reader should keep in mind that, in general and not just for functors to $\mathbf{1}$, Σ-operations are built from colimits in **Set**, and Π-operations are built from limits in **Set**.

3.5   Bonus: An Introduction to Limits and Colimits

What do products of sets, the results of $\Pi_!$-operations on database instances, and meets in a preorder all have in common? The answer, as we shall see, is that they are all examples of limits. Similarly, disjoint unions of sets, the results of $\Sigma_!$-operations on database instances, and joins in a preorder are all colimits. Let's begin with limits.

Recall that $\Pi_!$ takes a database instance $I: \mathcal{C} \to \mathbf{Set}$ and turns it into a set $\Pi_!(I)$. More generally, a limit turns a functor $F: \mathcal{C} \to \mathcal{D}$ into an object of $\mathcal{D}$.

3.5.1   Terminal Objects and Products

Terminal objects and products are each a sort of limit. Let's discuss them in turn.

Terminal objects

The most basic limit is a terminal object.

Definition 3.64.   Let $\mathcal{C}$ be a category. Then an object Z in $\mathcal{C}$ is a *terminal object* if, for each object C of $\mathcal{C}$, there exists a unique morphism $!: C \to Z$.

Since this *unique* morphism exists *for all* objects in $\mathcal{C}$, we say that terminal objects have a *universal property*.

Example 3.65.   In **Set**, any set with exactly one element is a terminal object. Why? Consider some such set $\{\bullet\}$. Then for any other set C we need to check that there is exactly one function $!: C \to \{\bullet\}$. This unique function is the one that does the only thing that can be done: it maps each element $c \in C$ to the element $\bullet \in \{\bullet\}$.

Exercise 3.66.   Let $(P, \leq)$ be a preorder, let $z \in P$ be an element, and let $\mathcal{P}$ be the corresponding category (see Section 3.2.3). Show that z is a terminal object in $\mathcal{P}$ if and only if it is a *top element* in P: that is, if and only if for all $c \in P$ we have $c \leq z$.   ◇

Exercise 3.67.   Name a terminal object in the category **Cat**. (Hint: recall Exercise 3.62.)   ◇

Exercise 3.68.   Not every category has a terminal object. Find one that doesn't.   ◇

Proposition 3.69.   All terminal objects in a category $\mathcal{C}$ are isomorphic.

Proof.   This is a simple, but powerful standard argument. Suppose Z and Z' are both terminal objects in some category $\mathcal{C}$. Then there exist (unique) maps $a: Z \to Z'$ and $b: Z' \to Z$. Composing these, we get a map $a \,\mathring{,}\, b: Z \to Z$. Now since Z is terminal, this map $Z \to Z$ must be unique. But id_Z is also such a map. So we must have $a \,\mathring{,}\, b = \mathrm{id}_Z$. Similarly, we find that $b \,\mathring{,}\, a = \mathrm{id}_{Z'}$. Thus a is an isomorphism, with inverse b.   □

Remark 3.70 ("The limit" vs. "a limit").   Not only are all terminal objects isomorphic, there is a unique isomorphism between any two. We hence say "terminal objects are unique up to unique isomorphism." To a category theorist, this is very nearly the same thing as saying "all terminal objects are equal." Thus we often abuse terminology and talk of "the" terminal object, rather than "a" terminal object. We will do the same for any sort of limit or colimit, e.g. speak of "the product" of two sets, rather than "a product." We saw a similar phenomenon in Definition 1.76.

Products

Products are slightly more complicated to formalize than terminal objects, but they are familiar in practice.

Definition 3.71. Let $\mathcal{C}$ be a category, and let X, Y be objects in $\mathcal{C}$. A *product* of X and Y is an object, denoted $X \times Y$, together with morphisms $p_X \colon X \times Y \to X$ and $p_Y \colon X \times Y \to Y$ such that for all objects C together with morphisms $f \colon C \to X$ and $g \colon C \to Y$, there exists a unique morphism $C \to X \times Y$, denoted $\langle f, g \rangle$, for which the following diagram commutes:

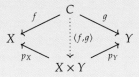

We will try to bring this down to earth in Example 3.72. Before we do, note that $X \times Y$ is an object equipped with morphisms to X and Y. Roughly speaking, it is like "the best object-equipped-with-morphisms-to-X-and-Y" in all of $\mathcal{C}$, in the sense that any other object-equipped-with-morphisms-to-X-and-Y maps to it uniquely. This is called a *universal property*. It's customary to use a dotted line to indicate the unique morphism that exists because of some universal property.

Example 3.72. In **Set**, a product of two sets X and Y is their usual cartesian product

$$X \times Y := \{(x, y) \mid x \in X, y \in Y\},$$

which comes with two projections $p_X \colon X \times Y \to X$ and $p_Y \colon X \times Y \to Y$, given by $p_X(x, y) := x$ and $p_Y(x, y) := y$.

Given any set C with functions $f \colon C \to X$ and $g \colon C \to Y$, the unique map from C to $X \times Y$ such that the required diagram commutes is given by $\langle f, g \rangle(c) := (f(c), g(c))$.

Here is a picture of the product $\underline{6} \times \underline{4}$ of sets $\underline{6}$ and $\underline{4}$.

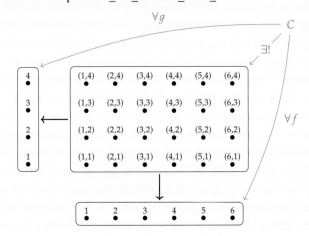

Exercise 3.73. Let $(P, \leq)$ be a preorder, let $x, y \in P$ be elements, and let $\mathcal{P}$ be the corresponding category. Show that the product $x \times y$ in $\mathcal{P}$ agrees with their meet $x \wedge y$ in P. ◇

Example 3.74. Given two categories $\mathcal{C}$ and $\mathcal{D}$, their product $\mathcal{C} \times \mathcal{D}$ may be given as follows. The objects of this category are pairs (c, d), where c is an object of $\mathcal{C}$ and d is an object of $\mathcal{D}$. Similarly, morphisms $(c, d) \to (c', d')$ are pairs (f, g) where $f : c \to c'$ is a morphism in $\mathcal{C}$ and $g : d \to d'$ is a morphism in $\mathcal{D}$. Composition of morphisms is simply given by composing each entry in the pair separately, so $(f, g) \mathbin{\raisebox{0.3ex}{\scriptsize ⨾}} (f', g') = (f \mathbin{\raisebox{0.3ex}{\scriptsize ⨾}} f', g \mathbin{\raisebox{0.3ex}{\scriptsize ⨾}} g')$.

Exercise 3.75.

1. What are the identity morphisms in a product category $\mathcal{C} \times \mathcal{D}$?
2. Why is composition in a product category associative?
3. What is the product category $\mathbf{1} \times \mathbf{2}$?
4. What is the product category $\mathcal{P} \times \mathcal{Q}$ when P and Q are preorders and $\mathcal{P}$ and $\mathcal{Q}$ are the corresponding categories? ◇

These two constructions, terminal objects and products, are subsumed by the notion of limit.

3.5.2   Limits

We'll get a little abstract. Consider the definition of product. This says that, given any pair of maps $X \xleftarrow{f} C \xrightarrow{g} Y$, there exists a unique map $C \to X \times Y$ such that certain diagrams commute. This has the flavor of being terminal – there is a unique map to $X \times Y$ – but it seems a bit more complicated. How are the two ideas related?

It turns out that products *are* terminal objects, but of a different category, which we'll call $\mathbf{Cone}(X, Y)$, *the category of cones over X and Y in $\mathcal{C}$*. We shall see in Exercise 3.76 that $X \xleftarrow{p_X} X \times Y \xrightarrow{p_Y} Y$ is a terminal object in $\mathbf{Cone}(X, Y)$.

An object of $\mathbf{Cone}(X, Y)$ is simply a pair of maps $X \xleftarrow{f} C \xrightarrow{g} Y$. A morphism from $X \xleftarrow{f} C \xrightarrow{g} Y$ to $X \xleftarrow{f'} C' \xrightarrow{g'} Y$ in $\mathbf{Cone}(X, Y)$ is a morphism $a : C \to C'$ in $\mathcal{C}$ such that the following diagram commutes:

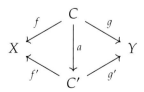

Exercise 3.76. Check that a product $X \xleftarrow{p_X} X \times Y \xrightarrow{p_Y} Y$ is exactly the same as a terminal object in $\mathbf{Cone}(X, Y)$. ◇

We're now ready for the abstract definition. Don't worry if the details are unclear; the main point is that it is possible to unify terminal objects, maximal elements, and meets, products of sets, preorders, and categories, and many other familiar friends under the scope of a single definition. In fact, they're all just terminal objects in different categories.

Recall from Definition 3.42 that, formally speaking, a diagram in $\mathcal{C}$ is just a functor $D \colon \mathcal{J} \to \mathcal{C}$. Here $\mathcal{J}$ is called the *indexing category* of the diagram D.

Definition 3.77.   Let $D \colon \mathcal{J} \to \mathcal{C}$ be a diagram. A *cone* (C, c_*) over D consists of

(i)   an object $C \in \mathcal{C}$;
(ii) for each object $j \in \mathcal{J}$, a morphism $c_j \colon C \to D(j)$.

To be a cone, these must satisfy the following property:

(a)   for each $f \colon j \to k$ in $\mathcal{J}$, we have $c_k = c_j \, \mathbin{\raise.2ex\hbox{$\circ$}} \, D(f)$.

A *morphism of cones* $(C, c_*) \to (C', c'_*)$ is a morphism $a \colon C \to C'$ in $\mathcal{C}$ such that for all $j \in \mathcal{J}$ we have $c_j = a \, \mathbin{\raise.2ex\hbox{$\circ$}} \, c'_j$. Cones over D, and their morphisms, form a category **Cone**(D).

The *limit* of D, denoted $\lim(D)$, is the terminal object in the category **Cone**(D). Say it is the cone $\lim(D) = (C, c_*)$; we refer to C as the *limit object* and the map c_j for any $j \in \mathcal{J}$ as the jth *projection map*.

For visualization purposes, if $\mathcal{J}$ is the free category on the graph

with five objects and five non-identity morphisms, then we may draw a diagram $D \colon \mathcal{J} \to \mathcal{C}$ inside $\mathcal{C}$ as on the left below, and a cone on it as on the right:

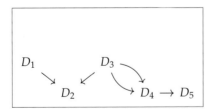 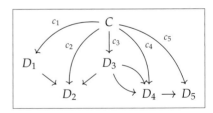

Here, any two parallel paths that start at C are considered the same. Note that both these diagrams depict a collection of objects and morphisms inside the category $\mathcal{C}$.

Example 3.78. Terminal objects are limits where the indexing category is empty, $\mathcal{J} = \varnothing$.

Example 3.79. Products are limits where the indexing category consists of two objects v, w and no arrows, $\mathcal{J} = \boxed{\begin{array}{cc} v & w \\ \bullet & \bullet \end{array}}$.

3.5.3   Finite Limits in **Set**

Recall that this discussion was inspired by wanting to understand Π-operations, and in particular $\Pi_!$. We can now see that a database instance $I \colon \mathcal{C} \to$ **Set** is a diagram in **Set**. The functor $\Pi_!$ takes the limit of this diagram. In this subsection we give a formula describing the result. This captures *all finite limits in* **Set**.

In database theory, we work with categories $\mathcal{C}$ that are presented by a finite graph plus equations. We won't explain the details, but in fact it's enough just to work with the graph part: as far as limits are concerned, the equations in $\mathcal{C}$ don't matter. For consistency with the rest of this section, let's denote the database schema by $\mathcal{J}$ instead of $\mathcal{C}$.

Theorem 3.80. Let $\mathcal{J}$ be a category presented by the finite graph (V, A, s, t) together with some equations, and let $D \colon \mathcal{J} \to$ **Set** be a set-valued functor. Write $V = \{v_1, \ldots, v_n\}$. The set

$$\lim_{\mathcal{J}} D := \big\{ (d_1, \ldots, d_n) \mid d_i \in D(v_i) \text{ for all } 1 \le i \le n \text{ and}$$
$$\text{for all } a \colon v_i \to v_j \in A, \text{ we have } D(a)(d_i) = d_j \big\}$$

together with the projection maps $p_i \colon (\lim_{\mathcal{J}} D) \to D(v_i)$ given by $p_i(d_1, \ldots, d_n) := d_i$, is a limit of D.

Example 3.81. If J is the empty graph $\square$, then $n = 0$: there are no vertices. There is exactly one empty tuple ( ), which vacuously satisfies the properties, so we've constructed the limit as the singleton set $\{( )\}$ consisting of just the empty tuple. Thus the limit of the empty diagram, i.e. the terminal object in **Set**, is the singleton set. See Remark 3.70.

Exercise 3.82. Show that the limit formula in Theorem 3.80 works for products. See Example 3.79. $\diamond$

Exercise 3.83. If $D \colon \mathbf{1} \to$ **Set** is a functor, what is the limit of D? Compute it using Theorem 3.80, and check your answer against Definition 3.77. $\diamond$

Pullbacks

In particular, the condition that the limit of $D \colon \mathcal{J} \to$ **Set** selects tuples $(d_1, \ldots, d_n)$ such that $D(a)(d_i) = d_j$ for each morphism $a \colon i \to j$ in $\mathcal{J}$ allows us to use limits to select data that satisfies certain equations or constraints. This is what allows us to express queries in terms of limits. Here is an example.

Example 3.84. If $\mathcal{J}$ is presented by the *cospan* graph $\boxed{\overset{x}{\bullet} \overset{f}{\longrightarrow} \overset{a}{\bullet} \overset{g}{\longleftarrow} \overset{y}{\bullet}}$, then its limit is known as a *pullback*. Given the diagram $X \overset{f}{\to} A \overset{g}{\leftarrow} Y$, the pullback is the cone shown on the left below:

$$
\begin{array}{ccc}
C & \overset{c_y}{\longrightarrow} & Y \\
{\scriptstyle c_x}\downarrow & \searrow{\scriptstyle c_a} & \downarrow{\scriptstyle g} \\
X & \underset{f}{\longrightarrow} & A
\end{array}
\qquad
\begin{array}{ccc}
X \times_A Y & \overset{c_y}{\longrightarrow} & Y \\
{\scriptstyle c_x}\downarrow & \lrcorner & \downarrow{\scriptstyle g} \\
X & \underset{f}{\longrightarrow} & A
\end{array}
$$

The fact that the diagram commutes means that the diagonal arrow c_a is in some sense superfluous, so one generally denotes pullbacks by dropping the diagonal arrow, naming the cone point $X \times_A Y$, and adding the $\lrcorner$ symbol, as shown to the right above.

Here is a picture to help us unpack the definition in **Set**. We take $X = \underline{6}$, $Y = \underline{4}$, and A to be the set of colors {red, blue, black}.

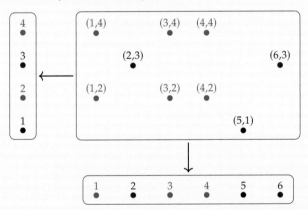

The functions $f \colon \underline{6} \to A$ and $g \colon \underline{4} \to A$ are expressed in the coloring of the dots: for example, $g(2) = g(4) = $ red, while $f(5) = $ black. The pullback selects pairs $(i, j) \in \underline{6} \times \underline{4}$ such that $f(i)$ and $g(j)$ have the same color.

Remark 3.85. As mentioned following Definition 3.56, this definition of pullback is not to be confused with the pullback of a set-valued functor along a functor; they are for now best thought of as different concepts which accidentally have the same name. Owing to the power of the primordial ooze, however, the pullback along a functor is a special case of pullback as the limit of a cospan: it can be understood as the pullback of a certain cospan in **Cat**. To unpack this requires the notions of category of elements and discrete opfibration; ask your friendly neighborhood category theorist.

3.5.4 A Brief Note on Colimits

Just like upper bounds have a dual concept – namely that of lower bounds – so limits have a dual concept: colimits. To expose the reader to this concept, we provide a succinct definition of these using opposite categories and opposite functors. The point, however, is just exposure; we will return to explore colimits in detail in Chapter 6.

Exercise 3.86. Recall from Example 3.20 that every category $\mathcal{C}$ has an opposite $\mathcal{C}^{op}$. Let $F: \mathcal{C} \rightarrow \mathcal{D}$ be a functor. How should we define its opposite, $F^{op}: \mathcal{C}^{op} \rightarrow \mathcal{D}^{op}$? That is, how should F^{op} act on objects, and how should it act on morphisms? ◇

Definition 3.87. Given a category $\mathcal{C}$ we say that a *cocone* in $\mathcal{C}$ is a cone in $\mathcal{C}^{op}$.
Given a diagram $D: \mathcal{J} \rightarrow \mathcal{C}$, we may take the limit of the functor $D^{op}: \mathcal{J}^{op} \rightarrow \mathcal{C}^{op}$.
   This is a cone in $\mathcal{C}^{op}$, and so by definition a cocone in $\mathcal{C}$. The *colimit* of D is this cocone.

Definition 3.87 is like a compressed file: useful for transmitting quickly, but completely useless for working with, unless you can successfully unpack it. We will unpack it later in Chapter 6 when we discuss electric circuits.

3.6 Summary and Further Reading

Congratulations on making it through one of the longest chapters in the book! We apologize for the length, but this chapter had a lot of work to do. Namely it introduced the "big three" of category theory – categories, functors, and natural transformations – as well as discussing adjunctions, limits, and very briefly colimits.

That's really quite a bit of material. For more on all these subjects, one can consult any standard book on category theory, of which there are many. The bible (old, important, seminal, and requires a priest to explain it) is [Mac98]; another thorough introduction is [Bor94]; a logical perspective is given in [Awo10]; a computer science perspective is given in [BW90], [Pie91], and [Wal92]; math students should probably read [Lei14], [Rie17], or [Gra18]; a general audience might start with [Spi14a].

We presented categories from a database perspective, because data is pretty ubiquitous in our world. A database schema – i.e. a system of interlocking tables – can be captured by a category $\mathcal{C}$, and filling it with data corresponds to a functor $\mathcal{C} \rightarrow \mathbf{Set}$. Here $\mathbf{Set}$ is the category of sets, perhaps the most important category to mathematicians.

The perspective of using category theory to model databases has been rediscovered several times. It seems to have first been discussed by various authors around the mid 1990s [IP94; CD95; PS95; TG96]. Bob Rosebrugh and collaborators took it much further in a series of papers including [FGR03; JR02; RW92]. Most of these authors tend to focus on sketches, which are more expressive categories. Spivak rediscovered the idea again quite a bit later, but focused on categories rather than sketches, so as to have all three data migration functors Δ, Σ, Π; see [Spi12; SW15b]. The version of this story presented in the chapter, including the white and black nodes in schemas, is part of a larger theory of algebraic databases, where a programming language such as Java or Haskell is attached to a database. The technical details are worked out in [Sch+17], and its use in database integration projects can be found in [SW15a; Wis+15].

Before we leave this chapter, we want to emphasize two things: coherence conditions and universal constructions.

Coherence conditions

In the definitions of category, functor, and natural transformations, we have data (indexed by (i)) that is required to satisfy certain properties (indexed by (a)). Indeed, for categories it was about associativity and unitality of composition, for functors it was about respecting composition and identities, and for natural transformations it was the naturality condition. These conditions are often called *coherence conditions*: we want the various structures to cohere, to work well together, rather than to flop around unattached.

Understanding why these particular structures and coherence conditions are "the right ones" is more science than mathematics: we empirically observe that certain combinations result in ideas that are both widely applicable and also strongly compositional. That is, we become satisfied with coherence conditions when they result in beautiful mathematics down the road.

Universal constructions

Universal constructions are one of the most important themes of category theory. Roughly speaking, one gives some specified shape in a category and says "find me the best solution!" And category theory comes back and says "do you want me to approximate from the left or the right (colimit or limit)?" You respond, and either there is a best solution or there is not. If there is, it's called the (co)limit; if there's not we say "the (co)limit does not exist."

Even data migration fits this form. We say "find me the closest thing in $\mathcal{D}$ that matches my $\mathcal{C}$-instance using my functor $F : \mathcal{C} \to \mathcal{D}$." In fact this approach – known as Kan extensions – subsumes the others. One of the two founders of category theory, Saunders Mac Lane, has a section in his book [Mac98] called "All concepts are Kan extensions," a big statement, no?

4 Collaborative Design: Profunctors, Categorification, and Monoidal Categories

4.1 Can We Build It?

When designing a large-scale system, many different fields of expertise are joined to work on a single project. Thus the whole project team is divided into multiple sub-teams, each of which is working on a sub-project. And we recurse downward: the sub-project is again factored into sub-sub-projects, each with their own team. One could refer to this sort of hierarchical design process as *collaborative design*, or co-design. In this chapter, we discuss a mathematical theory of co-design, due to Andrea Censi [Cen15].

Consider just one level of this hierarchy: a project and a set of teams working on it. Each team is supposed to *provide* resources – sometimes called "functionalities" – to the project, but the team also *requires* resources in order to do so. Different design teams must be allowed to plan and work independently from one another in order for progress to be made. Yet the design decisions made by one group affect the design decisions others can make: if A wants more space in order to provide a better radio speaker, then B must use less space. So these teams – though ostensibly working independently – are dependent on each other after all.

The combination of dependence and independence is crucial for progress to be made, and yet it can cause major problems. When a team requires more resources than it originally expected to require, or if it cannot provide the resources that it originally claimed it could provide, the usual response is for the team to issue a design-change notice. But these affect neighboring teams: if team A now requires more than originally claimed, team B may have to change their design, which can in turn affect team C. Thus these design-change notices can ripple through the system through feedback loops and can cause whole projects to fail [SLG+15].

As an example, consider the design problem of creating a robot to carry some load at some velocity. The top-level planner breaks the problem into three design teams: chassis team, motor team, and battery team. Each of these teams could break up into multiple parts and the process repeated, but let's remain at the top level and consider the resources produced and the resources required by each of our three teams.

The chassis in some sense provides all the functionality – it carries the load at the velocity – but it requires some things in order to do so. It requires money to build, of course, but more to the point it requires a source of torque and speed. These are supplied by the motor, which in turn needs voltage and current from the battery. Both the motor and the battery cost money, but more importantly they need to be carried by the chassis:

they become part of the load. A feedback loop is created: the chassis must carry all the weight, even that of the parts that power the chassis. A heavier battery might provide more energy to power the chassis, but is the extra power worth the heavier load?

In the following picture, each part – chassis, motor, battery, and robot – is shown as a box with ports on the left and right. The functionalities, or resources produced by the part, are shown as ports on the left of the box, and the resources required by the part are shown as ports on its right.

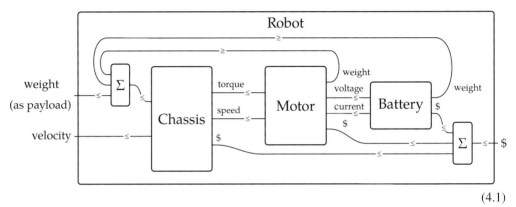

$$(4.1)$$

The boxes marked Σ correspond to summing inputs. These boxes are not to be designed, but we shall see later that they fit easily into the same conceptual framework. Note also the $\leq$s on each wire; they indicate that if box A requires a resource that box B produces, then A's requirement must be less-than-or-equal-to B's production. The chassis requires torque, and the motor must produce at least that much torque.

To formalize this a bit more, let's call diagrams like the one above *co-design diagrams*. Each of the wires in a co-design diagram represents a preorder of resources. For example, in Eq. (4.1) every wire corresponds to a resource type – weights, velocities, torques, speeds, costs, voltages, and currents – where resources of each type can be ordered from less useful to more useful. In general, these preorders do not have to be linear orders, though in the above cases each will likely correspond to a linear order: $\$10 \leq \$20$, 5W $\leq$ 6W, and so on.

Each of the boxes in a co-design diagram corresponds to what we call a *feasibility relation*. A feasibility relation matches resource production with requirements. For every pair $(p, r) \in P \times R$, where P is the preorder of resources to be produced and R is the preorder of resources to be required, the box says "true" or "false" – feasible or infeasible – for that pair. In other words, "yes I can provide p given r" or "no, I cannot provide p given r."

Feasibility relations hence define a function $\Phi\colon P \times R \to \mathbf{Bool}$. For a function $\Phi\colon P \times R \to \mathbf{Bool}$ to make sense as a feasibility relation, however, there are two conditions:

(a) If $\Phi(p, r) = \mathtt{true}$ and $p' \leq p$, then $\Phi(p', r) = \mathtt{true}$.
(b) If $\Phi(p, r) = \mathtt{true}$ and $r \leq r'$ then $\Phi(p, r') = \mathtt{true}$.

These conditions, which we shall see again in Definition 4.1, say that if you can produce p given resources r, you can (a) also produce less $p' \leq p$ with the same resources r, and (b) also produce p given more resources $r' \geq r$. We shall see that these two conditions are formalized by requiring Φ to be a monotone map $P^{\mathrm{op}} \times R \to \mathbf{Bool}$.

A *co-design problem*, represented by a co-design diagram, asks us to find the composite of some feasibility relations. It asks, for example, given these capabilities of the chassis, motor, and battery teams, can we build a robot together? Indeed, a co-design diagram factors a problem – for example, that of designing a robot – into interconnected subproblems, as in Eq. (4.1). Once the feasibility relation is worked out for each of the subproblems, i.e. the inner boxes in the diagram, the mathematics provides an algorithm producing the feasibility relation of the whole outer box. This process can be recursed downward, from the largest problem to tiny sub-problems.

In this chapter, we will understand co-design problems in terms of enriched profunctors, in particular **Bool**-profunctors. A **Bool**-profunctor is like a bridge connecting one preorder to another. We will show how the co-design framework gives rise to a structure known as a compact closed category, and that any compact closed category can interpret the sorts of wiring diagrams we see in Eq. (4.1).

4.2   Enriched Profunctors

In this section we will understand how co-design problems form a category. Along the way we will develop some abstract machinery that will allow us to replace preorder design spaces with other enriched categories.

4.2.1   Feasibility Relationships as **Bool**-Profunctors

The theory of co-design is based on preorders: each resource – e.g. velocity, torque, or $ – is structured as a preorder. The order $x \leq y$ represents the *availability of x given y*, i.e. that whenever you have y, you also have x. For example, in our preorder of wattage, if $5\mathrm{W} \leq 10\mathrm{W}$, it means that whenever we are provided 10W, we implicitly also have 5W. Above we referred to this as an order from less useful to more useful: if x is always available given y, then x is less useful than y.

We know from Section 2.3.2 that a preorder $\mathcal{X}$ can be conceived of as a **Bool**-category. Given $x, y \in X$, we have $\mathcal{X}(x, y) \in \mathbb{B}$; this value responds to the assertion "x is available given y," marking it either `true` or `false`.

Our goal is to see feasibility relations as **Bool**-profunctors, which are a special case of something called enriched profunctors. Indeed, we hope that this chapter will give you some intuition for profunctors, arising from the table

Bool-category	preorder
Bool-functor	monotone map
Bool-profunctor	feasibility relation

Because enriched profunctors are a bit abstract, we first concretely discuss **Bool**-profunctors as feasibility relations. Recall that if $\mathcal{X} = (X, \leq)$ is a preorder, then its opposite $\mathcal{X}^{op} = (X, \geq)$ has $x \geq y$ iff $y \leq x$.

Definition 4.1. Let $\mathcal{X} = (X, \leq_X)$ and $\mathcal{Y} = (Y, \leq_Y)$ be preorders. A *feasibility relation* for $\mathcal{X}$ given $\mathcal{Y}$ is a monotone map

$$\Phi: \mathcal{X}^{op} \times \mathcal{Y} \to \textbf{Bool}. \tag{4.2}$$

We denote this by $\Phi: \mathcal{X} \nrightarrow \mathcal{Y}$.

Given $x \in X$ and $y \in Y$, if $\Phi(x, y) = \texttt{true}$ we say x *can be obtained given* y.

As mentioned in the introduction, the requirement that Φ is monotone says that if $x' \leq_X x$ and $y \leq_Y y'$ then $\Phi(x, y) \leq_{\textbf{Bool}} \Phi(x', y')$. In other words, if x can be obtained given y, and if x' is available given x, then x' can be obtained given y. And if furthermore y is available given y', then x' can also be obtained given y'.

Exercise 4.2. Suppose we have the preorders

1. Draw the Hasse diagram for the preorder $\mathcal{X}^{op} \times \mathcal{Y}$.
2. Write down a profunctor $\Lambda: \mathcal{X} \nrightarrow \mathcal{Y}$ and, reading $\Lambda(x, y) = \texttt{true}$ as "my aunt can explain an x given y," give an interpretation of the fact that the preimage of $\texttt{true}$ forms an upper set in $\mathcal{X}^{op} \times \mathcal{Y}$.   ◇

To generalize the notion of feasibility relation, we must notice that the symmetric monoidal preorder **Bool** has more structure than just that of a symmetric monoidal preorder: as mentioned in Exercise 2.69, **Bool** is a quantale. That means it has all joins $\vee$, and a hom-operation, which we'll write $\Rightarrow: \mathbb{B} \times \mathbb{B} \to \mathbb{B}$. By definition, this operation satisfies the property that for all $b, c, d \in \mathbb{B}$ one has

$$b \wedge c \leq d \quad \text{iff} \quad b \leq (c \Rightarrow d). \tag{4.3}$$

The operation $\Rightarrow$ is given by the following table:

c	d	$c \Rightarrow d$
true	true	true
true	false	false
false	true	true
false	false	true

(4.4)

Exercise 4.3. Show that $\Rightarrow$ as defined in Eq. (4.4) indeed satisfies Eq. (4.3).   ◇

On an abstract level, it is the fact that **Bool** is a quantale which makes everything in this chapter work; any other (unital commutative) quantale also defines a way to interpret co-design diagrams. For example, we could use the quantale **Cost**, which would describe not *whether* x is available given y but the *cost* of obtaining x given y; see Example 2.21 and 2.30.

4.2.2 $\mathcal{V}$-Profunctors

We are now ready to recast Eq. (4.2) in abstract terms. Recall the notions of enriched product (Definition 2.53), enriched functor (Definition 2.48), and quantale (Definition 2.57).

Definition 4.4. Let $\mathcal{V} = (V, \leq, I, \otimes)$ be a (unital commutative) quantale,[1] and let $\mathcal{X}$ and $\mathcal{Y}$ be $\mathcal{V}$-categories. A $\mathcal{V}$-*profunctor* from $\mathcal{X}$ to $\mathcal{Y}$, denoted $\Phi \colon \mathcal{X} \nrightarrow \mathcal{Y}$, is a $\mathcal{V}$-functor

$$\Phi \colon \mathcal{X}^{\mathrm{op}} \times \mathcal{Y} \to \mathcal{V}.$$

Note that a $\mathcal{V}$-functor must have $\mathcal{V}$-categories for domain and codomain, so here we are considering $\mathcal{V}$ as enriched in itself; see Remark 2.65.

Exercise 4.5. Show that a $\mathcal{V}$-profunctor (Definition 4.4) is the same as a function $\Phi \colon \mathrm{Ob}(\mathcal{X}) \times \mathrm{Ob}(\mathcal{Y}) \to V$ such that for any $x, x' \in \mathcal{X}$ and $y, y' \in \mathcal{Y}$ the following inequality holds in $\mathcal{V}$:

$$\mathcal{X}(x', x) \otimes \Phi(x, y) \otimes \mathcal{Y}(y, y') \leq \Phi(x', y'). \qquad \diamond$$

Exercise 4.6. Is it true that a **Bool**-profunctor, as in Definition 4.4, is exactly the same as a feasibility relation, as in Definition 4.1, once you peel back all the jargon? Or is there some subtle difference? $\qquad \diamond$

We know that Definition 4.4 is quite abstract. But have no fear, we will take you through it in pictures.

Example 4.7 (**Bool**-profunctors and their interpretation as bridges). Let's discuss Definition 4.4 in the case $\mathcal{V} = $ **Bool**. One way to imagine a **Bool**-profunctor $\Phi \colon X \nrightarrow Y$ is in terms of building bridges between two cities. Recall that a preorder (a **Bool**-category) can be drawn using a Hasse diagram. We'll think of the preorder as a city, and each vertex in it as some point of interest. An arrow $A \to B$ in the Hasse diagram means that there exists a way to get from point A to point B in the city. So what's a profunctor?

A profunctor is just a bunch of bridges connecting points in one city to points in another. Let's see a specific example. Here is a picture of a **Bool**-profunctor $\Phi \colon X \nrightarrow Y$:

---

[1] From here on, as in Chapter 2, whenever we speak of quantales we mean unital commutative quantales.

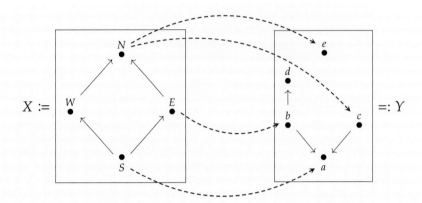

Both X and Y are preorders, e.g. with $W \leq N$ and $b \leq a$. With bridges coming from the profunctor in blue, one can now use both paths within the cities and the bridges to get from points in city X to points in city Y. For example, since there is a path from N to e and E to a, we have $\Phi(N, e) = $ true and $\Phi(E, a) = $ true. On the other hand, since there is no path from W to d, we have $\Phi(W, d) = $ false.

In fact, one could put a box around this entire picture and see a new preorder with $W \leq N \leq c \leq a$, etc. This is called the *collage* of Φ; we'll explore this in more detail later; see Definition 4.31.

Exercise 4.8.   We can express Φ as a matrix where the (m, n)th entry is the value of $\Phi(m, n) \in \mathbb{B}$. Copy and complete the **Bool**-matrix:

Φ	a	b	c	d	e
N	?	?	?	?	true
E	true	?	?	?	?
W	?	?	?	false	?
S	?	?	?	?	?

We'll call this the *feasibility matrix* of Φ.   ◇

Example 4.9 (Cost-profunctors and their interpretation as bridges). Let's now consider **Cost**-profunctors. Again we can view these as bridges, but this time our bridges are labeled by their length. Recall from Definition 2.36 and Eq. (2.18) that **Cost**-categories are Lawvere metric spaces, and can be depicted using weighted graphs. We'll think of such a weighted graph as a chart of distances between points in a city, and generate a **Cost**-profunctor by building a few bridges between the cities.

Here is a depiction of a **Cost**-profunctor $\Phi \colon X \nrightarrow Y$:

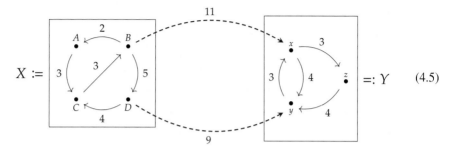

$$X := \qquad\qquad\qquad\qquad\qquad\qquad\qquad\qquad\qquad =: Y \qquad (4.5)$$

The distance from a point x in city X to a point y in city Y is given by the shortest path that runs from x through X, then across one of the bridges, and then through Y to the destination y. So for example

$$\Phi(B, x) = 11, \quad \Phi(A, z) = 20, \quad \Phi(C, y) = 17.$$

Exercise 4.10. Copy and complete the **Cost**-matrix:

Φ	x	y	z
A	?	?	20
B	11	?	?
C	?	17	?
D	?	?	?

$\diamond$

Remark 4.11 (Computing profunctors via matrix multiplication). We can give an algorithm for computing the above distance matrix using matrix multiplication. First, just like in Eq. (2.20), we can begin with the labeled graphs in Eq. (4.5) and read off the matrices of arrow labels for X, Y, and Φ:

M_X	A	B	C	D
A	0	∞	3	∞
B	2	0	∞	5
C	∞	3	0	∞
D	∞	∞	4	0

M_Φ	x	y	z
A	∞	∞	∞
B	11	∞	∞
C	∞	∞	∞
D	∞	9	∞

M_Y	x	y	z
x	0	4	3
y	3	0	∞
z	∞	4	0

Recall from Section 2.5.3 that the matrix of distances d_Y for **Cost**-category X can be obtained by taking the matrix power of M_X with smallest entries, and similarly for Y. The matrix of distances for the profunctor Φ will be equal to $d_X * M_\Phi * d_Y$. In fact, since X has four elements and Y has three, we also know that $\Phi = M_X^3 * M_\Phi * M_Y^2$.

Exercise 4.12.   Calculate $M_X^3 * M_\Phi * M_Y^2$, remembering to do matrix multiplication according to the (min, +)-formula for matrix multiplication in the quantale **Cost**; see Eq. (2.26).

Your answer should agree with what you got in Exercise 4.10; does it?   ◇

4.2.3   Back to Co-design Diagrams

Each box in a co-design diagram has a left-hand and a right-hand side, which in turn consist of a collection of ports, which in turn are labeled by preorders. For example, consider the chassis box below:

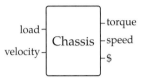

Its left side consists of two ports – one for load and one for velocity – and these are the functionality that the chassis produces. Its right side consists of three ports – one for torque, one for speed, and one for $ – and these are the resources that the chassis requires. Each of these resources is to be taken as a preorder. For example, load might be the preorder $([0, \infty], \leq)$, where an element $x \in [0, \infty]$ represents the idea "I can handle any load up to x," while $ might be the two-element preorder $\{up\_to\_\$100, more\_than\_\$100\}$, where the first element of this set is less than the second.

We then multiply – i.e. we take the product preorder – of all preorders on the left, and similarly for those on the right. The box then represents a feasibility relation between the results. For example, the chassis box above represents a feasibility relation

$$\text{Chassis}: \big(\text{load} \times \text{velocity}\big) \nrightarrow \big(\text{torque} \times \text{speed} \times \$\big).$$

Let's walk through this a bit more concretely. Consider the design problem of filming a movie, where you must pit the tone and entertainment value against the cost. A feasibility relation describing this situation details what tone and entertainment value can be obtained at each cost; as such, it is described by a feasibility relation $\Phi: (T \times E) \nrightarrow \$$. We represent this by the box

$$T\text{-}\boxed{\Phi}\text{-}\$$$
$$E\text{-}$$

where T, E, and $ are the preorders drawn below:

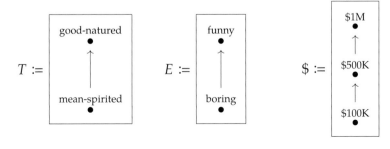

A possible feasibility relation is then described by the profunctor

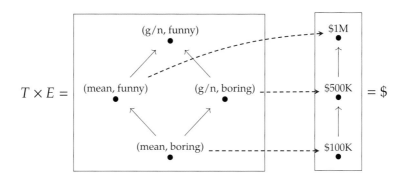

$$T \times E =$$

This says, for example, that a good-natured but boring movie costs $500K to produce (of course, the producers would also be happy to get $1M).

To elaborate, each arrow in the above diagram is to be interpreted as saying, "I can provide the source given the target." For example, there are arrows witnessing each of "I can provide $500K given $1M," "I can provide a good-natured but boring movie given $500K," and "I can provide a mean and boring movie given a good-natured but boring movie." Moreover, this relationship is transitive, so the path from (mean, boring) to $1M indicates also that "I can provide a mean and boring movie given $1M."

Note the similarity and difference with the bridge interpretation of profunctors in Example 4.7: the arrows still indicate the possibility of moving between source and target, but in this co-design driven interpretation we understand them as indicating that it is possible to get *to* the source *from* the target.

Exercise 4.13. In the above diagram, the node (g/n, funny) has no dashed blue arrow emerging from it. Is this valid? If so, what does it mean?   ◇

4.3   Categories of Profunctors

There is a category **Feas** whose objects are preorders and whose morphisms are feasibility relations. In order to describe it, we must give the composition formula and the identities, and prove that they satisfy the properties of being a category: unitality and associativity.

4.3.1   Composing Profunctors

If feasibility relations are to be morphisms, we need to give a formula for composing two of them in series. Imagine you have cities $\mathcal{P}$, $\mathcal{Q}$, and $\mathcal{R}$ and you have bridges – and hence feasibility matrices – connecting these cities, say $\Phi \colon \mathcal{P} \nrightarrow \mathcal{Q}$ and $\Psi \colon \mathcal{Q} \nrightarrow \mathcal{R}$.

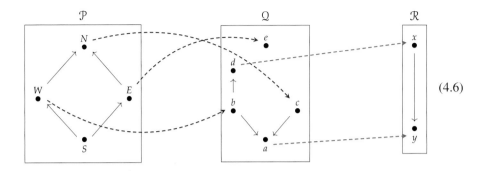

(4.6)

The feasibility matrices for Φ (in blue) and Ψ (in red) are:

Φ	a	b	c	d	e
N	true	false	true	false	false
E	true	false	true	false	true
W	true	true	true	true	false
S	true	true	true	true	true

Ψ	x	y
a	false	true
b	true	true
c	false	true
d	true	true
e	false	false

As in Remark 2.71, we personify a quantale as a navigator. So imagine a navigator is trying to give a feasibility matrix $\Phi \, \mathring{,} \, \Psi$ for getting from $\mathcal{P}$ to $\mathcal{R}$. How should this be done? Basically, for every pair $p \in \mathcal{P}$ and $r \in \mathcal{R}$, the navigator searches through $\mathcal{Q}$ for a way-point q, somewhere both to which we can get from p AND from which we can get to r. It is true that we can navigate from p to r iff there is a way-point q through which to travel; this is a big OR over all possible q. The composition formula is thus:

$$(\Phi \, \mathring{,} \, \Psi)(p, r) := \bigvee_{q \in \mathcal{Q}} \Phi(p, q) \wedge \Psi(q, r). \tag{4.7}$$

But as we said in Eq. (2.26), this can be thought of as matrix multiplication. In our example, the result is

$\Phi \, \mathring{,} \, \Psi$	x	y
N	false	true
E	false	true
W	true	true
S	true	true

and one can check that this answers the question, "can you get from here to there" in Eq. (4.6): you can't get from N to x but you can get from N to y.

The formula (4.7) is written in terms of the quantale **Bool**, but it works for arbitrary (unital commutative) quantales. We give the following definition.

Definition 4.14. Let $\mathcal{V}$ be a quantale, let $\mathcal{X}$, $\mathcal{Y}$, and $\mathcal{Z}$ be $\mathcal{V}$-categories, and let $\Phi \colon \mathcal{X} \nrightarrow \mathcal{Y}$ and $\Psi \colon \mathcal{Y} \nrightarrow \mathcal{Z}$ be $\mathcal{V}$-profunctors. We define their *composite*, denoted $\Phi \mathbin{\mathring{,}} \Psi \colon \mathcal{X} \nrightarrow \mathcal{Z}$ by the formula

$$(\Phi \mathbin{\mathring{,}} \Psi)(p, r) = \bigvee_{q \in Q} \big(\Phi(p, q) \otimes \Psi(q, r) \big).$$

Exercise 4.15. Consider the **Cost**-profunctors $\Phi \colon \mathcal{X} \nrightarrow \mathcal{Y}$ and $\Psi \colon \mathcal{Y} \nrightarrow \mathcal{Z}$ shown below:

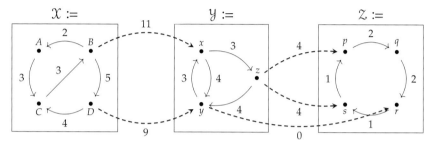

Copy and complete the matrix for the composite profunctor:

$\Phi \mathbin{\mathring{,}} \Psi$	p	q	r	s
A	?	24	?	?
B	?	?	?	?
C	?	?	?	?
D	?	?	9	?

$\diamond$

4.3.2   The Categories $\mathcal{V}$-Prof and Feas

A composition rule suggests a category, and there is indeed a category where the objects are **Bool**-categories and the morphisms are **Bool**-profunctors. To make this work more generally, however, we need to add one technical condition.

Recall from Remark 1.30 that a preorder is a skeletal preorder if whenever $x \le y$ and $y \le x$, we have $x = y$. Skeletal preorders are also known as posets. We say a quantale is skeletal if its underlying preorder is skeletal; **Bool** and **Cost** are skeletal quantales.

Theorem 4.16. For any skeletal quantale $\mathcal{V}$, there is a category $\mathbf{Prof}_{\mathcal{V}}$ whose objects are $\mathcal{V}$-categories $\mathcal{X}$, whose morphisms are $\mathcal{V}$-profunctors $\mathcal{X} \nrightarrow \mathcal{Y}$, and with composition defined as in Definition 4.14.

Definition 4.17. We define $\mathbf{Feas} := \mathbf{Prof}_{\mathbf{Bool}}$.

At this point perhaps you have two questions in mind. What are the identity morphisms? And why did we need to specialize to skeletal quantales? It turns out these two questions are closely related.

Define the *unit profunctor* $U_X : X \nrightarrow X$ on a V-category X by the formula

$$U_X(x, y) := X(x, y). \tag{4.8}$$

How do we interpret this? Recall that, by Definition 2.30, X already assigns to each pair of elements $x, y \in X$ a hom-object $X(x, y) \in V$. The unit profunctor U_X just assigns each pair (x, y) that same object.

In the **Bool** case the unit profunctor on some preorder X can be drawn like this:

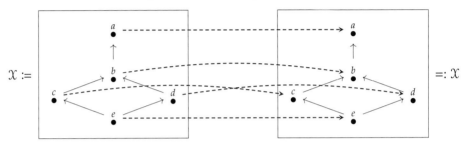

Obviously, composing a feasibility relation with with the unit leaves it unchanged; this is the content of Lemma 4.19.

Exercise 4.18.   Choose a not-too-simple **Cost**-category X. Give a bridge-style diagram for the unit profunctor $U_X : X \nrightarrow X$.   ◇

Lemma 4.19.   Composing any profunctor $\Phi : \mathcal{P} \nrightarrow \mathcal{Q}$ with either unit profunctor, $U_{\mathcal{P}}$ or $U_{\mathcal{Q}}$, returns Φ:

$$U_{\mathcal{P}} \, \mathring{,} \, \Phi = \Phi = \Phi \, \mathring{,} \, U_{\mathcal{Q}}.$$

Proof.   We show that $U_{\mathcal{P}} \, \mathring{,} \, \Phi = \Phi$ holds; proving $\Phi = \Phi \, \mathring{,} \, U_{\mathcal{Q}}$ is similar. Fix $p \in P$ and $q \in Q$. Since V is skeletal, to prove the equality it's enough to show $\Phi \leq U_{\mathcal{P}} \, \mathring{,} \, \Phi$ and $U_{\mathcal{P}} \, \mathring{,} \, \Phi \leq \Phi$. We have one direction:

$$\Phi(p, q) = I \otimes \Phi(p, q) \leq \mathcal{P}(p, p) \otimes \Phi(p, q)$$
$$\leq \bigvee_{p_1 \in P} \left(\mathcal{P}(p, p_1) \otimes \Phi(p_1, q) \right) = (U_{\mathcal{P}} \, \mathring{,} \, \Phi)(p, q). \tag{4.9}$$

For the other direction, we must show $\bigvee_{p_1 \in P} \left(\mathcal{P}(p, p_1) \otimes \Phi(p_1, q) \right) \leq \Phi(p, q)$. But by definition of join, this holds iff $\mathcal{P}(p, p_1) \otimes \Phi(p_1, q) \leq \Phi(p, q)$ is true for each $p_1 \in \mathcal{P}$. This follows from Definitions 2.30 and 4.4:

$$\mathcal{P}(p, p_1) \otimes \Phi(p_1, q) = \mathcal{P}(p, p_1) \otimes \Phi(p_1, q) \otimes I$$
$$\leq \mathcal{P}(p, p_1) \otimes \Phi(p_1, q) \otimes \mathcal{Q}(q, q) \leq \Phi(p, q). \tag{4.10}$$

□

Exercise 4.20.

1. Justify each of the four steps ($=, \leq, \leq, =$) in Eq. (4.9).
2. In the case $\mathcal{V} = \textbf{Bool}$, we can directly show each of the four steps in Eq. (4.9) is actually an equality. How?
3. Justify each of the three steps ($=, \leq, \leq$) in Eq. (4.10).   ◇

Composition of profunctors is also associative; we leave the proof to you.

Lemma 4.21.   Serial composition of profunctors is associative. That is, given profunctors $\Phi \colon \mathcal{P} \to \mathcal{Q}$, $\Psi \colon \mathcal{Q} \to \mathcal{R}$, and $\Upsilon \colon \mathcal{R} \to \mathcal{S}$, we have

$$(\Phi \, \mathbin{\mathring{,}} \, \Psi) \, \mathbin{\mathring{,}} \, \Upsilon = \Phi \, \mathbin{\mathring{,}} \, (\Psi \, \mathbin{\mathring{,}} \, \Upsilon).$$

Exercise 4.22.   Prove Lemma 4.21. (Hint: remember to use the fact that $\mathcal{V}$ is skeletal.)
                    ◇

So, feasibility relations form a category. Since this is the case, we can describe feasibility relations using wiring diagrams for categories (see also Section 4.4.2), which are very simple. Indeed, each box can only have one input and one output, and they're connected in a line:

On the other hand, we have seen that feasibility relations are the building blocks of co-design problems, and we know that co-design problems can be depicted with a much richer wiring diagram, for example:

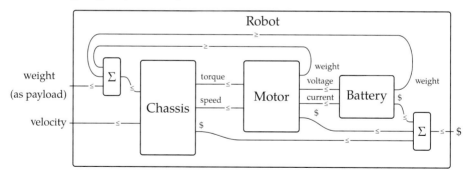

This hints that the category **Feas** has more structure. We have seen wiring diagrams where boxes can have multiple inputs and outputs before, in Chapter 2; there they depicted morphisms in a monoidal preorder. On other hand the boxes in the wiring diagrams of Chapter 2 could not have distinct labels, like the boxes in a co-design problem: all boxes in a wiring diagram for monoidal preorders indicate the order $\leq$, while above we see boxes labeled by "Chassis," "Motor," and so on. Similarly, we know that **Feas** is a proper category, not just a preorder. To understand these diagrams then, we must introduce a new structure, called a *monoidal category*. A monoidal category is a *categorified* monoidal preorder.

Remark 4.23. While we have chosen to define $\mathbf{Prof}_\mathcal{V}$ only for skeletal quantales in Theorem 4.16, it is not too hard to work with non-skeletal ones. There are two straightforward ways to do this. The first approach is to let the morphisms of $\mathbf{Prof}_\mathcal{V}$ be isomorphism classes of $\mathcal{V}$-profunctors. This is analogous to the trick we will use when defining the category $\mathbf{Cospan}_\mathcal{C}$ in Definition 6.39. An alternative approach is to relax what we mean by category, only requiring composition to be unital and associative "up to isomorphism." This is also a type of categorification, known as bicategory theory.

In the next section we'll discuss categorification and introduce monoidal categories. First, though, we finish this section by discussing why profunctors are called profunctors, and by formally introducing something called the *collage* of a profunctor.

4.3.3   Fun Profunctor Facts: Companions, Conjoints, Collages

Companions and conjoints

Recall that a preorder is a **Bool**-category and a monotone map is a **Bool**-functor. We said above that a profunctor is a generalization of a functor; how so?

In fact, every $\mathcal{V}$-functor gives rise to two $\mathcal{V}$-profunctors, called the companion and the conjoint.

Definition 4.24. Let $F: \mathcal{P} \to \mathcal{Q}$ be a $\mathcal{V}$-functor. The *companion of F*, denoted $\widehat{F}: \mathcal{P} \nrightarrow \mathcal{Q}$, and the *conjoint of F*, denoted $\check{F}: \mathcal{Q} \nrightarrow \mathcal{P}$, are defined to be the following $\mathcal{V}$-profunctors:

$$\widehat{F}(p, q) := \mathcal{Q}(F(p), q) \quad \text{and} \quad \check{F}(q, p) := \mathcal{Q}(q, F(p)).$$

Let's consider the **Bool** case again. One can think of a monotone map $F: \mathcal{P} \to \mathcal{Q}$ as a bunch of arrows, one coming out of each vertex $p \in P$ and landing at some vertex $F(p) \in Q$.

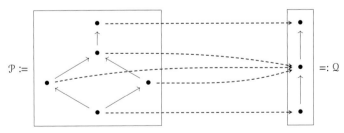

This looks like the pictures of bridges connecting cities, and if we look at the above picture in that light, we see the companion $\widehat{F}$. But now mentally reverse every dotted arrow, and the result would be bridges $\mathcal{Q}$ to $\mathcal{P}$! This is a profunctor $\mathcal{Q} \nrightarrow \mathcal{P}$! We call it $\check{F}$.

Example 4.25. For any preorder $\mathcal{P}$, there is an identity functor id: $\mathcal{P} \to \mathcal{P}$. Its companion and conjoint agree $\widehat{\text{id}} = \check{\text{id}}: \mathcal{P} \nrightarrow \mathcal{P}$. The resulting profunctor is in fact the unit profunctor, $U_\mathcal{P}$, as defined in Eq. (4.8).

Exercise 4.26. Check that the companion $\widehat{\mathrm{id}}$ of $\mathrm{id}\colon \mathcal{P} \to \mathcal{P}$ really has the unit profunctor formula given in Eq. (4.8). ◇

Example 4.27. Consider the function $+\colon \mathbb{R} \times \mathbb{R} \times \mathbb{R} \to \mathbb{R}$, sending a triple (a, b, c) of real numbers to $a + b + c \in \mathbb{R}$. This function is monotonic, because if $(a, b, c) \le (a', b', c')$ – i.e. if $a \le a', b \le b'$, and $c \le c'$ – then obviously $a + b + c \le a' + b' + c'$. Thus it has a companion and a conjoint.

Its companion $\widehat{+}\colon (\mathbb{R} \times \mathbb{R} \times \mathbb{R}) \nrightarrow \mathbb{R}$ is the function that sends (a, b, c, d) to `true` if $a + b + c \le d$ and to `false` otherwise.

Exercise 4.28. Let $+\colon \mathbb{R} \times \mathbb{R} \times \mathbb{R} \to \mathbb{R}$ be as in Example 4.27. What is its conjoint $\widecheck{+}$? ◇

Remark 4.29 ($\mathcal{V}$-Adjoints). Recall from Definition 1.90 the definition of Galois connection between preorders $\mathcal{P}$ and $\mathcal{Q}$. The definition of adjoint can be extended from the **Bool**-enriched setting (of preorders and monotone maps) to the $\mathcal{V}$-enriched setting for arbitrary monoidal preorders $\mathcal{V}$. In that case, the definition of a $\mathcal{V}$-adjunction is a pair of $\mathcal{V}$-functors $F\colon \mathcal{P} \to \mathcal{Q}$ and $G\colon \mathcal{Q} \to \mathcal{P}$ such that the following holds for all $p \in P$ and $q \in Q$:

$$\mathcal{P}(p, G(q)) \cong \mathcal{Q}(F(p), q). \tag{4.11}$$

Exercise 4.30. Let $\mathcal{V}$ be a skeletal quantale, let $\mathcal{P}$ and $\mathcal{Q}$ be $\mathcal{V}$-categories, and let $F\colon \mathcal{P} \to \mathcal{Q}$ and $G\colon \mathcal{Q} \to \mathcal{P}$ be $\mathcal{V}$-functors.

1. Show that F and G are $\mathcal{V}$-adjoints (as in Eq. (4.11)) if and only if the companion of the former equals the conjoint of the latter: $\widehat{F} = \widecheck{G}$.
2. Use this to prove that $\widehat{\mathrm{id}} = \widecheck{\mathrm{id}}$, as was stated in Example 4.25. ◇

Collage of a profunctor

We have been drawing profunctors as bridges connecting cities. One may get an inkling that, given a $\mathcal{V}$-profunctor $\Phi\colon X \nrightarrow Y$ between $\mathcal{V}$-categories X and Y, we have turned Φ into a some sort of new $\mathcal{V}$-category that has X on the left and Y on the right. This works for any $\mathcal{V}$ and profunctor Φ, and is called the collage construction.

Definition 4.31. Let $\mathcal{V}$ be a quantale, let X and Y be $\mathcal{V}$-categories, and let $\Phi\colon X \nrightarrow Y$ be a $\mathcal{V}$-profunctor. Recall the notation $\varnothing := \bigvee \varnothing$. The *collage of* Φ, denoted **Col**(Φ) is the $\mathcal{V}$-category defined as follows:

(i) $\mathrm{Ob}(\mathbf{Col}(\Phi)) := \mathrm{Ob}(X) \sqcup \mathrm{Ob}(Y)$.
(ii) For any $a, b \in \mathrm{Ob}(\mathbf{Col}(\Phi))$, define $\mathbf{Col}(\Phi)(a, b) \in \mathcal{V}$ to be

$$\mathrm{Col}(\Phi)(a, b) := \begin{cases} \mathcal{X}(a, b) & \text{if } a, b \in \mathcal{X}, \\ \Phi(a, b) & \text{if } a \in \mathcal{X}, b \in \mathcal{Y}, \\ \varnothing & \text{if } a \in \mathcal{Y}, b \in \mathcal{X}, \\ \mathcal{Y}(a, b) & \text{if } a, b \in \mathcal{Y}. \end{cases}$$

There are obvious functors $i_{\mathcal{X}} : \mathcal{X} \to \mathrm{Col}(\Phi)$ and $i_{\mathcal{Y}} : \mathcal{Y} \to \mathrm{Col}(\Phi)$, sending each object and morphism to "itself," called *collage inclusions*.

Some pictures will help clarify this.

Example 4.32. Consider the following picture of a **Cost**-profunctor $\Phi : \mathcal{X} \nrightarrow \mathcal{Y}$:

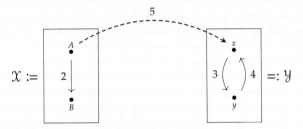

It corresponds to the following matrices

$\mathcal{X}$	A	B
A	0	2
B	∞	0

Φ	x	y
A	5	8
B	∞	∞

$\mathcal{Y}$	x	y
x	0	3
y	4	0

A generalized Hasse diagram of the collage can be obtained by simply taking the union of the Hasse diagrams for $\mathcal{X}$ and $\mathcal{Y}$, and adding in the bridges as arrows. Given the above profunctor Φ, we draw the Hasse diagram for $\mathrm{Col}(\Phi)$ below left, and the **Cost**-matrix representation of the resulting **Cost**-category on the right:

$\mathrm{Col}(\Phi) =$

$\mathrm{Col}(\Phi)$	A	B	x	y
A	0	2	5	8
B	∞	0	∞	∞
x	0	0	0	3
y	0	0	4	0

Exercise 4.33. Draw a Hasse diagram for the collage of the profunctor shown here:

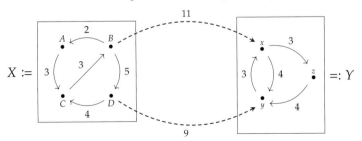

◇

4.4   Categorification

Here we switch gears, to discuss a general concept called *categorification*. We will begin
again with the basics, categorifying several of the notions we've encountered already.
The goal is to define compact closed categories and their feedback-style wiring diagrams. At that point we will return to the story of co-design, and $\mathcal{V}$-profunctors in
general, and show that they do in fact form a compact closed category, and thus interpret
the diagrams we've been drawing since Eq. (4.1).

4.4.1   The Basic Idea of Categorification

The general idea of categorification is that we take a thing we know and add structure to
it, so that what were formerly *properties* become *structures*. We do this in such a way
that we can recover the thing we categorified by forgetting this new structure. This is
rather vague; let's give an example.

Basic arithmetic concerns properties of the natural numbers $\mathbb{N}$, such as the fact
that $5 + 3 = 8$. One way to categorify $\mathbb{N}$ is to use the category **FinSet** of finite
sets and functions. To obtain a categorification, we replace the brute 5, 3, and 8 with
sets of that many elements, say $\overline{5} = \{$apple, banana, cherry, dragonfruit, elephant$\}$,
$\overline{3} = \{$apple, tomato, cantaloupe$\}$, and $\overline{8} = \{$Ali, Bob, Carl, Deb, Eli, Fritz, Gem, Helen$\}$
respectively. We also replace + with disjoint union of sets $\sqcup$, and the brute property
of equality with the structure of an isomorphism. What makes this a good categorification is that, having made these replacements, the analogue of $5 + 3 = 8$ is still true:
$\overline{5} \sqcup \overline{3} \cong \overline{8}$.

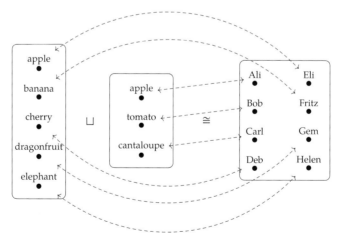

In this categorified world, we have more structure available to talk about the relationships between objects, so we can be more precise about how they relate to each other.
Thus it's not the case that $\overline{5} \sqcup \overline{3}$ is *equal* to our chosen eight-element set $\overline{8}$, but more
precisely that there exists an invertible function comparing the two, showing that they
are isomorphic in the *category* **FinSet**.

Note that in the above construction we made a number of choices; here we must
beware. Choosing a good categorification – like designing a good algebraic structure

such as that of preorders or quantales – is part of the *art* of mathematics. There is no prescribed way to categorify, and the success of a chosen categorification is often empirical: its richer structure should allow us more insights into the subject we want to model.

As another example, an empirically pleasing way to categorify preorders is to categorify them as, well, categories. In this case, rather than the brute property "there exists a morphism $a \to b$," denoted $a \leq b$ or $\mathcal{P}(a, b) = \texttt{true}$, we instead say "here is a set of morphisms $a \to b$." We get a hom-set rather than a hom-boolean. In fact – to state this in a way straight out of the primordial ooze – just as preorders are **Bool**-categories, ordinary categories are actually **Set**-categories.

4.4.2   A Reflection on Wiring Diagrams

Suppose we have a preorder. We introduced a very simple sort of wiring diagram in Section 2.2.2. These allowed us to draw a box

whenever $x_0 \leq x_1$. Chaining these together, we could prove facts in our preorder. For example

provides a proof that $x_0 \leq x_3$ (the exterior box) using three facts (the interior boxes), $x_0 \leq x_1$, $x_1 \leq x_2$, and $x_2 \leq x_3$.

As categorified preorders, categories have basically the same sort of wiring diagram as preorders – namely sequences of boxes inside a box. But since we have replaced the fact that $x_0 \leq x_1$ with the structure of a *set* of morphisms, we need to be able to label our boxes with morphism names:

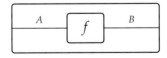

Suppose we are given additional morphisms $g \colon B \to C$, and $h \colon C \to D$. Representing these each as boxes like we did for f, we might be tempted to stick them together to form a new box:

Ideally this would also be a morphism in our category: after all, we have said that we can represent morphisms with boxes with one input and one output. But wait, you say! We don't know which morphism it is. Is it $f \mathbin{\fatsemi} (g \mathbin{\fatsemi} h)$? Or $(f \mathbin{\fatsemi} g) \mathbin{\fatsemi} h$? It's good that you are so careful. Luckily, we are saved by the properties that a category must have.

Associativity says $f \,\natural\, (g \,\natural\, h) = (f \,\natural\, g) \,\natural\, h$, so it doesn't matter which way we chose to try to decode the box.

Similarly, the identity morphism on an object x is drawn as on the left below, but we shall see that it is not harmful to draw id_x in any of the following three ways:

By Definition 3.2 the morphisms in a category satisfy two properties, called the unitality property and the associativity property. The unitality says that $\mathrm{id}_x \,\natural\, f = f = f \,\natural\, \mathrm{id}_y$ for any $f : x \to y$. In terms of diagrams this would say

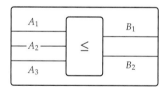

This means you can insert or discard any identity morphism you see in a wiring diagram. From this perspective, the coherence laws of a category – that is, the associativity law and the unitality law – are precisely what are needed to ensure we can lengthen and shorten wires without ambiguity.

In Section 2.2.2, we also saw wiring diagrams for monoidal preorders. Here we were allowed to draw boxes which can have multiple typed inputs and outputs, but with no choice of label (always $\leq$):

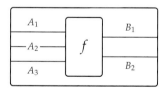

If we combine these ideas, we will obtain a categorification of symmetric monoidal preorders: symmetric monoidal categories. A symmetric monoidal category is an algebraic structure in which we have labeled boxes with multiple typed inputs and outputs:

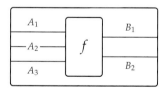

Furthermore, a symmetric monoidal category has a composition rule and a monoidal product, which permit us to combine these boxes to interpret diagrams like this:

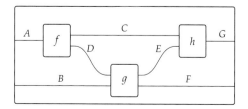

Finally, this structure must obey coherence laws, analogous to associativity and unitality in categories, that allow such diagrams to be unambiguously interpreted. In the next section we will be a bit more formal, but it is useful to keep in mind that, when we say our data must be "well behaved," this is all we mean.

4.4.3   Monoidal Categories

We defined $\mathcal{V}$-categories for a symmetric monoidal preorder $\mathcal{V}$ in Definition 2.30. Just as preorders turned out to be special kinds of categories (see Section 3.2.3), monoidal preorders are special kinds of monoidal categories. And just as we can consider $\mathcal{V}$-categories for a monoidal preorder, we can also consider $\mathcal{V}$-categories when $\mathcal{V}$ is a monoidal category. This is another sort of categorification.

We will soon meet the monoidal category $(\mathbf{Set}, \{1\}, \times)$. The monoidal product will take two sets, S and T, and return the set $S \times T = \{(s, t) \mid s \in S, t \in T\}$. But whereas for monoidal preorders we had the brute associative property $(p \otimes q) \otimes r = p \otimes (q \otimes r)$, the corresponding idea in **Set** is not quite true:

$$S \times (T \times U) := \big\{\big(s, (t, u)\big) \,\big|\, s \in S, t \in T, u \in U\big\}$$
$$=^? (S \times T) \times U := \big\{\big((s, t), u\big) \,\big|\, s \in S, t \in T, u \in U\big\}.$$

They are slightly different sets: the first contains pairs consisting of an elements in S and an element in $T \times U$, while the second contains pairs consisting of an element in $S \times T$ and an element in U. The sets are not equal, but they are clearly isomorphic, i.e. the difference between them is "just a matter of bookkeeping." We thus need a structure – a bookkeeping isomorphism – to keep track of the associativity:

$$\alpha_{s,t,u} \colon \{(s, (t, u)) \mid s \in S, t \in T, u \in U\} \xrightarrow{\cong} \{((s, t), u) \mid s \in S, t \in T, u \in U\}.$$

There are a couple of things to mention before we dive into these ideas. First, just because you replace brute things and properties with structures, it does not mean that you no longer have brute things and properties: new ones emerge! Not only that, but second, the new brute stuff tends to be more complex than what you started with. For example, above we replaced the associativity equation with an isomorphism $\alpha_{s,t,u}$, but now we need a more complex property to ensure that all these αs behave reasonably! The only way out of this morass is to add infinitely much structure, which leads one to "∞-categories," but we will not discuss that here.

Instead, we will continue with our categorification of monoidal preorders, starting with a rough definition of symmetric monoidal categories. It's rough in the sense that we suppress the technical bookkeeping, hiding it under the name "well behaved."

Rough Definition 4.34.   Let $\mathcal{C}$ be a category. A *symmetric monoidal structure* on $\mathcal{C}$ consists of the following constituents:

(i) an object $I \in \mathrm{Ob}(\mathcal{C})$ called the *monoidal unit*,
(ii) a functor $\otimes \colon \mathcal{C} \times \mathcal{C} \to \mathcal{C}$, called the *monoidal product*

subject to well-behaved, natural isomorphisms

(a) $\lambda_c : I \otimes c \cong c$ for every $c \in \mathrm{Ob}(\mathcal{C})$,
(b) $\rho_c : c \otimes I \cong c$ for every $c \in \mathrm{Ob}(\mathcal{C})$,
(c) $\alpha_{c,d,e} : (c \otimes d) \otimes e \cong c \otimes (d \otimes e)$ for every $c, d, e \in \mathrm{Ob}(\mathcal{C})$,
(d) $\sigma_{c,d} : c \otimes d \cong d \otimes c$ for every $c, d \in \mathrm{Ob}(\mathcal{C})$, called the *swap map*, such that $\sigma \circ \sigma = \mathrm{id}$.

A category equipped with a symmetric monoidal structure is called a *symmetric monoidal category*.

Remark 4.35. If the isomorphisms in (a), (b), and (c) – but *not* (d) – are replaced by equalities, then we say that the monoidal structure is *strict*, and this is a complete (non-rough) definition of *symmetric strict monoidal category*. In fact, symmetric strict monoidal categories are almost the same thing as symmetric monoidal categories, via a result known as Mac Lane's coherence theorem. An upshot of this theorem is that we can, when useful to us, pretend that our monoidal categories are strict: for example, we implicitly do this when we draw wiring diagrams. Ask your friendly neighborhood category theorist to explain how!

Remark 4.36. For those yet to find a friendly expert category theorist, we make the following remark. A complete (non-rough) definition of symmetric monoidal category is that a symmetric monoidal category is "a category equipped with an equivalence to (the underlying category of) a symmetric strict monoidal category". This can be unpacked, using Remark 4.35 and our comment about equivalence of categories in Remark 3.50, but we don't expect you to do so. Instead, we hope this gives you more incentive to ask a friendly expert category theorist!

Exercise 4.37. Check that monoidal categories indeed generalize monoidal preorders: a monoidal preorder is a monoidal category $(\mathcal{P}, I, \otimes)$ where, for every $p, q \in \mathcal{P}$, the set $\mathcal{P}(p, q)$ has at most one element.   ◇

Example 4.38. As we said above, there is a monoidal structure on **Set** where the monoidal unit is some choice of singleton set, say $I := \{1\}$, and the monoidal product is $\otimes := \times$. What it means that $\times$ is a functor is that:

- For any pair of objects, i.e. sets $(S, T) \in \mathrm{Ob}(\mathbf{Set} \times \mathbf{Set})$, one obtains a set $(S \times T) \in \mathrm{Ob}(\mathbf{Set})$. We know what it is: the set of pairs $\{(s, t) \mid s \in S, t \in T\}$.
- For any pair of morphisms, i.e. functions $f : S \to S'$ and $g : T \to T'$, one obtains a function $(f \times g) : (S \times T) \to (S' \times T')$. It works pointwise: $(f \times g)(s, t) := (f(s), g(t))$.
- These should preserve identities: $\mathrm{id}_S \times \mathrm{id}_T = \mathrm{id}_{S \times T}$ for any sets S, T.

- These should preserve composition: for any functions $S \xrightarrow{f} S' \xrightarrow{f'} S''$ and $T \xrightarrow{g}$ $T' \xrightarrow{g'} T''$, one has

$$(f \times g) \mathbin{\overset{\circ}{,}} (f' \times g') = (f \mathbin{\overset{\circ}{,}} g) \times (f' \mathbin{\overset{\circ}{,}} g').$$

The four conditions, (a), (b), (c), and (d) give isomorphisms $\{1\} \times S \cong S$, etc. These maps are obvious in the case of **Set**, e.g. the function $\{(1, s) \mid s \in S\} \to S$ sending $(1, s)$ to s. We have been calling such things bookkeeping.

Exercise 4.39.   Consider the monoidal category (**Set**, 1, $\times$), together with the diagram

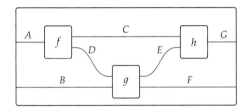

Suppose that $A = B = C = D = F = G = \mathbb{Z}$ and $E = \mathbb{B} = \{\texttt{true}, \texttt{false}\}$, and suppose that $f_C(a) = |a|$, $f_D(a) = a * 5$, $g_E(d, b) =$ "$d \leq b$," $g_F(d, b) = d - b$, and $h(c, e) = $ if e then c else $1 - c$.

1. What are $g_E(5, 3)$ and $g_F(5, 3)$?
2. What are $g_E(3, 5)$ and $g_F(3, 5)$?
3. What is $h(5, \texttt{true})$?
4. What is $h(-5, \texttt{true})$?
5. What is $h(-5, \texttt{false})$?

The whole diagram now defines a function $A \times B \to G \times F$; call it q.

6. What are $q_G(-2, 3)$ and $q_F(-2, 3)$?
7. What are $q_G(2, 3)$ and $q_F(2, 3)$?   ◇

We shall see more monoidal categories throughout the remainder of this book.

4.4.4   Categories Enriched in a Symmetric Monoidal Category

We will not need this again, but we once promised to explain why $\mathcal{V}$-categories, where $\mathcal{V}$ is a symmetric monoidal preorder, deserve to be seen as types of categories. The reason, as we have hinted, is that categories should really be called **Set**-categories. But wait, **Set** is not a preorder! We'll have to generalize – categorify – $\mathcal{V}$-categories.

We now give a rough definition of categories enriched in a symmetric monoidal category $\mathcal{V}$. As in Definition 4.34, we suppress some technical parts in this sketch, hiding them under the name "usual associative and unital laws."

Rough Definition 4.40. Let $\mathcal{V}$ be a symmetric monoidal category, as in Definition 4.34. To specify a *category enriched in* $\mathcal{V}$, or a $\mathcal{V}$-*category*, denoted $\mathcal{X}$,

(i) one specifies a collection $\mathrm{Ob}(\mathcal{X})$, elements of which are called *objects*;
(ii) for every pair $x, y \in \mathrm{Ob}(\mathcal{X})$, one specifies an object $\mathcal{X}(x, y) \in \mathcal{V}$, called the *hom-object* for x, y;
(iii) for every $x \in \mathrm{Ob}(\mathcal{X})$, one specifies a morphism $\mathrm{id}_x \colon I \to \mathcal{X}(x, x)$ in $\mathcal{V}$, called the *identity element*;
(iv) for each $x, y, z \in \mathrm{Ob}(\mathcal{X})$, one specifies a morphism $\stackrel{\circ}{,} \colon \mathcal{X}(x, y) \otimes \mathcal{X}(y, z) \to \mathcal{X}(x, z)$, called the *composition morphism*.

These constituents are required to satisfy the usual associative and unital laws.

The precise, non-rough, definition can be found in other sources, e.g. [nLa18], [Kel05], and Wikipedia.

Exercise 4.41. Recall from Example 4.38 that $\mathcal{V} = (\mathbf{Set}, \{1\}, \times)$ is a symmetric monoidal category. This means we can apply Definition 4.40. Does the (rough) definition roughly agree with the definition of category given in Definition 3.2? Or is there a subtle difference? ◇

Remark 4.42. We first defined $\mathcal{V}$-categories in Definition 2.30, where $\mathcal{V}$ was required to be a monoidal preorder. To check we're not abusing our terms, it's a good idea to make sure that $\mathcal{V}$-categories as per Definition 2.30 are still $\mathcal{V}$-categories as per Definition 4.40.

The first thing to observe is that every symmetric monoidal preorder is a symmetric monoidal category (Exercise 4.37). So, given a symmetric monoidal preorder $\mathcal{V}$, we can apply Definition 4.40. The required data (i) and (ii) then get us off to a good start: both definitions of $\mathcal{V}$-category require objects and hom-objects, and they are specified in the same way. On the other hand, Definition 4.40 requires two additional pieces of data: (iii) identity elements and (iv) composition morphisms. Where do these come from?

In the case of preorders, there is at most one morphism between any two objects, so we do not need to choose an identity element and a composition morphism. Instead, we just need to make sure that an identity element and a composition morphism exist. This is exactly what properties (a) and (b) of Definition 2.30 say.

For example, the requirement (iii) that a $\mathcal{V}$-category $\mathcal{X}$ has a chosen identity element $\mathrm{id}_x \colon I \to \mathcal{X}(x, x)$ for the object x simply becomes the requirement (a) that $I \leq \mathcal{X}(x, x)$ is true in $\mathcal{V}$. This is typical of the story of categorification: what were mere properties in Definition 2.30 have become structures in Definition 4.40.

Exercise 4.43. What are identity elements in Lawvere metric spaces (i.e. **Cost**-categories)? How do we interpret this in terms of distances? ◇

4.5   Profunctors Form a Compact Closed Category

In this section we will define compact closed categories and show that **Feas**, and more generally $\mathcal{V}$-profunctors, form such a thing. Compact closed categories are monoidal categories whose wiring diagrams allow feedback. The wiring diagrams look like this:

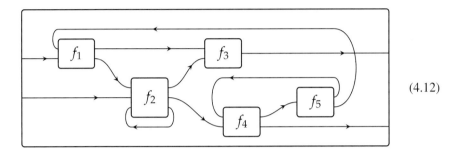

$$(4.12)$$

It's been a while since we thought about co-design, but these were the kinds of wiring diagrams we drew, e.g. connecting the chassis, the motor, and the battery in Eq. (4.1). Compact closed categories are symmetric monoidal categories, with a bit more structure that allow us to formally interpret the sorts of feedback that occur in co-design problems. This same structure shows up in many other fields, including quantum mechanics and dynamical systems.

In Eq. (2.6) and Section 2.2.3 we discussed various flavors of wiring diagrams, including those with icons for splitting and terminating wires. For compact closed categories, our additional icons allow us to bend outputs into inputs, and vice versa. To keep track of this, however, we draw arrows on our wire, which can point either forwards or backwards. For example, we can draw this

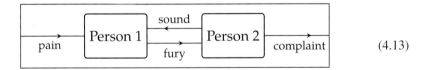

$$(4.13)$$

To accomplish this, we add icons – called a cap and a cup – allowing any wire to reverse direction from forwards to backwards and from backwards to forwards.

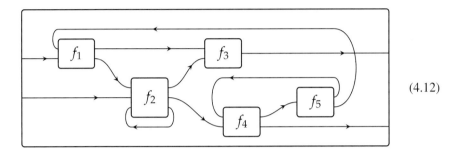

$$(4.14)$$

Thus we can draw the following

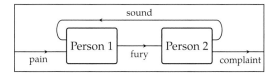

and its meaning is equivalent to that of Eq. (4.13).

We will begin by giving the axioms for a compact closed category. Then we will look again at feasibility relations in co-design – and more generally at enriched profunctors – and show that they indeed form a compact closed category.

4.5.1   Compact Closed Categories

As we said, compact closed categories are symmetric monoidal categories (see Definition 4.34) with extra structure.

Definition 4.44.   Let $(\mathcal{C}, I, \otimes)$ be a symmetric monoidal category, and $c \in \mathrm{Ob}(\mathcal{C})$ an object. A *dual for c* consists of three constituents

(i) an object $c^* \in \mathrm{Ob}(\mathcal{C})$, called the *dual of c*,
(ii) a morphism $\eta_c: I \to c^* \otimes c$, called the *unit for c*,
(iii) a morphism $\epsilon_c: c \otimes c^* \to I$, called the *counit for c*.

These are required to satisfy two equations for every $c \in \mathrm{Ob}(\mathcal{C})$, which we draw as commutative diagrams:

$$
\begin{array}{ccc}
c & =\!=\!=\!= & c \\
{\scriptstyle\cong}\downarrow & & \uparrow{\scriptstyle\cong} \\
c \otimes I & & I \otimes c \\
{\scriptstyle c\otimes\eta_c}\downarrow & & \uparrow{\scriptstyle\epsilon_c\otimes c} \\
c \otimes (c^* \otimes c) & \xrightarrow[\cong]{} & (c \otimes c^*) \otimes c
\end{array}
\qquad\qquad
\begin{array}{ccc}
c^* & =\!=\!=\!= & c^* \\
{\scriptstyle\cong}\downarrow & & \uparrow{\scriptstyle\cong} \\
I \otimes c^* & & c^* \otimes I \\
{\scriptstyle\eta_c\otimes c^*}\downarrow & & \uparrow{\scriptstyle c^*\otimes\epsilon_c} \\
(c^* \otimes c) \otimes c^* & \xrightarrow[\cong]{} & c^* \otimes (c \otimes c^*)
\end{array}
$$

$$(4.15)$$

These equations are sometimes called the *snake equations*.

If for every object $c \in \mathrm{Ob}(\mathcal{C})$ there exists a dual c^* for c, then we say that $(\mathcal{C}, I, \otimes)$ is *compact closed*.

In a compact closed category, each wire is equipped with a direction. For any object c, a forward-pointing wire labeled c is considered equivalent to a backward-pointing wire labeled c^*, i.e. $\xrightarrow{c}$ is the same as $\xleftarrow{c^*}$. The cup and cap discussed above are in fact the unit and counit morphisms; they are drawn as follows.

The snake equations (4.15) then say that the following are equal to the identity:

$$c \otimes \eta_c \qquad \epsilon_c \otimes c \qquad\qquad\qquad \eta_c \otimes c^* \qquad c^* \otimes \epsilon_c$$

Note that the pictures in Eq. (4.14) correspond to ϵ_{sound} and η_{sound^*} .

Recall the notion of monoidal closed preorder; a monoidal category can also be monoidal closed. This means that for every pair of objects $c, d \in \text{Ob}(\mathcal{C})$ there is an object $c \multimap d$ and an isomorphism $\mathcal{C}(b \otimes c, d) \cong \mathcal{C}(b, c \multimap d)$, natural in b. While we will not provide a full proof here, compact closed categories are so named because they are a special type of monoidal closed category.

> **Proposition 4.45.** If $\mathcal{C}$ is a compact closed category, then
>
> 1. $\mathcal{C}$ is monoidal closed;
>
> and for any object $c \in \text{Ob}(\mathcal{C})$,
>
> 2. if c^* and c' are both duals to c then there is an isomorphism $c^* \cong c'$; and
> 3. there is an isomorphism between c and its double-dual, $c \cong c^{**}$.

To prove part 1, the key idea is that, for any c and d, the object $c \multimap d$ is given by $c^* \otimes d$, and the natural isomorphism $\mathcal{C}(b \otimes c, d) \cong \mathcal{C}(b, c \multimap d)$ is given by precomposing with $\text{id}_b \otimes \eta_c$.

Before returning to co-design, we give another example of a compact closed category, called **Corel**, which we'll see again in the chapters to come.

Example 4.46. Recall, from Definition 1.13, that an equivalence relation on a set A is a reflexive, symmetric, and transitive binary relation on A. Given two finite sets, A and B, a *corelation* $A \to B$ is an equivalence relation on $A \sqcup B$.

So, for example, here is a corelation from a set A having five elements to a set B having six elements; two elements are equivalent if they are encircled by the same dashed line.

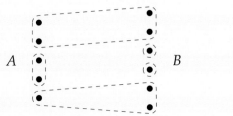

There exists a category, denoted **Corel**, where the objects are finite sets, and where a morphism from $A \to B$ is a corelation $A \to B$. The composition rule is simpler to look at than to write down formally.[2] If in addition to the corelation $\alpha \colon A \to B$ above we have another corelation $\beta \colon B \to C$

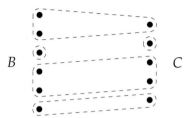

then the composite $\beta \circ \alpha$ of our two corelations is given by

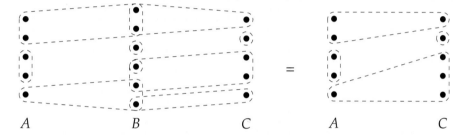

That is, two elements are equivalent in the composite corelation if we may travel from one to the other staying within equivalence classes of either α or β.

The category **Corel** may be equipped with the symmetric monoidal structure $(\varnothing, \sqcup)$. This monoidal category is compact closed, with every finite set its own dual. Indeed, note that for any finite set A there is an equivalence relation on $A \sqcup A :=$ $\{(a, 1), (a, 2) \mid a \in A\}$ where each part simply consists of the two elements $(a, 1)$ and $(a, 2)$ for each $a \in A$. The unit on a finite set A is the corelation $\eta_A \colon \varnothing \to A \sqcup A$ specified by this equivalence relation; similarly the counit on A is the corelation $\epsilon_A \colon A \sqcup A \to \varnothing$ specifed by this same equivalence relation.

Exercise 4.47. Consider the set $\underline{3} = \{1, 2, 3\}$.

1. Draw a picture of the unit corelation $\varnothing \to \underline{3} \sqcup \underline{3}$.
2. Draw a picture of the counit corelation $\underline{3} \sqcup \underline{3} \to \varnothing$.

---

[2] To compose corelations $\alpha \colon A \to B$ and $\beta \colon B \to C$, we need to construct an equivalence relation $\alpha \,\fatsemi\, \beta$ on $A \sqcup C$. To do so requires three steps: (i) consider α and β as relations on $A \sqcup B \sqcup C$, (ii) take the transitive closure of their union, and then (iii) restrict to an equivalence relation on $A \sqcup C$. Here is the formal description. Note that, as binary relations, we have $\alpha \subseteq (A \sqcup B) \times (A \sqcup B)$, and $\beta \subseteq (B \sqcup C) \times (B \sqcup C)$. We also have three inclusions: $\iota_{A \sqcup B} \colon A \sqcup B \to A \sqcup B \sqcup C$, $\iota_{B \sqcup C} \colon B \sqcup C \to A \sqcup B \sqcup C$, and $\iota_{A \sqcup C} \colon A \sqcup C \to A \sqcup B \sqcup C$. Recalling our notation from Section 1.4, we define

$$\alpha \,\fatsemi\, \beta := \iota_{A \sqcup C}^{*}((\iota_{A \sqcup B})_!(\alpha) \vee (\iota_{B \sqcup C})_!(\beta)).$$

3. Check that the snake equations (4.15) hold. (Since every object is its own dual, you only need to check one of them.)   ◇

4.5.2   Feas as a Compact Closed Category

We close the chapter by returning to co-design and showing that **Feas** has a compact closed structure. This is what allows us to draw the kinds of wiring diagrams we saw in Eqs. (4.1), (4.12), and (4.13): it is what puts actual mathematics behind these pictures.

Instead of just detailing this compact closed structure for **Feas** = **Prof**$_{\text{Bool}}$, it's no extra work to prove that for any skeletal (unital, commutative) quantale $(\mathcal{V}, I, \otimes)$ the profunctor category **Prof**$_\mathcal{V}$ of Theorem 4.16 is compact closed, so we'll discuss this general fact.

Theorem 4.48.   Let $\mathcal{V}$ be a skeletal quantale. The category **Prof**$_\mathcal{V}$ can be given the structure of a compact closed category, with monoidal product given by the product of $\mathcal{V}$-categories.

Indeed, all we need to do is construct the monoidal structure and duals for objects. Let's sketch how this goes.

Monoidal products in Prof$_\mathcal{V}$ are just product categories

In terms of wiring diagrams, the monoidal structure looks like stacking wires or boxes on top of one another, with no new interaction.

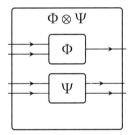

We take our monoidal product on **Prof**$_\mathcal{V}$ to be that given by the product of $\mathcal{V}$-categories; the definition was given in Definition 2.53, and we worked out several examples there. To recall, the formula for the hom-sets in $\mathcal{X} \times \mathcal{Y}$ is given by

$$(\mathcal{X} \times \mathcal{Y})((x, y), (x', y')) := \mathcal{X}(x, x') \otimes \mathcal{Y}(y, y').$$

But monoidal products need to be given on morphisms also, and the morphisms in **Prof**$_\mathcal{V}$ are $\mathcal{V}$-profunctors. So, given $\mathcal{V}$-profunctors $\Phi \colon \mathcal{X}_1 \nrightarrow \mathcal{X}_2$ and $\Psi \colon \mathcal{Y}_1 \nrightarrow \mathcal{Y}_2$, one defines a $\mathcal{V}$-profunctor $(\Phi \times \Psi) \colon \mathcal{X}_1 \times \mathcal{Y}_1 \nrightarrow \mathcal{X}_2 \times \mathcal{Y}_2$ by

$$(\Phi \times \Psi)((x_1, y_1), (x_2, y_2)) := \Phi(x_1, x_2) \otimes \Psi(y_1, y_2).$$

Exercise 4.49. Interpret the monoidal products in **Prof**$_{\mathbf{Bool}}$ in terms of feasibility. That is, preorders represent resources ordered by availability ($x \leq x'$ means that x is available given x') and a profunctor is a feasibility relation. Explain why $\mathcal{X} \times \mathcal{Y}$ makes sense as the monoidal product of resource preorders $\mathcal{X}$ and $\mathcal{Y}$ and why $\Phi \times \Psi$ makes sense as the monoidal product of feasibility relations Φ and Ψ.   ◇

The monoidal unit in Prof$_\mathcal{V}$ is 1

To define a monoidal structure on **Prof**$_\mathcal{V}$, we need not only a monoidal product – as defined above – but also a monoidal unit. Recall the $\mathcal{V}$-category $\mathbf{1}$; it has one object, say 1, and $(1, 1) = I$ is the monoidal unit of $\mathcal{V}$. We take $\mathbf{1}$ to be the monoidal unit of **Prof**$_\mathcal{V}$.

Exercise 4.50. In order for $\mathbf{1}$ to be a monoidal unit, there are supposed to be isomorphisms $\mathcal{X} \times \mathbf{1} \nrightarrow \mathcal{X}$ and $\mathbf{1} \times \mathcal{X} \nrightarrow \mathcal{X}$ in **Prof**$_\mathcal{V}$, for any $\mathcal{V}$-category $\mathcal{X}$. What are they?   ◇

Duals in Prof$_\mathcal{V}$ are just opposite categories

In order to regard **Prof**$_\mathcal{V}$ as a compact closed category (Definition 4.44), it remains to specify duals and the corresponding cup and cap.

Duals are easy: for every $\mathcal{V}$-category $\mathcal{X}$, its dual is its opposite category $\mathcal{X}^{\mathrm{op}}$ (see Exercise 2.52). The unit and counit then look like the unit profunctor, Eq. (4.8). To elaborate, the unit is a $\mathcal{V}$-profunctor $\eta_\mathcal{X} : \mathbf{1} \nrightarrow \mathcal{X}^{\mathrm{op}} \times \mathcal{X}$. By definition, this is a $\mathcal{V}$-functor

$$\eta_\mathcal{X} : \mathbf{1} \times \mathcal{X}^{\mathrm{op}} \times \mathcal{X} \to \mathcal{V};$$

we define it by $\eta_\mathcal{X}(1, x, x') := \mathcal{X}(x, x')$. Similarly, the counit is the profunctor $\epsilon_\mathcal{X} : (\mathcal{X} \times \mathcal{X}^{\mathrm{op}}) \nrightarrow \mathbf{1}$, defined by $\epsilon_\mathcal{X}(x, x', 1) := \mathcal{X}(x, x')$.

Exercise 4.51. Check these proposed units and counits do indeed obey the snake equations (4.15).   ◇

4.6   Summary and Further Reading

This chapter introduced three important ideas in category theory: profunctors, categorification, and monoidal categories. Let's talk about them in turn.

Profunctors generalize binary relations. In particular, we saw that the idea of profunctor over a monoidal preorder gave us the additional power necessary to formalize the idea of a feasibility relation between resource preorders. The idea of a feasibility relation is due to Andrea Censi; he called them *monotone co-design problems*. The basic idea is explained in [Cen15], where he also gives a programming language to specify and solve co-design problems. In [Cen17], Censi further discusses how to use estimation to make solving co-design problems computationally efficient.

We also saw profunctors over the preorder **Cost**, and how to think of these as bridges between Lawvere metric space. We referred earlier to Lawvere's paper [Law73]; plenty more on **Cost**-profunctors can be found there.

Profunctors, however, are vastly more general than the two examples we have discussed; $\mathcal{V}$-profunctors can be defined not only when $\mathcal{V}$ is a preorder, but for any symmetric monoidal category. A detailed exposition of profunctors and related concepts such as equipments, companions and conjoints, and symmetric monoidal bicategories can be found in [Shu08; Shu10].

We have not defined symmetric monoidal bicategories, but you would be correct if you guessed this is a sort of categorification of symmetric monoidal categories. Baez and Dolan tell the subtle story of categorifying categories to get ever *higher* categories in [BD98]. Crane and Yetter give a number of examples of categorification in [CY96].

Finally, we talked about monoidal categories and compact closed categories. Monoidal categories are a classic, central topic in category theory, and a quick introduction can be found in [Mac98]. Wiring diagrams play a huge role in this book and in applied category theory in general; while informally used for years, these were first formalized in the case of monoidal categories. You can find the details here [JS93; JSV96].

Compact closed categories are a special type of structured monoidal category; there are many others. For a broad introduction to the different flavors of monoidal category, detailed through their various styles of wiring diagram, see [Sel10].

5　Signal Flow Graphs: Props, Presentations, and Proofs

5.1　Comparing Systems as Interacting Signal Processors

Cyber-physical systems are systems that involve tightly interacting physical and computational parts. An example is an autonomous car: sensors inform a decision system that controls a steering unit that drives a car, whose movement changes the sensory input. While such systems involve complex interactions of many different subsystems – both physical ones, such as the driving of a wheel by a motor, or a voltage placed across a wire, and computational ones, such as a program that takes a measured velocity and returns a desired acceleration – it is often useful to model the system behavior as simply the passing around and processing of signals. For this illustrative sketch, we will just think of signals as things which we can add and multiply, such as real numbers.

Interaction in cyber-physical systems can often be understood as variable sharing; i.e. when two systems are linked, certain variables become shared. For example, when we connect two train carriages by a physical coupling, the train carriages must have the same velocity, and their positions differ by a constant. Similarly, when we connect two electrical ports, the electric potentials at these two ports now must be the same, and the current flowing into one must equal the current flowing out of the other. Of course, the way the shared variable is actually used may be very different for the different subsystems using it, but sharing the variable serves to couple those systems nonetheless.

Note that both the above examples involve the physical joining of two systems; more figuratively, we might express the interconnection by drawing a line connecting the boxes that represent the systems. In its simplest form, this is captured by the formalism of signal flow graphs, due to Claude Shannon in the 1940s. Here is an example of a signal flow graph:

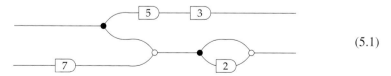

$$(5.1)$$

We consider the dangling wires on the left as inputs, and those on the right as outputs. In Eq. (5.1) we see three types of signal processing units, which we interpret as follows:

- Each unit labeled by a number a takes an input and multiplies it by a.
- Each black dot takes an input and produces two copies of it.

● Each white dot takes two inputs and produces their sum.

Thus the above signal flow graph takes in two input signals, say x (on the upper left wire) and y (on the lower left wire), and – going from left to right as described above – produces two output signals: $u = 15x$ (upper right) and $v = 3x + 21y$ (lower right). Let's show some steps from this computation (leaving others off to avoid clutter):

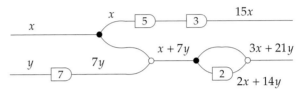

In words, the signal flow graph first multiplies y by 7, then splits x into two copies, adds the second copy of x to the lower signal to get $x + 7y$, and so on.

A signal flow graph might describe an existing system, or it might specify a system to be built. In either case, it is important to be able to analyze these diagrams to understand how the composite system converts inputs to outputs. This is reminiscent of a co-design problem from Chapter 4, which asks how to evaluate the composite feasibility relation from a diagram of simpler feasibility relations. We can use this process of evaluation to determine whether two different signal flow graphs in fact specify the same composite system, and hence to validate that a system meets a given specification.

In this chapter, however, we introduce categorical tools – props and their presentations – for reasoning more directly with the diagrams. Recall from Chapter 2 that symmetric monoidal preorders are a type of symmetric monoidal category where the *morphisms* are constrained to be very simple: there can be at most one morphism between any two objects. Here we shall see that signal flow graphs represent morphisms in a different, complementary simplification of the symmetric monoidal category concept, known as a *prop*.[1] A prop is a symmetric monoidal category where the *objects* are constrained to be very simple: they are generated, using the monoidal product, by just a single object. Just as the wiring diagrams for symmetric monoidal preorders did not require labels on the boxes, wiring diagrams for props do not require labels on the wires. This makes props particularly suited for describing diagrammatic formalisms such as signal flow graphs, which only have wires of a single type.

Finally, many systems behave in what is called a *linear* way, and linear systems form a foundational part of control theory, a branch of engineering that works on cyber-physical systems. Similarly, linear algebra is a foundational part of modern mathematics, both pure and applied, which includes not only control theory, but also the practice of computing, physics, statistics, and many others. As we analyze signal flow graphs, we shall see that they are in fact a way of recasting linear algebra – more specifically, matrix operations – in graphical terms. More formally, we shall say that signal flow graphs have *functorial semantics* as matrices.

---

[1] Historically, the word "prop" was written in all caps, "PROP," standing for "products and permutations category." However, we find "PROP" a bit loud, so like many modern authors we opt for writing it as "prop."

5.2   Props and Presentations

Signal flow graphs as in Eq. (5.1) are easily seen to be wiring diagrams of some sort. However they have the property that, unlike for monoidal preorders and monoidal categories, there is no need to label the wires. This corresponds to a form of symmetric monoidal category, known as a prop, which has a very particular set of objects.

5.2.1   Props: Definition and First Examples

Recall the definition of symmetric strict monoidal category from Definition 4.34 and Remark 4.35.

Definition 5.1.   A *prop* is a symmetric strict monoidal category $(\mathcal{C}, 0, +)$ for which $\mathrm{Ob}(\mathcal{C}) = \mathbb{N}$, the monoidal unit is $0 \in \mathbb{N}$, and the monoidal product on objects is given by addition.

Note that each object n is the n-fold monoidal product of the object 1; we call 1 the *generating object*. Since the objects of a prop are always the natural numbers, to specify a prop P it is enough to specify five things:

(i)   a set $\mathcal{C}(m, n)$ of morphisms $m \to n$, for $m, n \in \mathbb{N}$;
(ii)   for all $n \in \mathbb{N}$, an identity map $\mathrm{id}_n \colon n \to n$;
(iii)   for all $m, n \in \mathbb{N}$, a symmetry map $\sigma_{m,n} \colon m + n \to n + m$;
(iv)   a composition rule: given $f \colon m \to n$ and $g \colon n \to p$; a map $(f \,\mathring{,}\, g) \colon m \to p$;
(v)   a monoidal product on morphisms: given $f \colon m \to m'$ and $g \colon n \to n'$, a map $(f + g) \colon m + n \to m' + n'$.

Once one specifies the above data, one should check that one's specifications satisfy the rules of symmetric monoidal categories (see Definition 4.34).

Example 5.2.   There is a prop **FinSet** where the morphisms $f \colon m \to n$ are functions from $\underline{m} = \{1, \ldots, m\}$ to $\underline{n} = \{1, \ldots, n\}$. (The identities, symmetries, and composition rule are obvious.) The monoidal product on functions is given by the disjoint union of functions: that is, given $f \colon m \to m'$ and $g \colon n \to n'$, we define $f + g \colon m + n \longrightarrow m' + n'$ by

$$i \longmapsto \begin{cases} f(i) & \text{if } 1 \le i \le m, \\ m' + g(i) & \text{if } m + 1 \le i \le m + n. \end{cases} \tag{5.2}$$

Exercise 5.3.   In Example 5.2 we said that the identities, symmetries, and composition rule in **FinSet** "are obvious." In math lingo, this just means "we trust that the reader can figure them out, if she spends the time tracking down the definitions and fitting them together."

1. Draw a morphism $f : 3 \to 2$ and a morphism $g : 2 \to 4$ in **FinSet**.
2. Draw $f + g$.
3. What is the composition rule for morphisms $f : m \to n$ and $g : n \to p$ in **FinSet**?
4. What are the identities in **FinSet**? Draw some.
5. Choose $m, n \in \mathbb{N}$, and draw the symmetry map $\sigma_{m,n}$ in **FinSet**?   ◇

Example 5.4. Recall from Definition 1.17 that a bijection is a function that is both surjective and injective. There is a prop **Bij** where the morphisms $f : m \to n$ are bijections $\underline{m} \to \underline{n}$. Note that in this case morphisms $m \to n$ only exist when $m = n$; when $m \neq n$ the homset **Bij**(m, n) is empty. Since **Bij** is a subcategory of **FinSet**, we can define the monoidal product to be as in Eq. (5.2).

Example 5.5. The compact closed category **Corel**, in which the morphisms $f : m \to n$ are partitions on $\underline{m} \sqcup \underline{n}$ (see Example 4.46), is a prop.

Example 5.6. There is a prop **Rel** for which morphisms $m \to n$ are relations, $R \subseteq \underline{m} \times \underline{n}$. The composition of R with $S \subseteq \underline{n} \times \underline{p}$ is

$$R \,\mathring{9}\, S := \{(i, k) \in \underline{m} \times \underline{p} \mid \exists (j \in \underline{n}). (i, j) \in R \text{ and } (j, k) \in S\}.$$

The monoidal product is relatively easy to formalize using universal properties,[2] but one might get better intuition from pictures:

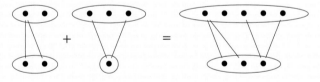

Exercise 5.7. A posetal prop is a prop that is also a poset. That is, a posetal prop is a symmetric monoidal preorder of the form $(\mathbb{N}, \preceq)$, for some poset relation $\preceq$ on $\mathbb{N}$, where the monoidal product on objects is addition. We've spent a lot of time discussing order structures on the natural numbers. Give three examples of a posetal prop.   ◇

Exercise 5.8. Choose one of Examples 5.4 to 5.6 and explicitly provide the five aspects of props discussed below Definition 5.1.   ◇

Definition 5.9. Let $\mathcal{C}$ and $\mathcal{D}$ be props. A *functor* $F : \mathcal{C} \to \mathcal{D}$ is called a *prop functor* if

---

[2] The monoidal product $R_1 + R_2$ of relations $R_1 \subseteq \underline{m_1} \times \underline{n_1}$ and $R_2 \subseteq \underline{m_2} \times \underline{n_2}$ is given by
$R_1 \sqcup R_2 \subseteq (\underline{m_1} \times \underline{n_1}) \sqcup (\underline{m_2} \times \underline{n_2}) \subseteq (\underline{m_1} \sqcup \underline{m_2}) \times (\underline{n_1} \sqcup \underline{n_2})$.

(a) F is identity-on-objects, i.e. $F(n) = n$ for all $n \in \mathrm{Ob}(\mathcal{C}) = \mathrm{Ob}(\mathcal{D}) = \mathbb{N}$, and
(b) for all $f: m_1 \to m_2$ and $g: n_1 \to n_2$ in $\mathcal{C}$, we have $F(f) + F(g) = F(f + g)$ in $\mathcal{D}$.

Example 5.10. The inclusion $i: \mathbf{Bij} \to \mathbf{FinSet}$ is a prop functor. Perhaps more interestingly, there is a prop functor $F: \mathbf{FinSet} \to \mathbf{Rel_{Fin}}$. It sends a function $f: \underline{m} \to \underline{n}$ to the relation $F(f) := \{(i, j) \mid f(i) = j\} \subseteq \underline{m} \times \underline{n}$.

5.2.2 The Prop of Port Graphs

An important example of a prop is the one in which morphisms are open, directed, acyclic port graphs, as we next define. We will just call them port graphs.

Definition 5.11. For $m, n \in \mathbb{N}$, an (m, n)-*port graph* (V, in, out, ι) is specified by

(i) a set V, elements of which are called *vertices*,
(ii) functions $in, out: V \to \mathbb{N}$, where $in(v)$ and $out(v)$ are called the *in degree* and *out degree* of each $v \in V$, and
(iii) a bijection $\iota: \underline{m} \sqcup O \xrightarrow{\cong} I \sqcup \underline{n}$, where $I = \{(v, i) \mid v \in V, 1 \le i \le in(v)\}$ is the set of *vertex inputs*, and $O = \{(v, i) \mid v \in V, 1 \le i \le out(v)\}$ is the set of *vertex outputs*.

This data must obey the following acyclicity condition. First, use the bijection ι to construct the graph with vertices V and with an arrow $e_{v,j}^{u,i}: u \to v$ for every $i, j \in \mathbb{N}$ such that $\iota(u, i) = (v, j)$; call it the *internal flow graph*. If the internal flow graph is acyclic – that is, if the only path from any vertex v to itself is the trivial path – then we say that (V, in, out, ι) is a port graph.

This seems quite a technical construction, but it's quite intuitive once you unpack it a bit. Let's do this.

Example 5.12. Here is an example of a $(2, 3)$-port graph, i.e. with $m = 2$ and $n = 3$:

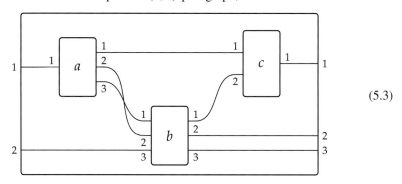

(5.3)

Since the port graph has type $(2, 3)$, we draw two ports on the left-hand side of the outer box, and three on the right. The vertex set is $V = \{a, b, c\}$ and, for example $in(a) = 1$ and $out(a) = 3$, so we draw one port on the left-hand side and three ports on the right-hand side of the box labeled a. The bijection ι is what tells us how the ports are connected by wires:

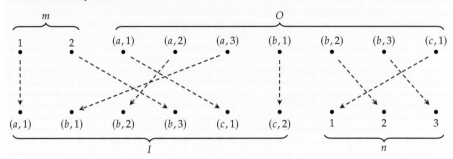

The internal flow graph – which one can see is acyclic – is shown below:

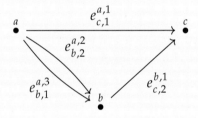

As you might guess from (5.3), port graphs are closely related to wiring diagrams for monoidal categories, and even more closely related to wiring diagrams for props.

A category PG whose morphisms are port graphs

Given an (m, n)-port graph (V, in, out, ι) and an (n, p)-port graph (V', in', out', ι'), we may compose them to produce an (m, p)-port graph $(V \sqcup V', [in, in'], [out, out'], \iota'')$. Here $[in, in']$ denotes the function $V \sqcup V' \to \mathbb{N}$ which maps elements of V according to in, and elements of V' according to in', and similarly for $[out, out']$. The bijection $\iota'' : \underline{m} \sqcup O \sqcup O' \to I \sqcup I' \sqcup \underline{p}$ is defined as follows:

$$\iota''(x) = \begin{cases} \iota(x) & \text{if } \iota(x) \in I, \\ \iota'(\iota(x)) & \text{if } \iota(x) \in \underline{n}, \\ \iota'(x) & \text{if } x \in O'. \end{cases}$$

Exercise 5.13. Describe how port graph composition looks, with respect to the visual representation of Example 5.12, and give a nontrivial example.   ◇

We thus have a category **PG**, whose objects are natural numbers $\mathrm{Ob}(\mathbf{PG}) = \mathbb{N}$, and whose morphisms are port graphs $\mathbf{PG}(m, n) = \{(V, in, out, \iota) \mid \text{as in Definition 5.11}\}$.

Composition of port graphs is as above, and the identity port graph on n is the (n, n)-port graph $(\varnothing, !, !, \mathrm{id}_{\underline{n}})$, where $! \colon \varnothing \to \mathbb{N}$ is the unique function. The identity on an object, say 3, is depicted as follows:

The monoidal structure on PG

This category **PG** is in fact a prop. The monoidal product of two port graphs $G :=$ (V, in, out, ι) and $G' := (V', in', out', \iota')$ is given by taking the disjoint union of ι and ι':

$$G + G' := \big((V \sqcup V'), [in, in'], [out, out'], (\iota \sqcup \iota')\big). \tag{5.4}$$

The monoidal unit is $(\varnothing, !, !, !)$.

Exercise 5.14. Draw the monoidal product of the morphism shown in Eq. (5.3) with itself. It will be a $(4, 6)$-port graph, i.e. a morphism $4 \to 6$ in **PG**. $\diamond$

5.2.3 Free Constructions and Universal Properties

Given some sort of categorical structure, such as a preorder, a category, or a prop, it is useful to be able to construct one according to your own specification. (This should not be surprising.) The minimally constrained structure that contains all the data you specify is called the *free structure* on your specification: it's free from unneccesary constraints! We have already seen some examples of free structures; let's recall and explore them.

Example 5.15 (The free preorder on a relation). For preorders, we saw the construction of taking the reflexive, transitive closure of a relation. That is, given a relation $R \subseteq P \times P$, the reflexive, transitive closure of R is the called the free preorder on R. Rather than specify all the inequalities in the preorder $(P, \leq)$, we can specify just a few inequalities $p \leq q$, and let our "closure machine" add in the minimum number of other inequalities necessary to make P a preorder. To obtain a preorder out of a graph, or Hasse diagram, we consider a graph (V, A, s, t) as defining a relation $\{(s(a), t(a)) \mid a \in A\} \subseteq V \times V$, and apply this closure machine.

But in what sense is the reflexive, transitive closure of a relation $R \subseteq P \times P$ really the *minimally constrained* preorder containing R? One way of understanding this is that the extra equalities impose no further constraints when defining a monotone map *out* of P. We are claiming that freeness has something to do with maps *out*! As strange as an asymmetry might seem here (one might ask, "why not maps in?"), the reader will have an opportunity to explore it for herself in Exercises 5.16 and 5.17.

A higher-level justification understands freeness as a left adjoint (see Example 3.61), but we will not discuss that here.

Exercise 5.16. Let P be a set, let $R \subseteq P \times P$ a relation, let $(P, \leq_P)$ be the preorder obtained by taking the reflexive, transitive closure of R, and let $(Q, \leq_Q)$ be an arbitrary preorder. Finally, let $f \colon P \to Q$ be a function, not assumed monotone.

1. Suppose that for every $x, y \in P$, if $R(x, y)$ then $f(x) \leq f(y)$. Show that f defines a monotone map $f \colon (P, \leq_P) \to (Q, \leq_Q)$.
2. Suppose that f defines a monotone map $f \colon (P, \leq_P) \to (Q, \leq_Q)$. Show that for every $x, y \in P$, if $R(x, y)$ then $f(x) \leq_Q f(y)$.

We call this the *universal property* of the free preorder $(P, \leq_P)$. ◇

Exercise 5.17. Let P, Q, R, etc. be as in Exercise 5.16. We want to see that the universal property is really about maps out of – and not maps into – the reflexive, transitive closure $(P, \leq)$. So let $g \colon Q \to P$ be a function.

1. Suppose that for every $a, b \in Q$, if $a \leq b$ then $(g(a), g(b)) \in R$. Is it automatically true that g defines a monotone map $g \colon (Q, \leq_Q) \to (P, \leq_P)$?
2. Suppose that g defines a monotone map $g \colon (Q, \leq_Q) \to (P, \leq_P)$. Is it automatically true that for every $a, b \in Q$, if $a \leq b$ then $(g(a), g(b)) \in R$?

The lesson is that maps between structured objects are defined to preserve constraints. This means the domain of a map must be somehow more constrained than the codomain. Thus having the fewest additional constraints coincides with having the most maps out – every function that respects our generating constraints should define a map. ◇

Example 5.18 (The free category on a graph). There is a similar story for categories. Indeed, we saw in Definition 3.3 the construction of the free category **Free**(G) on a graph G. The objects of **Free**(G) and the vertices of G are the same – nothing new here – but the morphisms of **Free**(G) are not just the arrows of G because morphisms in a category have stricter requirements: they must compose and there must be an identity. Thus morphisms in **Free**(G) are the *closure* of the set of arrows in G under these operations. Luckily (although this happens often in category theory), the result turns out to already be a relevant graph concept: the morphisms in **Free**(G) are exactly the paths in G. So **Free**(G) is a category that in a sense contains G and obeys no equations other than those that categories are forced to obey.

Exercise 5.19. Let $G = (V, A, s, t)$ be a graph, and let $\mathcal{G}$ be the free category on G. Let $\mathcal{C}$ be another category whose set of morphisms is denoted Mor$(\mathcal{C})$.

1. Someone tells you that there are "domain and codomain" functions dom, cod $\colon$ Mor$(\mathcal{C}) \to$ Ob$(\mathcal{C})$; interpret this statement.
2. Show that the set of functors $\mathcal{G} \to \mathcal{C}$ are in one-to-one correspondence with the set of pairs of functions (f, g), where $f \colon V \to$ Ob$(\mathcal{C})$ and $g \colon A \to$ Mor$(\mathcal{C})$, for which dom$(g(a)) = f(s(a))$ and cod$(g(a)) = f(t(a))$ for all a.

3. Is $(\mathrm{Mor}(\mathcal{C}), \mathrm{Ob}(\mathcal{C}), \mathrm{dom}, \mathrm{cod})$ a graph? If so, see if you can use the word "adjunction" in a sentence that describes the statement in part 2. If not, explain why not.   ◇

Exercise 5.20 (The free monoid on a set).   Recall from Example 3.7 that monoids are one-object categories. For any set A, there is a graph **Loop**(A) with one vertex and with one arrow from the vertex to itself for each $a \in A$. So if $A = \{a, b\}$ then **Loop**(A) looks like this:

$$a \rightleftarrows \bullet \rightleftarrows b$$

The free category on this graph is a one-object category, and hence a monoid; it's called the free monoid on A.

1. What are the elements of the free monoid on the set $A = \{a\}$?
2. Can you find a well-known monoid that is isomorphic to the free monoid on $\{a\}$?
3. What are the elements of the free monoid on the set $A = \{a, b\}$?   ◇

5.2.4   The Free Prop on a Signature

We have been discussing free constructions, in particular for preorders and categories. A similar construction exists for props. Since we already know what the objects of the prop will be – the natural numbers – all we need to specify is a set G of *generating morphisms*, together with the arities,[3] that we want to be in our prop. This information will be called a *signature*. Just as we can generate the free category from a graph, so too can we generate the free prop from a signature.

We now give an explicit construction of the free prop in terms of port graphs (see Definition 5.11).

Definition 5.21.   A *prop signature* is a tuple (G, s, t), where G is a set and $s, t \colon G \to \mathbb{N}$ are functions; each element $g \in G$ is called a *generator* and $s(g), t(g) \in \mathbb{N}$ are called its *in-arity and out-arity*. We often denote (G, s, t) simply by G, taking s, t to be implicit.

A *G-labeling* of a port graph $\Gamma = (V, in, out, \iota)$ is a function $\ell \colon V \to G$ such that the arities agree: $s(\ell(v)) = in(v)$ and $t(\ell(v)) = out(v)$ for each $v \in V$.

Define the *free prop on G*, denoted **Free**(G), to have as morphisms $m \to n$ all G-labeled (m, n)-port graphs. The composition and monoidal structure are just those for port graphs **PG** (see Eq. (5.4)); the labelings (the ℓ's) are just carried along.

The morphisms in **Free**(G) are port graphs (V, in, out, ι) as in Definition 5.11, that are equipped with a G-labeling. To draw a port graph, just as in Example 5.12, we draw each vertex $v \in V$ as a box with $in(v)$-many ports on the left and $out(v)$-many ports on the right. In wiring diagrams, we depict the labeling function $\ell \colon V \to G$ by using ℓ

---

[3] The arity of a prop morphism is a pair $(m, n) \in \mathbb{N} \times \mathbb{N}$, where m is the number of inputs and n is the number of outputs.

to add labels (in the usual sense) to our boxes. Note that multiple boxes can be labeled with the same generator. For example, if $G = \{f: 1 \to 1, g: 2 \to 2, h: 2 \to 1\}$, then the following is a morphism $3 \to 2$ in **Free**(G):

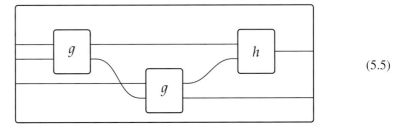

$$(5.5)$$

Note that the generator g is used twice, while the generator f is not used at all in Eq. (5.5). This is perfectly fine.

Example 5.22. The free prop on the empty set $\varnothing$ is **Bij**. This is because each morphism must have a labeling function of the form $V \to \varnothing$, and hence we must have $V = \varnothing$; see Exercise 1.20. Thus the only morphisms (n, m) are those given by port graphs $(\varnothing, !, !, \sigma)$, where $\sigma: n \to m$ is a bijection.

Exercise 5.23.   Consider the following prop signature:

$$G := \{\rho_{m,n} \mid m, n \in \mathbb{N}\}, \qquad s(\rho_{m,n}) := m, \quad t(\rho_{m,n}) := n,$$

i.e. having one generating morphism for each $(m, n) \in \mathbb{N}^2$. Show that **Free**(G) is the prop **PG** of port graphs from Section 5.2.2.   ◇

Just like free preorders and free categories, the free prop is characterized by a universal property in terms of maps out. The following can be proved in a manner similar to Exercise 5.19.

Proposition 5.24.   The free prop **Free**(G) on a signature (G, s, t) has the property that, for any prop $\mathcal{C}$, the prop functors **Free**$(G) \to \mathcal{C}$ are in one-to-one correspondence with functions $G \to \mathcal{C}$ that send each $g \in G$ to a morphism $s(g) \to t(g)$ in $\mathcal{C}$.

An alternate way to describe morphisms in Free(G)

Port graphs provide a convenient formalism of thinking about morphisms in the free prop on a signature G, but there is another approach which is also useful. It is syntactic, in the sense that we start with a small stock of basic morphisms, including elements of G, and then we inductively build new morphisms from them using the basic operations of props: namely composition and monoidal product. Sometimes the conditions of monoidal categories – e.g. associativity, unitality, functoriality; see Definition 4.34 – force two such morphisms to be equal, and so we dutifully equate them.

When we are done, the result is again the free prop **Free**(G). Let's make this more formal.

First, we need the notion of a prop expression. Just as prop signatures are the analogue of the graphs used to present categories, prop expressions are the analogue of paths in these graphs.

Definition 5.25. Suppose we have a set G and functions $s, t: G \to \mathbb{N}$. We define a *G-generated prop expression*, or simply *expression* $e: m \to n$, where $m, n \in \mathbb{N}$, inductively as follows:

- The empty morphism $\mathrm{id}_0: 0 \to 0$, the identity morphism $\mathrm{id}_1: 1 \to 1$, and the symmetry $\sigma: 2 \to 2$ are expressions.[4]
- The generators $g \in G$ are expressions $g: s(g) \to t(g)$.
- If $\alpha: m \to n$ and $\beta: p \to q$ are expressions, then $\alpha + \beta: m + p \to n + q$ is an expression.
- If $\alpha: m \to n$ and $\beta: n \to p$ are expressions, then $\alpha \,\mathring{\mathrm{s}}\, \beta: m \to p$ is an expression.

We write $\mathrm{Expr}(G)$ for the set of expressions in G. If $e: m \to n$ is an expression, we refer to (m, n) as its *arity*.

Example 5.26. Let $G = \{f: 1 \to 1, g: 2 \to 2, h: 2 \to 1\}$. Then

- $\mathrm{id}_1: 1 \to 1$,
- $f: 1 \to 1$,
- $f \,\mathring{\mathrm{s}}\, \mathrm{id}_1: 1 \to 1$,
- $h + \mathrm{id}_1: 3 \to 2$, and
- $(h + \mathrm{id}_1) \,\mathring{\mathrm{s}}\, \sigma \,\mathring{\mathrm{s}}\, g \,\mathring{\mathrm{s}}\, \sigma: 3 \to 2$

are all G-generated prop expressions.

Both G-labeled port graphs and G-generated prop expressions are ways to describe morphisms in the free prop **Free**(G). Note, however, that unlike for G-labeled port graphs, there may be two G-generated prop expressions that represent the same morphism. For example, we want to consider $f \,\mathring{\mathrm{s}}\, \mathrm{id}_1$ and f to be the same morphism, since the unitality axiom for categories says $f \,\mathring{\mathrm{s}}\, \mathrm{id}_1 = f$. Nonetheless, we only consider two G-generated prop expressions equal when some axiom from the definition of prop requires that they be so; again, the free prop is the *minimally constrained* way to take G and obtain a prop.

Since both port graphs and prop expressions describe morphisms in **Free**(G), you might be wondering how to translate between them. Here's how to turn a port graph into a prop expression: imagine a vertical line moving through the port graph from left to right. Whenever you see "action" – either a box or wires crossing – write down the sum

---

[4] One can think of σ as the "swap" icon $\times: 2 \to 2$.

(using $+$) of all the boxes g, all the symmetries σ, and all the wires id_1 in that column. Finally, compose all of those action columns. For example, in the picture below we see four action columns:

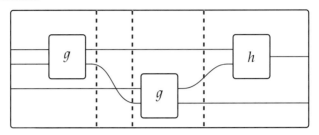

Here the result is $(g + \mathrm{id}_1) \mathbin{\substack{\circ \\ 9}} (\mathrm{id}_1 + \sigma) \mathbin{\substack{\circ \\ 9}} (\mathrm{id}_1 + g) \mathbin{\substack{\circ \\ 9}} (h + \mathrm{id}_1)$.

Exercise 5.27. Consider again the free prop on generators $G = \{f: 1 \to 1, g: 2 \to 2, h: 2 \to 1\}$. Draw a picture of $(f + \mathrm{id}_1 + \mathrm{id}_1) \mathbin{\substack{\circ \\ 9}} (\sigma + \mathrm{id}_1) \mathbin{\substack{\circ \\ 9}} (\mathrm{id}_1 + h) \mathbin{\substack{\circ \\ 9}} \sigma \mathbin{\substack{\circ \\ 9}} g$, where $\sigma: 2 \to 2$ is the symmetry map. ◇

Another way of describing when we should consider two prop expressions equal is to say that they are equal if and only if they represent the same port graph. In either case, these notions induce an equivalence relation on the set of prop expressions. To say that we consider these certain prop expressions equal is to say that the morphisms of the free prop on G are the G-generated prop expressions *quotiented* by this equivalence relation (see Definition 1.16).

5.2.5   Props via Presentations

In Section 3.2.2 we saw that a presentation for a category, or database schema, consists of a graph together with imposed equations between paths. Similarly here, sometimes we want to construct a prop whose morphisms obey specific equations. But rather than mere paths, the things we want to equate are prop expressions as in Definition 5.25.

Rough Definition 5.28. A *presentation* (G, s, t, E) for a prop is a set G, functions $s, t: G \to \mathbb{N}$, and a set $E \subseteq \mathrm{Expr}(G) \times \mathrm{Expr}(G)$ of pairs of G-generated prop expressions, such that e_1 and e_2 have the same arity for each $(e_1, e_2) \in E$. We refer to G as the set of generators and to E as the set of *equations* in the presentation.[5]

The prop $\mathcal{G}$ *presented* by the presentation (G, s, t, E) is the prop whose morphisms are elements in $\mathrm{Expr}(G)$, quotiented by both the equations $e_1 = e_2$ where $(e_1, e_2) \in E$, and by the axioms of symmetric strict monoidal categories.

Remark 5.29. Given a presentation (G, s, t, E), it can be shown that the prop $\mathcal{G}$ has a universal property in terms of "maps out." Namely prop functors from $\mathcal{G}$ to any

---

[5] Elements of E, which we call equations, are traditionally called "relations." We think of $(e_1, e_2) \in E$ as standing for the equation $e_1 = e_2$, as this will be forced soon.

other prop $\mathcal{C}$ are in one-to-one correspondence with functions f from G to the set of morphisms in $\mathcal{C}$ such that

- for all $g \in G$, $f(g)$ is a morphism $s(g) \to t(g)$, and
- for all $(e_1, e_2) \in E$, we have that $f(e_1) = f(e_2)$ in $\mathcal{C}$, where $f(e)$ denotes the morphism in $\mathcal{C}$ obtained by applying f to each generator in the expression e, and then composing the result in $\mathcal{C}$.

Exercise 5.30. Is it the case that the free prop on generators (G, s, t), defined in Definition 5.21, is the same thing as the prop presented by $(G, s, t, \varnothing)$, having no relations, as defined in Definition 5.28? Or is there a subtle difference somehow?   ◇

5.3   Simplified Signal Flow Graphs

We now return to signal flow graphs, expressing them in terms of props. We will discuss a simplified form without feedback (the only sort we have discussed so far), and then extend to the usual form of signal flow graphs in Section 5.4.3. But before we can do that, we must say what we mean by signals; this gets us into the algebraic structure of "rigs." We will get to signal flow graphs in Section 5.3.2.

5.3.1   Rigs

Signals can be amplified, and they can be added. Adding and amplification interact via a distributive law, as follows: if we add two signals, and then amplify them by some amount a, it should be the same as amplifying the two signals separately by a, then adding the results.

We can think of all the possible amplifications as forming a structure called a rig,[6] defined as follows.

Definition 5.31. A *rig* is a tuple $(R, 0, +, 1, *)$, where R is a set, $0, 1 \in R$ are elements, and $+, *: R \times R \to R$ are functions, such that

(a) $(R, +, 0)$ is a commutative monoid,
(b) $(R, *, 1)$ is a monoid,[7]
(c) $a * (b + c) = a * b + a * c$ and $(a + b) * c = a * c + b * c$ for all $a, b, c \in R$,
(d) $a * 0 = 0 = 0 * a$ for all $a \in R$.

We have already encountered many examples of rigs.

Example 5.32. The natural numbers form a rig $(\mathbb{N}, 0, +, 1, *)$.

---

[6] Rigs are also known as *semirings*.
[7] Note that we did not demand that $(R, *, 1)$ be commutative; we shall see a naturally arising example where it is not commutative in Example 5.35.

Example 5.33. The booleans form a rig $(\mathbb{B}, \mathtt{false}, \vee, \mathtt{true}, \wedge)$.

Example 5.34. Any quantale $\mathcal{V} = (V, \leq, I, \otimes)$ determines a rig $(V, 0, \vee, I, \otimes)$, where $0 = \bigvee \varnothing$ is the empty join. See Definition 2.57.

Example 5.35. If R is a rig and $n \in \mathbb{N}$ is any natural number, then the set $\mathrm{Mat}_n(R)$ of $(n \times n)$-matrices in R forms a rig. A matrix $M \in \mathrm{Mat}_n(R)$ is a function $M: \underline{n} \times \underline{n} \to R$. Addition $M + N$ of matrices is given by $(M + N)(i, j) := M(i, j) + N(i, j)$ and multiplication $M * N$ is given by $(M * N)(i, j) := \sum_{k \in \underline{n}} M(i, k) * N(k, j)$. The 0-matrix is $0(i, j) := 0$ for all $i, j \in \underline{n}$. Note that $\mathrm{Mat}_n(R)$ is generally not commutative.

Exercise 5.36.

1. We said in Example 5.35 that, for any rig R, the set $\mathrm{Mat}_n(R)$ forms a rig. What is its multiplicative identity $1 \in \mathrm{Mat}_n(R)$?
2. We also said that $\mathrm{Mat}_n(R)$ is generally not commutative. Pick an n and show that $\mathrm{Mat}_n(\mathbb{N})$ is not commutative, where $\mathbb{N}$ is as in Example 5.32.   ◇

The following is an example for readers who are familiar with the algebraic structure known as "rings."

Example 5.37. Any ring forms a rig. In particular, the real numbers $(\mathbb{R}, 0, +, 1, *)$ are a rig. The difference between a ring and rig is that a ring, in addition to all the properties of a rig, must also have additive inverses, or *negatives*. A common mnemonic is that a rig is a ring without negatives.

5.3.2   The Iconography of Signal Flow Graphs

A signal flow graph is supposed to keep track of the amplification, by elements of a rig R, to which signals are subjected. While not strictly necessary,[8] we will assume the signals themselves are elements of the same rig R. We refer to elements of R as *signals* for the time being.

Amplification of a signal by some value $a \in R$ is simply depicted like so:

(scalar mult.)

---

[8] The necessary requirement for the material below to make sense is that the signals take values in an *R-module M*. We will not discuss this here, keeping to the simpler requirement that $M = R$.

We interpret the above icon as depicting a system where a signal enters on the left-hand wire, is multiplied by a, and is output on the right-hand wire.

What is more interesting than just a single signal amplification, however, is the interaction of signals. There are four other important icons in signal flow graphs.

Let's go through them one by one. The first two are old friends from Chapter 2: copy and discard.

(copy)

We interpret this diagram as taking in an input signal on the left, and outputting that same value to both wires on the right. It is basically the "copy" operation from Section 2.2.3.

Next, we have the ability to discard signals.

(discard)

This takes in any signal, and outputs nothing. It is basically the "waste" operation from Section 2.2.3.

Next, we have the ability to add signals.

(add, +)

This takes the two input signals and adds them, to produce a single-output signal.

Finally, we have the zero signal.

(zero, 0)

This has no inputs, but always outputs the 0 element of the rig.

Using these icons, we can build more complex signal flow graphs. To compute the operation performed by a signal flow graph we simply trace the paths with the above interpretations, plugging outputs of one icon into the inputs of the next icon.

For example, consider the rig $R = \mathbb{N}$ from Example 5.32, where the scalars are the natural numbers. Recall the signal flow graph from Eq. (5.1) in the introduction:

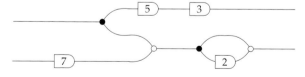

As we explained, this takes in two input signals x and y, and returns two output signals $a = 15x$ and $b = 3x + 21y$.

In addition to tracing the processing of the values as they move forward through the graph, we can also calculate these values by summing over paths. More explicitly, to get the contribution of a given input wire to a given output wire, we take the sum, over all paths p joining the wires, of the total amplification along that path.

So, for example, there is one path from the top input to the top output. On this path, the signal is first copied, which does not affect its value, then amplified by 5, and finally amplified by 3. Thus, if x is the first input signal, then this contributes $15x$ to the first output. Since there is no path from the bottom input to the top output (one is not allowed to traverse paths backwards), the signal at the first output is exactly $15x$. Both inputs contribute to the bottom output. In fact, each input contributes in two ways, as there are two paths to it from each input. The top input thus contributes $3x = x + 2x$, whereas the bottom input, passing through an additional $*7$ amplification, contributes $21y$.

Exercise 5.38.   The following flow graph takes in two natural numbers x and y

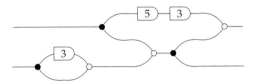

and produces two output signals. What are they?   ◇

Example 5.39. This example is for those who have some familiarity with differential equations. A linear system of differential equations provides a simple way to specify the movement of a particle. For example, consider a particle whose position (x, y, z) in three-dimensional space is determined by the following equations:

$$\dot{x} + 3\ddot{y} - 2z = 0,$$
$$\ddot{y} + 5\dot{z} = 0.$$

Using what is known as the Laplace transform, one can convert this into a linear system involving a formal variable D, which stands for "differentiate." Then the system becomes

$$Dx + 3D^2y - 2z = 0,$$
$$D^2y + 5Dz = 0,$$

which can be represented by the signal flow graph

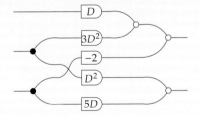

Signal flow graphs as morphisms in a free prop

We can formally define simplified signal flow graphs using props.

Definition 5.40. Let R be a rig (see Definition 5.31). Consider the set

$$G_R := \left\{ \, \supset\!\!-, \, \circ\!\!-, \, -\!\!\subset, \, \bullet\!\!- \, \right\} \cup \left\{ \, -\!\boxed{a}\!- \mid a \in R \right\},$$

and let $s, t \colon G_R \to \mathbb{N}$ be given by the number of dangling wires on the left and right of the generator icon, respectively. A *simplified signal flow graph* is a morphism in the free prop **Free**(G_R) on this set G_R of generators. We define **SFG**$_R :=$ **Free**(G_R).

For now we'll drop the term "simplified," since these are the only sort of signal flow graph we know. We'll return to signal flow graphs in their full glory – i.e. including feedback – in Section 5.4.3.

Example 5.41. To be more in line with our representations of both wiring diagrams and port graphs, morphisms in **Free**(G_R) should be drawn slightly differently. For example, technically the signal flow graph from Exercise 5.38 should be drawn as follows:

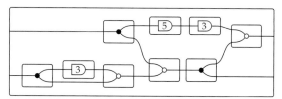

because we said we would label boxes with the elements of G. But it is easier on the eye to draw remove the boxes and just look at the icons inside as in Exercise 5.38, and so we'll draw our diagrams in that fashion.

More importantly, props provide language to understand the semantics of signal flow graphs. Although the signal flow graphs themselves are free props, their semantics – their meaning in our model of signals flowing – will arise when we add equations to our props, as in Definition 5.28. These equations will tell us when two signal flow graphs act the same way on signals. For example,

$$-\!\!\bullet\!\!\subset\!\!\!\!\!\!< \qquad \text{and} \qquad -\!\!\bullet\!\!\subset\!\!\!\!\!\!< \tag{5.6}$$

both express the same behavior: a single-input signal is copied twice so that three identical copies of the input signal are output.

If two signal flow graphs S, T are almost the same, with the one exception being that somewhere we replace the left-hand side of Eq. (5.6) with the right-hand side, then S and T have the same behavior. But there are other replacements we could make to a signal flow graph that do not change its behavior. Our next goal is to find a complete description of these replacements.

5.3.3   The Prop of Matrices over a Rig

Signal flow graphs are closely related to matrices. In previous chapters we showed how a matrix with values in a quantale $\mathcal{V}$ – a closed monoidal preorder with all joins – represents a system of interrelated points and connections between them, such as a profunctor. The quantale gave us the structure and axioms we needed in order for matrix multiplication to work properly. But we know from Example 5.34 that quantales are examples of rigs, and in fact matrix multiplication makes sense in any rig R. In Example 5.35, we explained that the set $\mathrm{Mat}_n(R)$ of $(n \times n)$-matrices in R can naturally be assembled into a rig, for any fixed choice of $n \in \mathbb{N}$. But what if we want to do better, and assemble *all* matrices into a single algebraic structure? The result is a prop!

An $(m \times n)$-*matrix M with values in R* is a function $M \colon (\underline{m} \times \underline{n}) \to R$. Given an $(m \times n)$-matrix M and an $(n \times p)$-matrix N, their *composite* is the $(m \times p)$-matrix $M \,\mathring{,}\, N$ defined as follows for any $a \in \underline{m}$ and $c \in \underline{p}$:

$$M \,\mathring{,}\, N(a, c) := \sum_{b \in \underline{n}} M(a, b) \times N(b, c). \tag{5.7}$$

Here the $\sum_{b \in \underline{n}}$ just means repeated addition (using the rig R's $+$ operation), as usual.

Remark 5.42.   Conventionally, one generally considers a matrix A acting on a vector v by multiplication in the order Av, where v is a column vector. In keeping with our composition convention, we use the opposite order, $v \,\mathring{,}\, A$, where v is a row vector. See for example Eq. (5.8) for when this is implicitly used.

Definition 5.43.   Let R be a rig. We define the *prop of R-matrices*, denoted **Mat**(R), to be the prop whose morphisms $m \to n$ are the $(m \times n)$-matrices with values in R. Composition of morphisms is given by matrix multiplication as in Eq. (5.7). The monoidal product is given by the direct sum of matrices: given matrices $A \colon m \to n$ and $b \colon p \to q$, we define $A + B \colon m + p \to n + q$ to be the block matrix

$$\begin{pmatrix} A & 0 \\ 0 & B \end{pmatrix}$$

where each 0 represents a matrix of zeros of the appropriate dimension ($m \times q$ and $n \times p$). We refer to any combination of multiplication and direct sum as an *interconnection* of matrices.

Exercise 5.44.   Let A and B be the following matrices with values in $\mathbb{N}$:

$$A = \begin{pmatrix} 3 & 3 & 1 \\ 2 & 0 & 4 \end{pmatrix}, \qquad\qquad B = \begin{pmatrix} 2 & 5 & 6 & 1 \end{pmatrix}.$$

What is the direct sum matrix $A + B$?   ◇

5.3.4   Turning Signal Flow Graphs into Matrices

Let's now consider more carefully what we mean when we talk about the meaning, or *semantics*, of each signal flow graph. We'll use matrices.

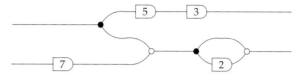

In the examples like the one above (copied from Eq. (5.1)), the signals emanating from output wires, say a and b, are given by certain sums of amplified input values, say x and y. If we can only measure the input and output signals, and care nothing for what happens in between, then each signal flow graph may as well be reduced to a matrix of amplifications. We can represent the signal flow graph of Eq. (5.1) by either the matrix on the left (for more detail) or the matrix on the right if the labels are clear from context:

$$
\begin{array}{c|cc}
 & a & b \\
\hline
x & 15 & 3 \\
y & 0 & 21
\end{array}
\qquad\qquad
\begin{pmatrix} 15 & 3 \\ 0 & 21 \end{pmatrix}
$$

Every signal flow graph can be interpreted as a matrix

The generators G_R from Definition 5.40 are shown again in the table below, where each is interpreted as a matrix. For example, we interpret amplification by $a \in R$ as the (1×1)-matrix $(a) \colon 1 \to 1$: it is an operation that takes an input $x \in R$ and returns $a * x$. Similarly, we can interpret $\rhd\!-$ as the (2×1)-matrix $\left(\begin{smallmatrix}1\\1\end{smallmatrix}\right)$: it is an operation that takes a row vector consisting of two inputs, x and y, and returns $x + y$. Here is a table showing the interpretation of each generator.

Generator	Icon	Matrix	Arity
amplify by $a \in R$	$-\boxed{a}-$	(a)	$1 \to 1$
add	$\rhd\!-$	$\left(\begin{smallmatrix}1\\1\end{smallmatrix}\right)$	$2 \to 1$
zero	$\circ\!-$	$()$	$0 \to 1$
copy	$-\!\!\lhd$	$(1 \quad 1)$	$1 \to 2$
discard	$-\!\!\bullet$	$()$	$1 \to 0$

(5.8)

Note that both zero and discard are represented by empty matrices, but of differing dimensions. In linear algebra it is unusual to consider matrices of the form $0 \times n$ or $n \times 0$ for various n to be different, but they can be kept distinct for bookkeeping purposes: you can multiply a (0×3)-matrix by a $(3 \times n)$-matrix for any n, but you can not multiply it by a $(2 \times n)$-matrix.

Since signal flow graphs are morphisms in a free prop, the table in (5.8) is enough to show that we can interpret any signal flow diagram as a matrix.

Theorem 5.45. There is a prop functor $S\colon \mathbf{SFG}_R \to \mathbf{Mat}(R)$ that sends the generators $g \in G$ icons to the matrices as described in Table 5.8.

Proof. This follows immediately from the universal property of free props, Remark 5.29. $\qquad\square$

We have now constructed a matrix $S(g)$ from any signal flow graph g. But how can we produce this matrix explicitly? Both for the example signal flow graph in Eq. (5.1) and for the generators in Definition 5.40, the associated matrix has dimension $m \times n$, where m is the number of inputs and n the number of outputs, with (i, j)th entry describing the amplification of the ith input that contributes to the jth output. This is how one would hope or expect the functor S to work in general; but does it? We have used a big hammer – the universal property of free constructions – to obtain our functor S. Our next goal is to check that it works in the expected way. Doing so is a matter of using induction over the set of prop expressions, as we now see.[9]

Proposition 5.46. Let g be a signal flow graph with m inputs and n outputs. The matrix $S(g)$ is the $(m \times n)$-matrix whose (i, j)-entry describes the amplification of the ith input that contributes to the jth output.

Proof. Recall from Definition 5.25 that an arbitrary G_R-generated prop expression is built from the morphisms $\mathrm{id}_0\colon 0 \to 0$, $\mathrm{id}_1\colon 1 \to 1$, $\sigma\colon 2 \to 2$, and the generators in G_R, using the following two rules:

- If $\alpha\colon m \to n$ and $\beta\colon p \to q$ are expressions, then $(\alpha + \beta)\colon (m + p) \to (n + q)$ is an expression.
- If $\alpha\colon m \to n$ and $\beta\colon n \to p$ are expressions, then $\alpha \,\fatsemi\, \beta\colon m \to p$ is an expression.

S is a prop functor by Theorem 5.45, which by Definition 5.9 must preserve identities, compositions, monoidal products, and symmetries. We first show that the proposition is true when g is equal to id_0, id_1, and σ.

The empty signal flow graph $\mathrm{id}_0\colon 0 \to 0$ must be sent to the unique (empty) matrix $()\colon 0 \to 0$. The morphisms id_1, σ, and $a \in R$ map to the identity matrix, the swap matrix, and the scalar matrix (a), respectively:

$$\underline{\qquad} \;\mapsto\; \begin{pmatrix}1\end{pmatrix} \qquad \text{and} \qquad \times \;\mapsto\; \begin{pmatrix}0 & 1 \\ 1 & 0\end{pmatrix} \qquad \text{and} \qquad \underset{\text{---}\boxed{a}\text{---}}{} \;\mapsto\; \begin{pmatrix}a\end{pmatrix}.$$

In each case, the (i, j)-entry gives the amplification of the ith input to the jth output.

It remains to show that if the proposition holds for $\alpha\colon m \to n$ and $\beta\colon p \to q$, then it holds for (i) $\alpha \,\fatsemi\, \beta$ (when $n = p$) and for (ii) $\alpha + \beta$ (in general).

---

[9] Mathematical induction is a formal proof technique that can be thought of like a domino rally: if you knock over all the starting dominoes, and you're sure that each domino will be knocked down if its predecessors are, then you're sure every domino will eventually fall. If you want more rigor, or you want to understand the proof of Proposition 5.46 as a genuine case of induction, ask a friendly neighborhood mathematician!

To prove (i), consider the following picture of $\alpha \,\mathbin{\raise1pt\hbox{$\circ$}\kern-2.5pt\raise-3pt\hbox{$\circ$}}\, \beta$:

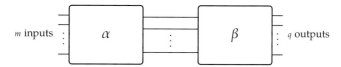

Here $\alpha : m \to n$ and $\beta : n \to q$ are signal flow graphs, assumed to obey the proposition. Consider the ith input and kth output of $\alpha \,\mathbin{\raise1pt\hbox{$\circ$}\kern-2.5pt\raise-3pt\hbox{$\circ$}}\, \beta$; we'll just call these i and k. We want to show that the amplification that i contributes to k is the sum – over all paths from i to k – of the amplification along that path. So let's also fix some $j \in \underline{n}$, and consider paths from i to k that run through j. By distributivity of the rig R, the total amplification from i to k through j is the total amplification over all paths from i to j times the total amplification over all paths from j to k. Since all paths from i to k must run through some jth output of α/input of β, the amplification that i contributes to k is

$$\sum_{j \in \underline{n}} \alpha(i, j) * \beta(j, k).$$

This is exactly the formula for matrix multiplication, which is composition $S(\alpha) \,\mathbin{\raise1pt\hbox{$\circ$}\kern-2.5pt\raise-3pt\hbox{$\circ$}}\, S(\beta)$ in the prop **Mat**(R); see Definition 5.43. So $\alpha \,\mathbin{\raise1pt\hbox{$\circ$}\kern-2.5pt\raise-3pt\hbox{$\circ$}}\, \beta$ obeys the proposition when α and β do.

Proving (ii) is more straightforward. The monoidal product $\alpha + \beta$ of signal flow graphs looks like this:

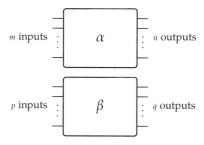

No new paths are created; the only change is to reindex the inputs and outputs. In particular, the ith input of α is the ith input of $\alpha + \beta$, the jth output of α is the jth output of $\alpha + \beta$, the ith input of β is the $(m + i)$th output of $\alpha + \beta$, and the jth output of β is the $(n + j)$th output of $\alpha + \beta$. This means that the matrix with (i, j)th entry describing the amplification of the ith input that contributes to the jth output is $S(\alpha) + S(\beta) = S(\alpha + \beta)$, as in Definition 5.43. This proves the proposition. $\square$

Exercise 5.47.

1. What matrix does the signal flow graph

   represent?

2. What about the following signal flow graph?

3. Are they equal?   ◇

5.3.5   The Idea of Functorial Semantics

Let's pause for a moment to reflect on what we have just learned. First, signal flow diagrams are the morphisms in a prop. This means we have two special operations we can do to form new signal flow diagrams from old, namely composition (combining in series) and monoidal product (combining in parallel). We might think of this as specifying a "grammar" or "syntax" for signal flow diagrams.

As a language, signal flow graphs have not only syntax but also semantics: each signal flow diagram can be interpreted as a matrix. Moreover, matrices have the same grammatical structure: they form a prop, and we can construct new matrices from old using composition and monoidal product. In Theorem 5.45 we completed this picture by showing that semantic interpretation is a prop functor between the prop of signal flow graphs and the prop of matrices. Thus we say that matrices give *functorial semantics* for signal flow diagrams.

Functorial semantics is a key manifestation of compositionality. It says that the matrix meaning $S(g)$ for a big signal flow graph g can be computed by:

1. splitting g up into little pieces,
2. computing the very simple matrices for each piece,
3. using matrix multiplication and direct sum to put the pieces back together to obtain the desired meaning, $S(g)$.

This functoriality is useful in practice, for example in speeding up computation of the semantics of signal flow graphs: for large signal flow graphs, composing matrices is much faster than tracing paths.

5.4   Graphical Linear Algebra

In this section we will begin to develop something called graphical linear algebra, which extends the ideas above. This formalism is actually quite powerful. For example, with it we can easily and *graphically* prove certain conjectures from control theory that, although they were eventually solved, required fairly elaborate matrix algebra arguments [FSR16].

5.4.1   A Presentation of Mat(R)

Let R be a rig, as defined in Definition 5.31. The main theorem of the previous section, Theorem 5.45, provided a functor $S \colon \mathbf{SFG}_R \to \mathbf{Mat}(R)$ that converts any signal flow graph into a matrix. Next we show that S is "full": that any matrix can be represented by a signal flow graph.

Proposition 5.48. Given any matrix $M \in \mathbf{Mat}(R)$, there exists a signal flow graph $g \in \mathbf{SFG}_R$ such that such that $S(g) = M$.

Proof sketch. Let $M \in \mathbf{Mat}(R)$ be an $(m \times n)$-matrix. We want a signal flow graph g such that $S(g) = M$. In particular, to compute $S(g)(i, j)$, we know that we can simply compute the amplification that the ith input contributes to the jth output. The key idea then is to construct g so that there is exactly one path from ith input to the jth output, and that this path has exactly one scalar multiplication icon, namely $M(i, j)$.

The general construction is a little technical (see Exercise 5.50), but the idea is clear from just considering the case of (2×2)-matrices. Suppose M is the (2×2)-matrix $\left(\begin{smallmatrix} a & b \\ c & d \end{smallmatrix}\right)$. Then we define g to be the signal flow graph

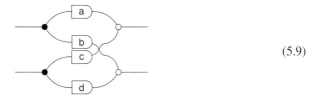

$$(5.9)$$

Tracing paths, it is easy to see that $S(g) = M$. Note that g is the composite of four layers, each layer respectively a monoidal product of (i) copy and discard maps, (ii) scalar multiplications, (iii) swaps and identities, (iv) addition and zero maps.

For the general case, see Exercise 5.50. □

Exercise 5.49. Draw signal flow graphs that represent the following matrices:

$$1. \begin{pmatrix} 0 \\ 1 \\ 2 \end{pmatrix} \qquad 2. \begin{pmatrix} 0 & 0 \\ 0 & 0 \end{pmatrix} \qquad 3. \begin{pmatrix} 1 & 2 & 3 \\ 4 & 5 & 6 \end{pmatrix}$$

◇

Exercise 5.50. Write down a detailed proof of Proposition 5.48. Suppose M is an $(m \times n)$-matrix. Follow the idea of the (2×2)-case in Eq. (5.9), and construct the signal flow graph g – having m inputs and n outputs – as the composite of four layers, respectively comprising (i) copy and discard maps, (ii) scalars, (iii) swaps and identities, (iv) addition and zero maps.

◇

We can also use Proposition 5.48 and its proof to give a presentation of $\mathbf{Mat}(R)$, which was defined in Definition 5.43.

Theorem 5.51.   The prop $\mathbf{Mat}(R)$ is isomorphic to the prop with the following presentation. The set of generators is the set

$$G_R := \left\{ \rhd\!\!\text{-},\; \text{-}\!\!\lhd,\; \text{-}\!\!\bullet\; \right\} \cup \left\{ \text{-}\boxed{a}\text{-} \mid a \in R \right\},$$

the same as the set of generators for $\mathbf{SFG}_R$; see Definition 5.40.
   We have the following equations for any $a, b \in R$:

Proof.   The key idea is that these equations are sufficient to rewrite any G_R-generated prop expression into a normal form – the one used in the proof of Proposition 5.48 – with all the black nodes to the left, all the white nodes to the right, and all the scalars in the middle. This is enough to show the equality of any two expressions that represent the same matrix. Details can be found in [BE15] or [BSZ17]. $\square$

Sound and complete presentation of matrices

Once you get used to it, Theorem 5.51 provides an intuitive, visual way to reason about matrices. Indeed, the theorem implies that two signal flow graphs represent the same matrix if and only if one can be turned into the other by local application of the above equations and the prop axioms.

   The fact that you can prove two SFGs to be the same by using only graphical rules can be stated in the jargon of logic: we say that the graphical rules provide a *sound and complete reasoning system*. To be more specific, *sound* refers to the forward direction of the above statement: two signal flow graphs represent the same matrix if one can be turned into the other using the given rules. *Complete* refers to the reverse direction: if two signal flow graphs represent the same matrix, then we can convert one into the other using the equations of Theorem 5.51.

Example 5.52. Both of the signal flow graphs below represent the same matrix, $\binom{0}{6}$:

and

This means that one can be transformed into the other by using only the equations from Theorem 5.51. Indeed, here

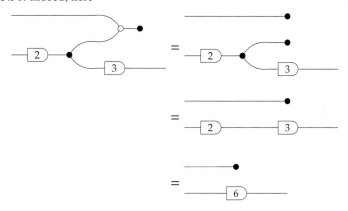

Exercise 5.53.

1. For each matrix in Exercise 5.49, draw another signal flow graph that represents that matrix.
2. Using the above equations and the prop axioms, prove that the two signal flow graphs represent the same matrix.   ◇

Exercise 5.54. Consider the signal flow graphs

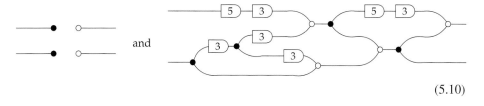

and

(5.10)

1. Let $R = (\mathbb{N}, 0, +, 1, *)$. By examining the presentation of **Mat**(R) in Theorem 5.51, and without computing the matrices that the two signal flow graphs in Eq. (5.10) represent, prove that they do *not* represent the same matrix.
2. Now suppose the rig is $R = \mathbb{N}/3\mathbb{N}$; if you do not know what this means, just replace all 3's with 0's in the right-hand diagram of Eq. (5.10). Find what you would call a minimal representation of this diagram, using the presentation in Theorem 5.51.   ◇

5.4.2   Aside: Monoid Objects in a Monoidal Category

Various subsets of the equations in Theorem 5.51 encode structures that are familiar from many other parts of mathematics, e.g. representation theory. For example one can find the axioms for (co)monoids, (co)monoid homomorphisms, Frobenius algebras, and (with a little rearranging) Hopf algebras, sitting inside this collection. The first example, the notion of monoids, is particularly familiar to us by now, so we briefly discuss it below, both in algebraic terms (Definition 5.55) and in diagrammatic terms (Example 5.58).

Definition 5.55.   A *monoid object* (M, μ, η) in a symmetric monoidal category $(\mathcal{C}, I, \otimes)$ is an object M of $\mathcal{C}$ together with morphisms $\mu \colon M \otimes M \to M$ and $\eta \colon I \to M$ such that

(a) $(\mu \otimes \mathrm{id}) \,\mathring{,}\, \mu = (\mathrm{id} \otimes \mu) \,\mathring{,}\, \mu$,
(b) $(\eta \otimes \mathrm{id}) \,\mathring{,}\, \mu = \mathrm{id} = (\mathrm{id} \otimes \eta) \,\mathring{,}\, \mu$.

A *commutative monoid object* is a monoid object that further obeys

(c) $\sigma_{M,M} \,\mathring{,}\, \mu = \mu$,

where $\sigma_{M,M}$ is the swap map on M in $\mathcal{C}$. We often denote it simply by σ.

Monoid objects are so named because they are an abstraction of the usual concept of monoid.

Example 5.56. A monoid object in $(\mathbf{Set}, 1, \times)$ is just a regular old monoid, as defined in Example 2.5; see also Example 3.7. That is, it is a set M, a function $\mu \colon M \times M \to M$, which we denote by infix notation $*$, and an element $\eta(1) \in M$, which we denote by e, satisfying $(a * b) * c = a * (b * c)$ and $a * e = a = e * a$.

Exercise 5.57.   Consider the set $\mathbb{R}$ of real numbers.

1. Show that if $\mu \colon \mathbb{R} \times \mathbb{R} \to \mathbb{R}$ is defined by $\mu(a, b) = a * b$ and if $\eta \in \mathbb{R}$ is defined to be $\eta = 1$, then $(\mathbb{R}, *, 1)$ satisfies all three conditions of Definition 5.55.
2. Show that if $\mu \colon \mathbb{R} \times \mathbb{R} \to \mathbb{R}$ is defined by $\mu(a, b) = a + b$ and if $\eta \in \mathbb{R}$ is defined to be $\eta = 0$, then $(\mathbb{R}, +, 0)$ satisfies all three conditions of Definition 5.55. $\diamond$

Example 5.58. Graphically, we can depict $\mu = \,\rightfork\,$ and $\eta = \,\multimap\,$. Then axioms (a), (b), and (c) from Definition 5.55 become:

(a) $=$

(b) $=$ ———

(c) $=$

All three of these are found in Theorem 5.51. Thus we can immediately conclude the following: the triple $(1, \rhd\text{-}, \text{-}\circ)$ is a commutative monoid object in the prop **Mat**(R).

Exercise 5.59. For any rig R, there is a functor $U : \textbf{Mat}(R) \to \textbf{Set}$, sending the object $n \in \mathbb{N}$ to the set R^n, and sending a morphism (matrix) $M : m \to n$ to the function $R^m \to R^n$ given by vector-matrix multiplication.

Recall that, in **Mat**(R), the monoidal unit is 0 and the monoidal product is $+$, because it is a prop. Recall also that, in (the usual monoidal structure on) **Set**, the monoidal unit is $\{1\}$, a set with one element, and the monoidal product is $\times$ (see Example 4.38).

1. Check that the functor $U : \textbf{Mat}(R) \to \textbf{Set}$, defined above, preserves the monoidal unit and the monoidal product.
2. Show that if (M, μ, η) is a monoid object in **Mat**(R) then $(U(M), U(\mu), U(\eta))$ is a monoid object in **Set**. (This works for any monoidal functor – which we will define in Definition 6.58 – not just for U in particular.)
3. In Example 5.58, we said that the triple $(1, \rhd\text{-}, \text{-}\circ)$ is a commutative monoid object in the prop **Mat**(R). If $R = \mathbb{R}$ is the rig of real numbers, this means that we have a monoid structure on the set $\mathbb{R}$. But in Exercise 5.57 we gave two such monoid structures. Which one is it? $\diamond$

Example 5.60. The triple $(1, \text{-}\lhd, \circ\text{-})$ in **Mat**(R) forms a commutative monoid object in **Mat**$(R)^{\text{op}}$. We hence also say that $(1, \text{-}\lhd, \circ\text{-})$ forms a *co-commutative comonoid object* in **Mat**(R).

Example 5.61. A *symmetric strict monoidal category*, is just a commutative monoid object in $(\textbf{Cat}, \times, \textbf{1})$. We will unpack this in Section 6.4.1.

Example 5.62. A symmetric monoidal preorder, which we defined in Definition 2.1, is just a commutative monoid object in the symmetric monoidal category $(\textbf{Preord}, \times, \textbf{1})$ of preorders and monotone maps.

Example 5.63. For those who know what tensor products of commutative monoids are (or can guess): a rig is a monoid object in the symmetric monoidal category $(\textbf{CMon}, \otimes, \mathbb{N})$ of commutative monoids with tensor product.

Remark 5.64. If we present a prop $\mathcal{M}$ using two generators $\mu : 2 \to 1$ and $\eta : 0 \to 1$, and the three equations from Definition 5.55, we could call it "the theory of monoids in monoidal categories." This means that, in any monoidal category $\mathcal{C}$, the monoid objects in $\mathcal{C}$ correspond to strict monoidal functors $\mathcal{M} \to \mathcal{C}$. This sort of idea leads to the study of algebraic theories, due to Bill Lawvere and extended by many others; see Section 5.5.

5.4.3 Signal Flow Graphs: Feedback and More

At this point in the story, we have seen that every signal flow graph represents a matrix, and this gives us a new way of reasoning about matrices. This is just the beginning of a beautiful tale, one not only of graphical matrices, but of graphical *linear algebra*. We close this chapter with some brief hints at how the story continues.

The pictoral nature of signal flow graphs invites us to play with them. While we normally draw the copy icon like so, $\prec$, we could just as easily reverse it and draw an icon $\succ$. What might it mean? Let's think again about the semantics of flow graphs.

The behavioral approach

A signal flow graph $g: m \to n$ takes an input $x \in R^m$ and gives an output $y \in R^n$. In fact, since this is all we care about, we might just think about representing a signal flow graph g as describing a set of input and output pairs (x, y). We'll call this set the *behavior* of g and denote it $\mathsf{B}(g) \subseteq R^m \times R^n$. For example, the "copy" flow graph

sends the input 1 to the output $(1, 1)$, so we consider $(1, (1, 1))$ to be an element of copy behavior. Similarly, $(x, (x, x))$ is copy behavior for every $x \in R$, thus we have

$$\mathsf{B}(\prec) = \{(x, (x, x)) \mid x \in R\}.$$

In the abstract, the signal flow graph $g: m \to n$ has the behavior

$$\mathsf{B}(g) = \left\{ \left(x, S(g)(x)\right) \mid x \in R^m \right\} \subseteq R^m \times R^n. \tag{5.11}$$

The mirror image of an icon

The above behavioral perspective provides a clue about how to interpret the mirror images of the diagrams discussed above. Reversing an icon $g: m \to n$ exchanges the inputs with the outputs, so if we denote this reversed icon by g^{op}, we must have $g^{\mathrm{op}}: n \to m$. Thus if $\mathsf{B}(g) \subseteq R^m \times R^n$ then we need $\mathsf{B}(g^{\mathrm{op}}) \subseteq R^n \times R^m$. One simple way to do this is to replace each (a, b) with (b, a), so we would have

$$\mathsf{B}(g^{\mathrm{op}}) := \left\{ \left(S(g)(x), x\right) \mid x \in R^m \right\} \subseteq R^n \times R^m. \tag{5.12}$$

This is called the *transposed relation*.

Exercise 5.65.

1. What is the behavior $\mathsf{B}(\prec)$ of the reversed addition icon $\prec: 1 \to 2$?
2. What is the behavior $\mathsf{B}(\succ)$ of the reversed copy icon $\succ: 2 \to 1$? $\diamond$

Eqs. (5.11) and (5.12) give us formulas for interpreting signal flow graphs and their mirror images. But this would easily lead to disappointment, if we couldn't combine the two directions behaviorally; luckily we can.

Combining directions

What should the behavior be for a diagram such as the following:

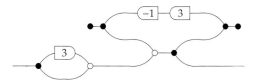

Let's formalize our thoughts a bit and begin by thinking about behaviors. The behavior of a signal flow graph $m \to n$ is a subset $B \subseteq R^m \times R^n$, i.e. a relation. Why not try to construct a prop where the morphisms $m \to n$ are relations?

We'll need to know how to compose and take monoidal products of relations. And if we want this prop of relations to contain the old prop $\mathbf{Mat}(R)$, we need the new compositions and monoidal products to generalize the old ones in $\mathbf{Mat}(R)$. Given signal flow graphs with matrices $M : m \to n$ and $N : n \to p$, we see that their behaviors are the relations $B_1 := \{(x, Mx) \mid x \in R^m\}$ and $B_2 := \{(y, Ny) \mid y \in R^n\}$, while the behavior of $M \,\mathring{,}\, N$ is the relation $\{(x, x \,\mathring{,}\, M \,\mathring{,}\, N) \mid x \in R^m\}$. This is a case of relation composition. Given relations $B_1 \subseteq R^m \times R^n$ and $B_2 \subseteq R^n \times R^p$, their composite $B_1 \,\mathring{,}\, B_2 \subseteq R^m \times R^p$ is given by

$$B_1 \,\mathring{,}\, B_2 := \{(x, z) \mid \text{there exists } y \in R^n \text{ such that } (x, y) \in B_1 \text{ and } (y, z) \in B_2\}. \quad (5.13)$$

We will use this as the general definition for composing two behaviors.

Definition 5.66.   Let R be a rig. We define the prop $\mathbf{Rel}_R$ of R-relations to have subsets $B \subseteq R^m \times R^n$ as morphisms. These are composed by the composition rule from Eq. (5.13), and we take the product of two sets to form their monoidal product.

Exercise 5.67.   In Definition 5.66 we went quickly through monoidal products $+$ in the prop $\mathbf{Rel}_R$. If $B \subseteq R^m \times R^n$ and $C \subseteq R^p \times R^q$ are morphisms in $\mathbf{Rel}_R$, write down $B + C$ in set notation.   ◇

(No-longer simplified) signal flow graphs

Recall that above, e.g. in Definition 5.40, we wrote G_R for the set of generators of signal flow graphs. In Section 5.4.3, we wrote g^{op} for the mirror image of g, for each $g \in G_R$. So let's write $G_R^{\mathrm{op}} := \{g^{\mathrm{op}} \mid g \in G_R\}$ for the set of all the mirror images of generators. We define a prop

$$\mathbf{SFG}_R^+ := \mathbf{Free}\left(G_R \sqcup G_R^{\mathrm{op}}\right). \quad (5.14)$$

We call a morphism in the prop $\mathbf{SFG}_R^+$ a *(non-simplified) signal flow graph*: these extend our simplified signal flow graphs from Definition 5.40 because now we can also use the mirrored icons. By the universal property of free props, since we have said what the behavior of the generators is (the behavior of a reversed icon is the transposed relation; see Eq. (5.12)), we have specified the behavior of any signal flow graph.

The following two exercises help us understand what this behavior is.

Exercise 5.68. Let $g: m \to n$, $h: \ell \to n$ be signal flow graphs. Note that $h^{\mathrm{op}}: n \to \ell$ is a signal flow graph, and we can form the composite $g \,\mathring{,}\, (h^{\mathrm{op}})$:

Show that the behavior of $g \,\mathring{,}\, (h^{\mathrm{op}}) \subseteq R^m \times R^\ell$ is equal to

$$\mathrm{B}(g \,\mathring{,}\, (h^{\mathrm{op}})) = \{(x, y) \mid S(g)(x) = S(h)(y)\}. \qquad \diamond$$

Exercise 5.69. Let $g: m \to n$, $h: m \to p$ be signal flow graphs. Note that $(g^{\mathrm{op}}): n \to m$ is a signal flow graph, and we can form the composite $g^{\mathrm{op}} \,\mathring{,}\, h$

Show that the behavior of $g^{\mathrm{op}} \,\mathring{,}\, h$ is equal to

$$\mathrm{B}((g^{\mathrm{op}}) \,\mathring{,}\, h) = \{(S(g)(x), S(h)(x)) \mid x \in R^m\}. \qquad \diamond$$

Linear algebra via signal flow graphs

In Eq. (5.11) we see that every matrix, or linear map, can be represented as the behavior of a signal flow graph, and in Exercise 5.68 we see that solution sets of linear equations can also be represented. This includes central concepts in linear algebra, like kernels and images.

Exercise 5.70. Here is an exercise for those who know linear algebra, in particular kernels and images. Let R be a field, let $g: m \to n$ be a signal flow graph, and let $S(g) \in \mathbf{Mat}(R)$ be the associated $(m \times n)$-matrix (see Theorem 5.45).

1. Show that the composite of g with 0-reverses, shown here

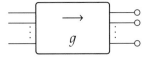

   is equal to the kernel of the matrix $S(g)$.

2. Show that the composite of discard-reverses with g, shown here

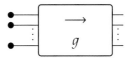

   is equal to the image of the matrix $S(g)$.
3. Show that, for any signal flow graph g, the subset $\mathsf{B}(g) \subseteq R^m \times R^n$ is a linear subspace. That is, if $b_1, b_2 \in \mathsf{B}(g)$ then so are $b_1 + b_2$ and $r * b_1$, for any $r \in R$. ◇

We have thus seen that signal flow graphs provide a uniform, compositional language to talk about many concepts in linear algebra. Moreover, in Exercise 5.70 we showed that the behavior of a signal flow graph is a linear relation, i.e. a relation whose elements can be added and multiplied by scalars $r \in R$. In fact the converse is true too: any linear relation $B \subseteq R^m \times R^n$ can be represented by a signal flow graph.

Exercise 5.71. One might want to show that linear relations on R form a prop, $\mathbf{LinRel}_R$. That is, one might want to show that there is a sub-prop of the prop $\mathbf{Rel}_R$ from Definition 5.66, where the morphisms $m \to n$ are the subsets $B \subseteq R^m \times R^n$ such that B is linear. In other words, where for any $(x, y) \in B$ and $r \in R$ the element $(r * x, r * y) \in R^m \times R^n$ is in B, and for any $(x', y') \in B$, the element $(x + x', y + y')$ is in B.

This is certainly doable, but for this exercise we only ask that you prove that the composite of two linear relations is linear. ◇

Just as we gave a sound and complete presentation for the prop of matrices in Theorem 5.51, it is possible to give a sound and complete presentation for linear relations on R. Moreover, it is possible to give such a presentation whose generating set is $G_R \sqcup G_R^{\mathrm{op}}$ as in Eq. (5.14) and whose equations include those from Theorem 5.51, plus a few more. This presentation gives a graphical method for doing linear algebra: an equation between linear subspaces is true *if and only if* it can be proved using the equations from the presentation.

Although not difficult, we leave the full presentation to further reading (Section 5.5). Instead, we'll conclude our exploration of the prop of linear relations by noting that some of these "few more" equations state that relations – just like co-design problems in Chapter 4 – form a compact closed category.

Compact closed structure
Using the icons available to us for signal flow graphs, we can build morphisms that look like the "cup" and "cap" from Definition 4.44:

$$\qquad\qquad \text{and} \qquad\qquad\qquad \tag{5.15}$$

The behaviors of these graphs are respectively

$$\{(0, (x, x)) \mid x \in R\} \subseteq R^0 \times R^2 \quad \text{and} \quad \{((x, x), 0) \mid x \in R\} \subseteq R^2 \times R^0.$$

In fact, these show that the object 1 in the prop $\mathbf{Rel}_R$ is dual to itself: the morphisms from Eq. (5.15) serve as the η_1 and ϵ_1 from Definition 4.44. Using monoidal products of these morphisms, one can show that any object in $\mathbf{Rel}_R$ is dual to itself.

Graphically, this means that the three signal flow graphs

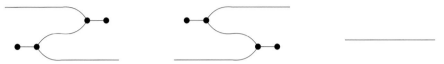

all represent the same relation.

Using these relations, it is straightforward to check the following result.

Theorem 5.72. The prop $\mathbf{Rel}_R$ is a compact closed category in which every object $n \in \mathbb{N}$ is dual to itself, $n = n^*$.

To make our signal flow graphs simpler, we define new icons cup and cap by the equations

$$\text{cup} := \quad \text{and} \quad \text{cap} :=$$

Back to control theory

Let's close by thinking about how to represent a simple control theory problem in this setting. Suppose we want to design a system to maintain the speed of a car at a desired speed u. We'll work in signal flow diagrams over the rig $\mathbb{R}[s, s^{-1}]$ of polynomials in s and s^{-1} with coefficients in $\mathbb{R}$ and where $ss^{-1} = s^{-1}s = 1$. This is standard in control theory: we think of s as integration, and s^{-1} as differentiation.

There are three factors that contribute to the actual speed v. First, there is the actual speed v. Second, there are external forces F. Third, we have our control system: this will take some linear combination $a * u + b * v$ of the desired speed and actual speed, amplify it by some factor p to give a (possibly negative) acceleration. We can represent this system as follows, where m is the mass of the car.

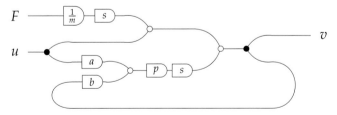

This can be read as the following equation, where one notes that v occurs twice:

$$v = \int \frac{1}{m} F(t)dt + u(t) + p \int au(t) + bv(t)dt.$$

Our control problem then asks: how do we choose a and b to make the behavior of this signal flow graph close to the relation $\{(F, u, v) \mid u = v\}$? By phrasing problems in this way, we can use extensions of the logic we have discussed above to reason about such complex real-world problems.

5.5   Summary and Further Reading

The goal of this chapter was to explain how props formalize signal flow graphs, and to provide a new perspective on linear algebra. To do this, we examined the idea of free and presented structures in terms of universal properties. This allowed us to build props that exactly suited our needs.

Paweł Sobociński's Graphical Linear Algebra blog is an accessible and fun exploration of the key themes of this chapter, which goes on to describe how concepts such as determinants, eigenvectors, and division by zero can be expressed using signal flow graphs [Sob]. For the technical details, one could start with Baez and Erbele [BE15], or Zanasi's thesis [Zan15] and its related series of papers [BSZ14; BSZ15; BSZ17]. For details about applications to control theory, see [FSR16]. From the control-theoretic perspective, the ideas and philosophy of this chapter are heavily influenced by Willems' behavioral approach [Wil07].

For the reader who has not studied abstract algebra, we mention that rings, monoids, and matrices are standard fare in abstract algebra, and can be found in any standard introduction, such as [Fra67]. Rigs, also known as semirings, are a bit less well known, but no less interesting; a comprehensive survey of the literature can be found in [Gla13].

Perhaps the most significant idea in this chapter is the separation of structure into syntax and semantics, related by a functor. This is not only present in the running theme of studying signal flow graphs, but in our aside Section 5.4.2, where we talk, for example, about monoid objects in monoidal categories. The idea of functorial semantics is yet another due to Lawvere, first appearing in his thesis [Law04].

6 Electric Circuits: Hypergraph Categories and Operads

6.1 The Ubiquity of Network Languages

Electric circuits, chemical reaction networks, finite-state automata, Markov processes: these are all models of physical or computational systems that are commonly described using network diagrams. Here, for example, we draw a diagram that models a flip-flop, an electric circuit – important in computer memory – that can store a bit of information:

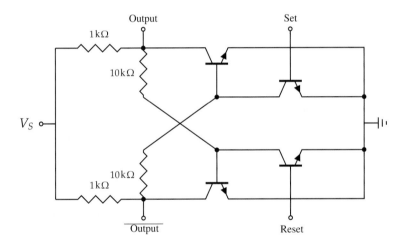

Network diagrams have time-tested utility. In this chapter, we are interested in understanding the common mathematical structure that they share, for the purposes of translating between and unifying them; for example certain types of Markov processes can be simulated and hence solved using circuits of resistors. When we understand the underlying structures that are shared by network diagram languages, we can make comparisons between the corresponding mathematical models easily.

At first glance network diagrams appear quite different from the wiring diagrams we have seen so far. For example, the wires are undirected in the case above, whereas in a category – including monoidal categories seen in resource theories or co-design – every morphism has a domain and codomain, giving it a sense of direction. Nonetheless, we shall see how to use categorical constructions such as universal properties to create categorical models that precisely capture the above type of "network" compositionality, i.e. allowing us to effectively drop directedness when convenient.

In particular we'll return to the idea of a colimit, which we sketched for you at the end of Chapter 3, and show how to use colimits in the category of sets to formalize ideas of connection. Here's the key idea.

Connections via colimits

Let's say we want to install some lights: we want to create a circuit so that when we flick a switch, a light turns on or off. To start, we have a bunch of circuit components: a power source, a switch, and a lamp connected to a resistor:

We want to connect them together, but there are many ways to do so. How should we describe the particular way that will form a light switch?

First, we claim that circuits should really be thought of as open circuits: each carries the additional structure of an "interface" exposing it to the rest of the electrical world. Here by *interface* we mean a certain set of locations, or *ports*, at which we are able to connect them with other components.[1] As is so common in category theory, we begin by making this more-or-less obvious fact explicit. Let's depict the available ports using a bold •. If we say that in the each of the three drawings above, the ports are simply the dangling end points of the wires, they would be redrawn as follows:

Next, we have to describe which ports should be connected. We'll do this by drawing empty circles ∘ connected by arrows to two ports •. Each will be a witness-to-connection, saying 'connect these two!'

Looking at this picture, it is clear what we need to do: just identify – i.e. *merge* or *make equal* – the ports as indicated, to get the following circuit:

---

[1] If your circuit has no such ports, it still falls within our purview, by taking its interface to be the empty set.

But mathematics doesn't have a visual cortex with which to generate the intuitions we can count on with a human reader such as yourself.[2] Thus we need to specify formally what "identifying ports as indicated" means mathematically. As it turns out, we can do this using finite colimits in a given category $\mathcal{C}$.

Colimits are diagrams with certain universal properties, which is kind of an epiphenomenon of the category $\mathcal{C}$. Our goal is to obtain $\mathcal{C}$'s colimits more directly, as a kind of operation in some context, so that we can think of them as telling us how to connect circuit parts together. To that end, we produce a certain monoidal category – namely that of *cospans in* $\mathcal{C}$, denoted **Cospan**$_{\mathcal{C}}$ – that can conveniently package $\mathcal{C}$'s colimits in terms of its own basic operations: composition and monoidal product.

In summary, the first part of this chapter is devoted to the slogan "colimits model interconnection." In addition to universal constructions such as colimits, however, another way to describe interconnection is to use wiring diagrams. We go full circle when we find that these wiring diagrams are strongly connected to cospans, and hence colimits.

Composition operations and wiring diagrams

In this book we have seen the utility of defining syntactic or algebraic structures that describe the sort of composition operations that make sense and can be performed in a given application area. Examples include monoidal preorders with discarding, props, and compact closed categories. Each of these has an associated sort of wiring diagram style, so that any wiring diagram of that style represents a composition operation that makes sense in the given area: the first makes sense in manufacturing, the second in signal flow, and the third in collaborative design. So our second goal is to answer the question, "how do we describe the compositional structure of network-style wiring diagrams?"

Network-type interconnection can be described using something called a hypergraph category. Roughly speaking, these are categories whose wiring diagrams are those of symmetric monoidal categories together with, for each pair of natural numbers (m, n), an icon $s_{m,n} \colon m \to n$. These icons, known as *spiders*,[3] are drawn as follows:

Two spiders can share a leg, and when they do, we can fuse them into one spider. The intuition is that spiders are connection points for a number of wires, and when two connection points are connected, they fuse to form an even more "connect-y" connection point. Here is an example:

![spider fusion diagram]

---

[2] Unless the future has arrived since the writing of this book.

[3] Our spiders have any number of legs.

A hypergraph category may have many species of spiders with the rule that spiders of different species cannot share a leg – and hence not fuse – but two spiders of the same species can share legs and fuse. We add spider diagrams to the iconography of hypergraph categories.

As we shall see, the ideas of describing network interconnection using colimits and hypergraph categories come together in the notion of a theory. We first introduced the idea of a theory in Section 5.4.2, but here we explore it more thoroughly, starting with the idea that, approximately speaking, cospans in the category **FinSet** form the theory of hypergraph categories.

We can assemble all cospans in **FinSet** into something called an "operad." Throughout this book we have talked about using free structures and presentations to create instances of algebraic structures such as preorders, categories, and props, tailored to the needs of a particular situation. Operads can be used to tailor the algebraic structures *themselves* to the needs of a particular situation. We will discuss how this works, in particular how operads encode various sorts of wiring diagram languages and corresponding algebraic structures, at the end of the chapter.

6.2   Colimits and Connection

Universal constructions are central to category theory. They allow us to define objects, at least up to isomorphism, by describing their relationship with other objects. So far we have seen this theme in a number of different forms: meets and joins (Section 1.3), Galois connections and adjunctions (Sections 1.4 and 3.4), limits (Section 3.5), and free and presented structures (Section 5.2.3–5.2.5). Here we turn our attention to colimits.

In this section, our main task is to have a concrete understanding of colimits in the category **FinSet** of finite sets and functions. The idea will be to take a bunch of sets – say two or fifteen or zero – use functions between them to designate that elements in one set "should be considered the same as" elements in another set, and then merge the sets together accordingly.

6.2.1   Initial Objects

Just as the simplest limit is a terminal object (see Section 3.5.1), the simplest colimit is an initial object. This is the case where you start with no objects and you merge them together.

Definition 6.1.   Let $\mathcal{C}$ be a category. An *initial object* in $\mathcal{C}$ is an object $\varnothing \in \mathcal{C}$ such that for each object T in $\mathcal{C}$ there exists a unique morphism $!_T : \varnothing \to T$.

The symbol $\varnothing$ is just a default name, a notation, intended to evoke the right idea; see Example 6.4 for the reason why we use the notation $\varnothing$, and Exercise 6.7 for a case when the default name $\varnothing$ would probably not be used.

Again, the hallmark of universality is the existence of a unique map to any other comparable object.

Example 6.2. An initial object of a preorder is a bottom element – i.e. an element that is less than every other element. For example, 0 is the initial object in $(\mathbb{N}, \leq)$, whereas $(\mathbb{R}, \leq)$ has no initial object.

Exercise 6.3.   Consider the set $A = \{a, b\}$. Find a preorder relation $\leq$ on A such that

1. $(A, \leq)$ has no initial object.
2. $(A, \leq)$ has exactly one initial object.
3. $(A, \leq)$ has two initial objects.   ◇

Example 6.4. The initial object in **FinSet** is the empty set. Given any finite set T, there is a unique function $\varnothing \to T$, since $\varnothing$ has no elements.

Example 6.5. As seen in Exercise 6.3, a category $\mathcal{C}$ need not have an initial object. As a different sort of example, consider the category shown here:

$$\mathcal{C} := \boxed{\; A \underset{g}{\overset{f}{\rightrightarrows}} B \;}$$

If there were to be an initial object $\varnothing$, it would be either A or B. Either way, we need to show that for each object $T \in \mathrm{Ob}(\mathcal{C})$ (i.e. for both $T = A$ and $T = B$) there is a unique morphism $\varnothing \to T$. Trying the case $\varnothing =^? A$ this condition fails when $T = B$: there are two morphisms $A \to B$, not one. And trying the case $\varnothing =^? B$ this condition fails when $T = A$: there are zero morphisms $B \to A$, not one.

Exercise 6.6.   For each of the graphs below, consider the free category on that graph, and say whether it has an initial object.

1. $\boxed{\overset{a}{\bullet}}$ 2.   3. $\boxed{\overset{a}{\bullet} \quad \overset{b}{\bullet}}$ 4.

◇

Exercise 6.7. Recall the notion of rig from Chapter 5. A *rig homomorphism* from $(R, 0_R, +_R, 1_R, *_R)$ to $(S, 0_S, +_S, 1_S, *_S)$ is a function $f : R \to S$ such that $f(0_R) = 0_S$, $f(r_1 +_R r_2) = f(r_1) +_S f(r_2)$, etc.

1. We said "etc." Guess the remaining conditions for f to be a rig homomorphism.
2. Let **Rig** denote the category whose objects are rigs and whose morphisms are rig homomorphisms. We claim **Rig** has an initial object. What is it? ◇

Exercise 6.8. Explain the statement "the hallmark of universality is the existence of a unique map to any other comparable object," in the context of Definition 6.1. In particular, what is being universal in Definition 6.1, and which is the "comparable object"? ◇

Remark 6.9. As mentioned in Remark 3.70, we often speak of "the" object that satisfies a universal property, such as "the initial object," even though many different objects could satisfy the initial object condition. Again, the reason is that initial objects are unique up to unique isomorphism: any two initial objects will have a canonical isomorphism between them, which one finds using various applications of the universal property.

Exercise 6.10. Let $\mathcal{C}$ be a category, and suppose that c_1 and c_2 are initial objects. Find an isomorphism between them, using the universal property from Definition 6.1. ◇

6.2.2 Coproducts

Coproducts generalize both joins in a preorder and disjoint unions of sets.

Definition 6.11. Let A and B be objects in a category $\mathcal{C}$. A *coproduct* of A and B is an object, which we denote $A + B$, together with a pair of morphisms $(\iota_A : A \to A + B, \iota_B : B \to A + B)$ such that, for all objects T and pairs of morphisms $(f : A \to T, g : B \to T)$, there exists a unique morphism $[f, g] : A + B \to T$ such that the following diagram commutes:

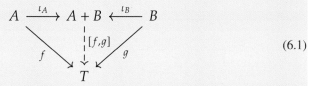

$$(6.1)$$

We call $[f, g]$ the *copairing* of f and g.

Exercise 6.12. Explain why, in a preorder, coproducts are the same as joins. ◇

Example 6.13. Coproducts in the categories **FinSet** and **Set** are disjoint unions. More precisely, suppose A and B are sets. Then the coproduct of A and B is given by the disjoint union $A \sqcup B$ together with the inclusion functions $\iota_A : A \longrightarrow A \sqcup B$ and $\iota_B : B \to A \sqcup B$.

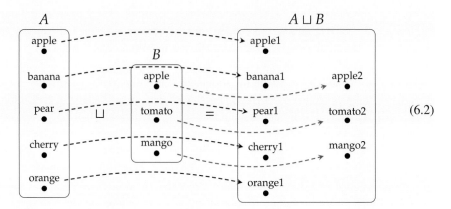

$$(6.2)$$

Suppose we have functions $f : A \to T$ and $g : B \to T$ for some other set T, unpictured. The universal property of coproducts says there is a unique function $[f, g] : A \sqcup B \to T$ such that $\iota_A \, \mathring{,} \, [f, g] = f$ and $\iota_B \, \mathring{,} \, [f, g] = g$. What is it? Any element $x \in A \sqcup B$ is either "from A" or "from B," i.e. either there is some $a \in A$ with $x = \iota_A(a)$ or there is some $b \in B$ with $x = \iota_B(b)$. By Eq. (6.1), we must have:

$$[f, g](x) = \begin{cases} f(x) & \text{if } x = \iota_A(a) \text{ for some } a \in A, \\ g(x) & \text{if } x = \iota_B(b) \text{ for some } b \in B. \end{cases}$$

Exercise 6.14.   Suppose $T = \{a, b, c, \dots, z\}$ is the set of letters in the alphabet, and let A and B be the sets from Eq. (6.2). Consider the function $f : A \to T$ sending each element of A to the first letter of its label, e.g. $f(\text{apple}) = a$. Let $g : B \to T$ be the function sending each element of B to the last letter of its label, e.g. $g(\text{apple}) = e$. Write down the function $[f, g](x)$ for all eight elements of $A \sqcup B$. ◇

Exercise 6.15.   Let $f : A \to C$, $g : B \to C$, and $h : C \to D$ be morphisms in a category $\mathcal{C}$ with coproducts. Show that

1. $\iota_A \, \mathring{,} \, [f, g] = f$,
2. $\iota_B \, \mathring{,} \, [f, g] = g$,
3. $[f, g] \, \mathring{,} \, h = [f \, \mathring{,} \, h, g \, \mathring{,} \, h]$,
4. $[\iota_A, \iota_B] = \text{id}_{A+B}$. ◇

Exercise 6.16.   Suppose a category $\mathcal{C}$ has coproducts, denoted $+$, and an initial object, denoted $\varnothing$. Then $(\mathcal{C}, +, \varnothing)$ is a symmetric monoidal category (recall Definition 4.34). In this exercise we develop the data relevant to this fact:

1. Show that $+$ extends to a functor $\mathcal{C} \times \mathcal{C} \to \mathcal{C}$. In particular, how does it act on morphisms in $\mathcal{C} \times \mathcal{C}$?

2. Using the universal properties of the initial object and coproduct, show that there are isomorphisms $A + \varnothing \to A$ and $\varnothing + A \to A$.
3. Using the universal property of the coproduct, write down morphisms
   (a) $(A + B) + C \to A + (B + C)$,
   (b) $A + B \to B + A$.
   If you like, check that these are isomorphisms.

It can then be checked that this data obeys the axioms of a symmetric monoidal category, but we'll end the exercise here. ◇

6.2.3 Pushouts

Pushouts are a way of combining sets. Like a union of subsets, a pushout can combine two sets in a nondisjoint way: elements of one set may be identified with elements of the other. The pushout construction, however, is much more general: it allows (and requires) the user to specify exactly which elements will be identified. We'll see a demonstration of this additional generality in Example 6.25.

> **Definition 6.17.** Let $\mathcal{C}$ be a category and let $f \colon A \to X$ and $g \colon A \to Y$ be morphisms in $\mathcal{C}$ that have a common domain. The *pushout* $X +_A Y$ is the colimit of the diagram
>
> $$A \xrightarrow{f} X$$
> $$\Big\downarrow{\scriptstyle g}$$
> $$Y$$

In more detail, a pushout consists of (i) an object $X +_A Y$ and (ii) morphisms $\iota_X \colon X \to X +_A Y$ and $\iota_Y \colon Y \to X +_A Y$ satisfying (a) and (b) below.

(a) The diagram

$$
\begin{array}{ccc}
A & \xrightarrow{\;f\;} & X \\
{\scriptstyle g}\big\downarrow & \ulcorner & \big\downarrow{\scriptstyle \iota_X} \\
Y & \xrightarrow[\iota_Y]{} & X +_A Y
\end{array}
\tag{6.3}
$$

commutes. (We will explain the '$\ulcorner$' symbol below.)

(b) For all objects T and morphisms $x \colon X \to T$, $y \colon Y \to T$, if the diagram

$$
\begin{array}{ccc}
A & \xrightarrow{\;f\;} & X \\
{\scriptstyle g}\big\downarrow & & \big\downarrow{\scriptstyle x} \\
Y & \xrightarrow[y]{} & T
\end{array}
$$

commutes, then there exists a unique morphism $t \colon X +_A Y \to T$ such that

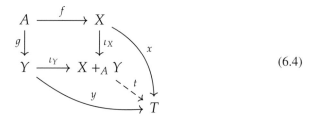

$$\tag{6.4}$$

commutes.

If $X +_A Y$ is a pushout, we denote that fact by drawing the commutative square Eq. (6.3), together with the $\ulcorner$ symbol as shown; we call it a *pushout square*.

We further call ι_X the *pushout of g along f*, and similarly ι_Y the *pushout of f along g*.

Example 6.18. In a preorder, pushouts and coproducts have a lot in common. The pushout of a diagram $B \leftarrow A \rightarrow C$ is equal to the coproduct $B \sqcup C$: namely, both are equal to the join $B \vee C$.

Example 6.19. Let $f \colon A \rightarrow X$ be a morphism in a category $\mathcal{C}$. For any isomorphisms $i \colon A \rightarrow A'$ and $j \colon X \rightarrow X'$, we can take X' to be the pushout $X +_A A'$, i.e. the following is a pushout square:

$$
\begin{array}{ccc}
A & \xrightarrow{\ f\ } & X \\
{\scriptstyle i}\downarrow & \ulcorner & \downarrow{\scriptstyle j} \\
A' & \xrightarrow{\ f'\ } & X'
\end{array}
$$

where $f' := i^{-1} \,\mathbin{\fatsemi}\, f \,\mathbin{\fatsemi}\, j$. To see this, observe that if there is any object T such that the following square commutes:

$$
\begin{array}{ccc}
A & \xrightarrow{\ f\ } & X \\
{\scriptstyle i}\downarrow & & \downarrow{\scriptstyle x} \\
A' & \xrightarrow{\ a\ } & T
\end{array}
$$

then $f \,\mathbin{\fatsemi}\, x = i \,\mathbin{\fatsemi}\, a$, and so we are forced to take $x' \colon X \rightarrow T$ to be $x' := j^{-1} \,\mathbin{\fatsemi}\, x$. This makes the following diagram commute:

$$
\begin{array}{ccc}
A & \xrightarrow{\ f\ } & X \\
{\scriptstyle i}\downarrow & & \downarrow{\scriptstyle j} \quad\searrow{\scriptstyle x} \\
A' & \xrightarrow{\ f'\ } & X' \\
& \searrow{\scriptstyle a} & \downarrow{\scriptstyle x'} \\
& & T
\end{array}
$$

because $f' \,\mathbin{\fatsemi}\, x' = i^{-1} \,\mathbin{\fatsemi}\, f \,\mathbin{\fatsemi}\, j \,\mathbin{\fatsemi}\, j^{-1} \,\mathbin{\fatsemi}\, x = i^{-1} \,\mathbin{\fatsemi}\, i \,\mathbin{\fatsemi}\, a = a$.

Exercise 6.20. For any set S, we have the discrete category $\mathbf{Disc}_S$, with S as objects and only identity morphisms.

1. Show that all pushouts exist in $\mathbf{Disc}_S$, for any set S.
2. For what sets S does $\mathbf{Disc}_S$ have an initial object?   ◇

Example 6.21. In the category **FinSet**, pushouts always exist. The pushout of functions $f : A \to X$ and $g : A \to Y$ is the set of equivalence classes of $X \sqcup Y$ under the equivalence relation generated by – i.e. the reflexive, transitive, symmetric closure of – the relation $\{f(a) \sim g(a) \mid a \in A\}$.

We can think of this in terms of interconnection too. Each element $a \in A$ provides a connection between $f(a)$ in X and $g(a)$ in Y. The pushout is the set of connected components of $X \sqcup Y$.

Exercise 6.22. What is the pushout of the functions $f : \underline{4} \to \underline{5}$ and $g : \underline{4} \to \underline{3}$ pictured below?

$$f : \underline{4} \to \underline{5} \qquad\qquad g : \underline{4} \to \underline{3}$$

Check your answer using the abstract description from Example 6.21.   ◇

Example 6.23. Suppose a category $\mathcal{C}$ has an initial object $\varnothing$. For any two objects $X, Y \in \mathrm{Ob}\,\mathcal{C}$, there is a unique morphism $f : \varnothing \to X$ and a unique morphism $g : \varnothing \to Y$; this is what it means for $\varnothing$ to be initial.

The diagram $X \xleftarrow{f} \varnothing \xrightarrow{g} Y$ has a pushout in $\mathcal{C}$ iff X and Y have a coproduct in $\mathcal{C}$, and the pushout and the coproduct will be the same. Indeed, suppose X and Y have a coproduct $X + Y$; then the diagram to the left below

$$
\begin{array}{ccc}
\varnothing & \xrightarrow{\;f\;} & X \\
{\scriptstyle g}\downarrow & & \downarrow{\scriptstyle \iota_X} \\
Y & \xrightarrow[\iota_Y]{} & X + Y
\end{array}
\qquad\qquad
\begin{array}{ccc}
\varnothing & \xrightarrow{\;f\;} & X \\
{\scriptstyle g}\downarrow & & \downarrow{\scriptstyle x} \\
Y & \xrightarrow[y]{} & T
\end{array}
$$

commutes (why?$_1$), and for any object T and commutative diagram as shown to the right, there is a unique map $X + Y \to T$ making the diagram as in Eq. (6.4) commute (why?$_2$). This shows that $X + Y$ is a pushout, $X +_\varnothing Y \cong X + Y$.

Similarly, if a pushout $X +_\varnothing Y$ exists, then it satisfies the universal property of the coproduct (why?$_3$).

Exercise 6.24. In Example 6.23 we asked "why?" three times.

1. Give a justification for "why?$_1$."

2. Give a justification for "why?$_2$."
3. Give a justification for "why?$_3$."   ◇

Example 6.25. Let $A = X = Y = \mathbb{N}$. Consider the functions $f : A \to X$ and $g : A \to Y$ given by the "floor" functions, $f(a) := \lfloor a/2 \rfloor$ and $g(a) := \lfloor (a+1)/2 \rfloor$.

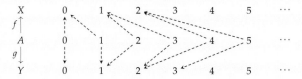

What is their pushout? Let's figure it out using the definition.

If T is any other set and we have maps $x : X \to T$ and $y : Y \to T$ that commute with f and g, i.e. $f \cong x = g \cong y$, then this commutativity implies that

$$y(0) = y(g(0)) = x(f(0)) = x(0).$$

In other words, Y's 0 and X's 0 go to the same place in T, say t. But since $f(1) = 0$ and $g(1) = 1$, we also have that $t = x(0) = x(f(1)) = y(g(1)) = y(1)$. This means Y's 1 goes to t also. But since $g(2) = 1$ and $f(2) = 1$, we also have that $t = g(1) = y(g(2)) = x(f(2)) = x(1)$, which means that X's 1 also goes to t. One can keep repeating this and find that every element of Y and every element of X go to t! Using mathematical induction, one can prove that the pushout is in fact a one-element set, $X \sqcup_A Y \cong \{1\}$.

6.2.4   Finite Colimits

Initial objects, coproducts, and pushouts are all types of colimits. We gave the general definition of colimit in Section 3.5.4. Just as a limit in $\mathcal{C}$ is a terminal object in a category of cones over a diagram $D : \mathcal{J} \to \mathcal{C}$, a colimit is an initial object in a category of cocones over some diagram $D : \mathcal{J} \to \mathcal{C}$. For our purposes it is enough to discuss finite colimits – i.e. when $\mathcal{J}$ is a finite category – which subsume initial objects, coproducts, and pushouts.[4]

In Definition 3.87, cocones in $\mathcal{C}$ are defined to be cones in $\mathcal{C}^{\mathrm{op}}$. For visualization purposes, if $D : \mathcal{J} \to \mathcal{C}$ looks like the diagram to the left, then a cocone on it is shown in the diagram to the right:

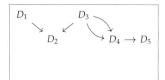   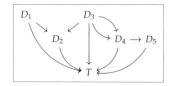

Here, any two parallel paths that end at T are equal in $\mathcal{C}$.

---

[4] If a category $\mathcal{J}$ has finitely many morphisms, we say that $\mathcal{J}$ is a *finite category*. Note that in this case it must have finitely many objects too, because each object $j \in \mathrm{Ob}\,\mathcal{J}$ has its own identity morphism id_j.

Definition 6.26.   We say that a category $\mathcal{C}$ *has finite colimits* if a colimit, $\mathrm{colim}_{\mathcal{J}} D$, exists whenever $\mathcal{J}$ is a finite category and $D \colon \mathcal{J} \to \mathcal{C}$ is a diagram.

Example 6.27.   The initial object in a category $\mathcal{C}$, if it exists, is the colimit of the functor $! \colon \mathbf{0} \to \mathcal{C}$, where $\mathbf{0}$ is the category with no objects and no morphisms, and $!$ is the unique such functor. Indeed, a cocone over $!$ is just an object of $\mathcal{C}$, and so the initial cocone over $!$ is just the initial object of $\mathcal{C}$.

   Note that $\mathbf{0}$ has finitely many objects (none); thus initial objects are finite colimits.

   We often want to know that a category $\mathcal{C}$ has *all* finite colimits (in which case, we often drop the "all" and just say "$\mathcal{C}$ has finite colimits"). To check that $\mathcal{C}$ has (all) finite colimits, it's enough to check that it has a few simpler forms of colimit, which generate all the rest.

Proposition 6.28.   Let $\mathcal{C}$ be a category. The following are equivalent:

1. $\mathcal{C}$ has all finite colimits.
2. $\mathcal{C}$ has an initial object and all pushouts.
3. $\mathcal{C}$ has all coequalizers and all finite coproducts.

Proof.   We will not give precise details here, but the key idea is an inductive one: one can build arbitrary finite diagrams using some basic building blocks. Full details can be found in [Bor94, Prop. 2.8.2]. □

Example 6.29.   Let $\mathcal{C}$ be a category with all pushouts, and suppose we want to take the colimit of the following diagram in $\mathcal{C}$:

$$
\begin{array}{ccc}
 & B & \longrightarrow Z \\
 & \downarrow & \\
A & \longrightarrow C & \\
\downarrow & & \\
D & &
\end{array}
\tag{6.5}
$$

In it we see two diagrams ready to be pushed out, and we know how to take pushouts. So suppose we do that; then we see another pushout diagram so we take the pushout again:

is the result – consisting of the object S, together with all the morphisms from the original diagram to S – the colimit of the original diagram? One can check that it indeed has the correct universal property and thus is a colimit.

Exercise 6.30. Check that the pushout of pushouts from Example 6.29 satisfies the universal property of the colimit for the original diagram, Eq. (6.5).   ◇

We have already seen that the categories **FinSet** and **Set** both have an initial object and pushouts. We thus have the following corollary.

Corollary 6.31. The categories **FinSet** and **Set** have (all) finite colimits.

In Theorem 3.80 we gave a general formula for computing finite limits in **Set**. It is also possible to give a formula for computing finite colimits. There is a duality between products and coproducts and between subobjects and quotient objects, so whereas a finite limit is given by a subset of a product, a finite colimit is given by a quotient of a coproduct.

Theorem 6.32. Let $\mathcal{J}$ be presented by the finite graph (V, A, s, t) and some equations, and let $D: \mathcal{J} \to \textbf{Set}$ be a diagram. Consider the set

$$\operatorname*{colim}_{\mathcal{J}} D := \big\{(v, d) \mid v \in V \text{ and } d \in D(v)\big\}/\sim$$

where this denotes the set of equivalence classes under the equivalence relation $\sim$ generated by putting $(v, d) \sim (w, e)$ if there is an arrow $a: v \to w$ in J such that $D(a)(d) = e$. Then this set, together with the functions $\iota_v: D(v) \to \operatorname{colim}_{\mathcal{J}} D$ given by sending $d \in D(v)$ to its equivalence class, constitutes a colimit of D.

Example 6.33. Recall that an initial object is the colimit on the empty graph. The formula thus says the initial object in **Set** is the empty set $\varnothing$: there are no $v \in V$.

Example 6.34. A coproduct is a colimit on the graph $\mathcal{J} = \boxed{\overset{v_1}{\bullet} \quad \overset{v_2}{\bullet}}$. A functor $D: \mathcal{J} \to \textbf{Set}$ can be identified with a choice of two sets, $X := D(v_1)$ and $Y := D(v_2)$. Since there are no arrows in $\mathcal{J}$, the equivalence relation $\sim$ is vacuous, so the formula in Theorem 6.32 says that a coproduct is given by

$$\{(v, d) \mid d \in D(v), \text{ where } v = v_1 \text{ or } v = v_2\}.$$

In other words, the coproduct of sets X and Y is their disjoint union $X \sqcup Y$, as expected.

Example 6.35. If $\mathcal{J}$ is the category $\mathbf{1} = \boxed{\overset{v}{\bullet}}$, the formula in Theorem 6.32 yields the set

$$\{(v, d) \mid d \in D(v)\}.$$

This is isomorphic to the set $D(v)$. In other words, if X is a set considered as a diagram $X \colon \mathbf{1} \to \mathbf{Set}$, then its colimit (like its limit) is just X again.

Exercise 6.36. Use the formula in Theorem 6.32 to show that pushouts – colimits on a diagram $X \xleftarrow{f} N \xrightarrow{g} Y$ – agree with the description we gave in Example 6.21. ◇

Example 6.37. Another important type of finite colimit is the *coequalizer*. These are colimits over the graph $\boxed{\bullet \rightrightarrows \bullet}$ consisting of two parallel arrows.

Consider some diagram $X \mathrel{\substack{f \\ \longrightarrow \\ g}} Y$ on this graph in $\mathbf{Set}$. The coequalizer of this diagram is the set of equivalence classes of Y under equivalence relation generated by declaring $y \sim y'$ whenever there exists x in X such that $f(x) = y$ and $g(x) = y'$.

Let's return to the example circuit in the introduction to hint at why colimits are useful for interconnection. Consider the following picture:

We've redrawn this picture with one change: some of the arrows are now red, and others are now blue. If we let X be the set of white circles ∘, and Y be the set of black circles •, the blue and red arrows respectively define functions $f, g \colon X \to Y$. Let's leave the actual circuit components out of the picture for now; we're just interested in the dots. What is the coequalizer?

It is a three-element set, consisting of one element for each newly connected pair of •s . Thus the colimit describes the set of terminals after performing the interconnection operation. In Section 6.4 we'll see how to keep track of the circuit components too.

6.2.5   Cospans

When a category $\mathcal{C}$ has finite colimits, an extremely useful way to package them is by considering the category of cospans in $\mathcal{C}$.

Definition 6.38. Let $\mathcal{C}$ be a category. A *cospan* in $\mathcal{C}$ is just a pair of morphisms to a common object $A \rightarrow N \leftarrow B$. The common object N is called the *apex* of the cospan and the other two objects A and B are called its *feet*.

If we want to say that cospans form a category, we should begin by saying how composition would work. So suppose we have two cospans in $\mathcal{C}$

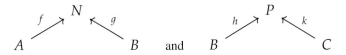

Since the right foot of the first is equal to the left foot of the second, we might stick them together into a diagram like this:

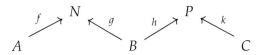

Then, if a pushout of $N \xleftarrow{g} B \xrightarrow{h} P$ exists in $\mathcal{C}$, as shown on the left, we can extract a new cospan in $\mathcal{C}$, as shown on the right:

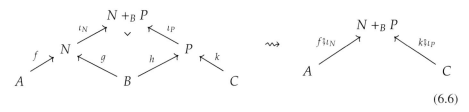

$$(6.6)$$

It might look like we have achieved our goal, but we're missing some things. First, we need an identity on every object $C \in \mathrm{Ob}\,\mathcal{C}$; but that's not hard: use $C \rightarrow C \leftarrow C$ where both maps are identities in $\mathcal{C}$. More importantly, we don't know that $\mathcal{C}$ has all pushouts, so we don't know that every two sequential morphisms $A \rightarrow B \rightarrow C$ can be composed. And beyond that, there is a technical condition that when we form pushouts, we only get an answer "up to isomorphism": anything isomorphic to a pushout counts as a pushout (check the definition to see why). We want all these different choices to count as the same thing, so we define two cospans to be equivalent iff there is an isomorphism between their respective apexes. That is, the cospan $A \rightarrow P \leftarrow B$ and $A \rightarrow P' \leftarrow B$ in the diagram shown left below are equivalent iff there is an isomorphism $P \cong P'$ making the diagram to the right commute:

Now we are getting somewhere. As long as our category $\mathcal{C}$ has pushouts, we are in business: **Cospan**$_{\mathcal{C}}$ will form a category. But in fact, we are very close to getting more.

If we also demand that $\mathcal{C}$ has an initial object $\varnothing$ as well, then we can upgrade **Cospan**$_{\mathcal{C}}$ to a symmetric monoidal category.

Recall from Proposition 6.28 that a category $\mathcal{C}$ has all finite colimits iff it has an initial object and all pushouts.

Definition 6.39. Let $\mathcal{C}$ be a category with finite colimits. Then there exists a category **Cospan**$_{\mathcal{C}}$ with the same objects as $\mathcal{C}$, i.e. $\mathrm{Ob}(\textbf{Cospan}_{\mathcal{C}}) = \mathrm{Ob}(\mathcal{C})$, where the morphisms $A \to B$ are the (equivalence classes of) cospans from A to B, and composition is given by the above pushout construction.

There is a symmetric monoidal structure on this category, denoted $(\textbf{Cospan}_{\mathcal{C}}, \varnothing, +)$. The monoidal unit is the initial object $\varnothing \in \mathcal{C}$ and the monoidal product is given by coproduct. The coherence isomorphisms, e.g. $A + \varnothing \cong A$, can be defined in a similar way to those in Exercise 6.16.

It is a straightforward but time-consuming exercise to verify that $(\textbf{Cospan}_{\mathcal{C}}, \varnothing, +)$ from Definition 6.39 really does satisfy all the axioms of a symmetric monoidal category, but it does.

Example 6.40. The category **FinSet** has finite colimits (see Corollary 6.31). So, we can define a symmetric monoidal category **Cospan**$_{\textbf{FinSet}}$. What does it look like? It looks a lot like wires connecting ports.

The objects of **Cospan**$_{\textbf{FinSet}}$ are finite sets; here let's draw them as collections of $\bullet$s. The morphisms are cospans of functions. Let A and N be five-element sets, and B be a six-element set. Below are two depictions of a cospan $A \xrightarrow{f} N \xleftarrow{g} B$.

In the depiction on the left, we simply represent the functions f and g by drawing arrows from each $a \in A$ to $f(a)$ and each $b \in B$ to $g(b)$. In the depiction on the right, we make this picture resemble wires a bit more, simply drawing a wire where before we had an arrow, and removing the unnecessary center dots. We also draw a dotted line around points that are connected, to emphasize an important perspective, that cospans establish that certain ports are connected, i.e. part of the same equivalence class.

The monoidal category **Cospan**$_{\textbf{FinSet}}$ then provides two operations for combining cospans: composition and monoidal product. Composition is given by taking the pushout of the maps coming from the common foot, as described in Definition 6.39. Here is an example of cospan composition, where all the functions are depicted with arrow notation:

$$\tag{6.7}$$

$$A \quad N \quad B \quad P \quad C \qquad A \quad N +_B P \quad C$$

The monoidal product is given simply by the disjoint union of two cospans; in pictures it is simply combining two cospans by stacking one above another.

Exercise 6.41. In Eq. (6.7) we showed morphisms $A \to B$ and $B \to C$ in $\mathbf{Cospan}_{\mathbf{FinSet}}$. Draw their monoidal product as a morphism $A + B \to B + C$ in $\mathbf{Cospan}_{\mathbf{FinSet}}$. ◇

Exercise 6.42. Depicting the composite of cospans in Eq. (6.7) with the wire notation gives

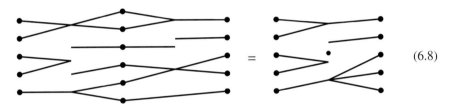

$$\tag{6.8}$$

Comparing Eq. (6.7) and Eq. (6.8), describe the composition rule in $\mathbf{Cospan}_{\mathbf{FinSet}}$ in terms of wires and connected components. ◇

6.3 Hypergraph Categories

A hypergraph category is a type of symmetric monoidal category whose wiring diagrams are networks. We shall soon see that electric circuits can be organized into a hypergraph category; this is what we've been building up to. But to define hypergraph categories, it is useful to first introduce Frobenius monoids.

6.3.1 Frobenius Monoids

The pictures of cospans we saw above, e.g. in Eq. (6.8), look something like icons in signal flow graphs (see Section 5.3.2): various wires merge and split, initialize and terminate. And these follow the same rules they did for linear relations, which we briefly discussed in Exercise 5.70. There's a lot of potential for confusion, so let's start from scratch and build back up.

In any symmetric monoidal category $(\mathcal{C}, I, \otimes)$, recall from Section 4.4.2 that objects can be drawn as wires and morphisms can be drawn as boxes. Particularly noteworthy morphisms might be iconified as dots rather than boxes, to indicate that the morphisms there are not arbitrary but notation-worthy. One case of this is when there is an object X with special "abilities," e.g. the ability to duplicate into two, or disappear into nothing.

To make this precise, recall from Definition 5.55 that a commutative monoid (X, μ, η) in symmetric monoidal category $(\mathcal{C}, I, \otimes)$ is an object X of $\mathcal{C}$ together with (noteworthy) morphisms

$$\mu\colon X \otimes X \to X \qquad\qquad \eta\colon I \to X$$

obeying

| (associativity) | (unitality) | (commutativity) | (6.9) |

where $\times$ is the symmetry on $X \otimes X$. A cocommutative comonoid (X, δ, ϵ) is an object X with maps $\delta\colon X \to X \otimes X$, $\epsilon\colon X \to I$, obeying the mirror images of the laws in Eq. (6.9).

Suppose X has both the structure of a commutative monoid and cocommutative comonoid, and consider a wiring diagram built only from the icons μ, η, δ, and ϵ, where every wire is labeled X. These diagrams have a left and right, and are pictures of how ports on the left are connected to ports on the right. The commutative monoid and cocommutative comonoid axioms thus both express when to consider two such connection pictures the same. For example, associativity says the order of connecting ports on the left doesn't matter; coassociativity (not drawn) says the same for the right.

If you want to go all the way and say "all I care about is which port is connected to which; I don't even care about left and right," then you need a few more axioms to say how the morphisms μ and δ, the merger and the splitter, interact.

Definition 6.43. Let X be an object in a symmetric monoidal category $(\mathcal{C}, \otimes, I)$. A *Frobenius structure* on X consists of a 4-tuple $(\mu, \eta, \delta, \epsilon)$ such that (X, μ, η) is a commutative monoid and (X, δ, ϵ) is a cocommutative comonoid, which satisfies the six equations above ((co)associativity, (co)unitality, (co)commutativity), as well as the following three equations:

| (The Frobenius law) | (The special law) | (6.10) |

We refer to an object X equipped with a Frobenius structure as a *special commutative Frobenius monoid*, or just *Frobenius monoid* for short.

With these two equations, it turns out that two morphisms $X^{\otimes m} \to X^{\otimes n}$ – defined by composing and tensoring identities on X and the noteworthy morphisms μ, δ, etc. – are equal if and only if their string diagrams connect the same ports. This link between connectivity and Frobenius monoids can be made precise as follows.

Definition 6.44.   Let $(X, \mu, \eta, \delta, \epsilon)$ be a Frobenius monoid in a monoidal category $(\mathcal{C}, I, \otimes)$. Let $m, n \in \mathbb{N}$. Define $s_{m,n} \colon X^{\otimes m} \to X^{\otimes n}$ to be the following morphism

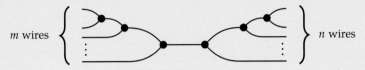

It can be written formally as $(m-1)$ μs followed by $(n-1)$ δs, with special cases when $m = 0$ or $n = 0$.

We call $s_{m,n}$ the *spider of type* (m, n), and can draw it more simply as the icon

$$m \text{ legs} \left\{ \ \rightthreetimes\kern-6pt\leftthreetimes \ \right\} n \text{ legs}$$

So a special commutative Frobenius monoid, aside from being a mouthful, is a "spiderable" wire. You agree that in any monoidal category wiring diagram language, wires represent objects and boxes represent morphisms? Well in our weird way of talking, if a wire is spiderable, it means that we have a bunch of morphisms $\mu, \eta, \delta, \epsilon, \sigma$ that we can combine without worrying about the order of doing so: the result is just "how many ins, and how many outs", a spider. Here's a formal statement.

Theorem 6.45.   Let $(X, \mu, \eta, \delta, \epsilon)$ be a Frobenius monoid in a monoidal category $(\mathcal{C}, I, \otimes)$. Suppose that we have a map $f \colon X^{\otimes m} \to X^{\otimes n}$ constructed from spiders and the symmetry map $\sigma \colon X^{\otimes 2} \to X^{\otimes 2}$ using composition and the monoidal product, and such that the string diagram of f has only one connected component. Then it is a spider: $f = s_{m,n}$.

Example 6.46.   As the following two morphisms both (i) have the same number of inputs and outputs, (ii) are constructed only from spiders, and (iii) are connected, Theorem 6.45 immediately implies they are equal:

Exercise 6.47.   Let X be an object equipped with a Frobenius structure. Which of the morphisms $X \otimes X \to X \otimes X \otimes X$ in the following list are necessarily equal?

1.

2.

3.

4.

5.

6.

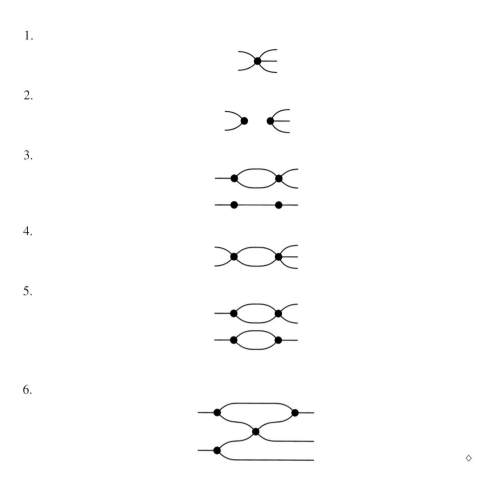

◇

Back to cospans

Another way of understanding Frobenius monoids is to relate them to cospans. Recall the notion of prop presentation from Definition 5.28.

Theorem 6.48. Consider the four-element set $G := \{\mu, \eta, \delta, \epsilon\}$ and define $in, out: G \to \mathbb{N}$ as follows:

$$in(\mu) := 2, \qquad in(\eta) := 0, \qquad in(\delta) := 1, \qquad in(\epsilon) := 1,$$
$$out(\mu) := 1, \qquad out(\eta) := 1, \qquad out(\delta) := 2, \qquad out(\epsilon) := 0.$$

Let E be the set of Frobenius axioms, i.e. the nine equations from Definition 6.43. Then the free prop on (G, E) is equivalent, as a symmetric monoidal category,[5] to $\mathbf{Cospan}_{\mathbf{FinSet}}$.

---

[5] We will not explain precisely what it means to be equivalent as a symmetric monoidal category, but you probably have some idea: "they are the same for all category-theoretic intents and purposes." The idea is similar to that of equivalence of categories, as explained in Remark 3.50.

Thus we see that ideal wires, connectivity, cospans, and objects with Frobenius structures are all intimately related. We use Frobenius structures (all that splitting, merging, initializing, and terminating stuff) as a way to capture the grammar of circuit diagrams.

6.3.2   Wiring Diagrams for Hypergraph Categories

We introduce hypergraph categories through their wiring diagrams. As for monoidal categories, the formal definition is just the structure required to unambiguously interpret these diagrams.

Indeed, our interest in hypergraph categories is best seen in their wiring diagrams. The key idea is that wiring diagrams for hypergraph categories are network diagrams. This means, in addition to drawing labeled boxes with inputs and outputs, as we can for monoidal categories, and in addition to bending these wires around as we can for compact closed categories, we are allowed to split, join, terminate, and initialize wires.

Here is an example of a wiring diagram that represents a composite of morphisms in a hypergraph category

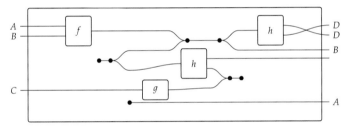

We have suppressed some of the object/wire labels for readability, since all types can be inferred from the labeled ones.

Exercise 6.49.

1. What label should be on the input to h?
2. What label should be on the output of g?
3. What label should be on the fourth output wire of the composite?   ◇

Thus hypergraph categories are general enough to talk about all network-style diagrammatic languages, like circuit diagrams.

6.3.3   Definition of Hypergraph Category

We are now ready to define hypergraph categories formally. Since the wiring diagrams for hypergraph categories are just those for symmetric monoidal categories with a few additional icons, the definition is relatively straightforward: we just want a Frobenius structure on every object. The only coherence condition is that these interact nicely with the monoidal product.

Definition 6.50. A *hypergraph category* is a symmetric monoidal category $(\mathcal{C}, I, \otimes)$ in which each object X is equipped with a Frobenius structure $(X, \mu_X, \delta_X, \eta_X, \epsilon_X)$ such that

for all objects X, Y, and such that $\eta_I = \mathrm{id}_I = \epsilon_I$.

A *hypergraph prop* is a hypergraph category that is also a prop, e.g. $\mathrm{Ob}(\mathcal{C}) = \mathbb{N}$, etc.

Example 6.51. For any $\mathcal{C}$ with finite colimits, $\mathbf{Cospan}_\mathcal{C}$ is a hypergraph category. The Frobenius morphisms $\mu_X, \delta_X, \eta_X, \epsilon_X$ for each object X are constructed using the universal properties of colimits:

$$\mu_X := \left(X + X \xrightarrow{[\mathrm{id}_X, \mathrm{id}_X]} X \xleftarrow{\mathrm{id}_X} X \right)$$

$$\eta_X := \left(\varnothing \xrightarrow{!_X} X \xleftarrow{\mathrm{id}_X} X \right)$$

$$\delta_X := \left(X \xrightarrow{\mathrm{id}_X} X \xleftarrow{[\mathrm{id}_X, \mathrm{id}_X]} X + X \right)$$

$$\epsilon_X := \left(X \xrightarrow{\mathrm{id}_X} X \xleftarrow{!_X} \varnothing \right)$$

Exercise 6.52. By Example 6.51, the category $\mathbf{Cospan}_{\mathbf{FinSet}}$ is a hypergraph category. (In fact, it is equivalent to a hypergraph prop.) Draw the Frobenius morphisms for the object 1 in $\mathbf{Cospan}_{\mathbf{FinSet}}$ using both the function and wiring depictions as in Example 6.40. ◇

Exercise 6.53. Using your knowledge of colimits, show that the maps defined in Example 6.51 do indeed obey the special law (see Definition 6.43). ◇

Example 6.54. Recall the monoidal category $(\mathbf{Corel}, \varnothing, \sqcup)$ from Example 4.46; its objects are finite sets and its morphisms are corelations. Given a finite set X, define the corelation $\mu_X : X \sqcup X \to X$ such that two elements of $X \sqcup X \sqcup X$ are equivalent if and only if they come from the same underlying element of X. Define $\delta_X : X \to X \sqcup X$ in the same way, and define $\eta_X : \varnothing \to X$ and $\epsilon_X : X \to \varnothing$ such that no two elements of $X = \varnothing \sqcup X = X \sqcup \varnothing$ are equivalent.

These maps define a special commutative Frobenius monoid $(X, \mu_X, \eta_X, \delta_X, \epsilon_X)$, and in fact give **Corel** the structure of a hypergraph category.

Example 6.55. The prop of linear relations, which we briefly mentioned in Exercise 5.70, is a hypergraph category. In fact, it is a hypergraph category in two ways, by choosing either the black "copy" and "discard" generators or the white "add" and "zero" generators as the Frobenius maps.

We can generalize the construction we gave in Theorem 5.72.

Proposition 6.56.   Hypergraph categories are self-dual compact closed categories, if we define the cup and cap to be

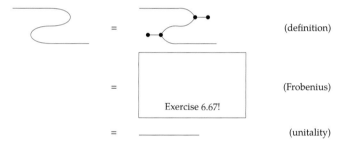

and

Proof.   The proof is a straightforward application of the Frobenius and unitality axioms:

	=	(definition)
	=	(Frobenius)
	Exercise 6.67!	
	=	(unitality)

$\square$

Exercise 6.57.   Copy and complete the missing diagram in the proof of Proposition 6.56 using the equations from Eq. (6.9), their opposites, and Eq. (6.10).   ◇

6.4   Decorated Cospans

The goal of this section is to show how we can construct a hypergraph category whose morphisms are electric circuits. To do this, we first must introduce the notion of structure-preserving map for symmetric monoidal categories, a generalization of monoidal monotones known as symmetric monoidal functors. Then we introduce a general method – that of decorated cospans – for producing hypergraph categories. Doing all this will tie up lots of loose ends: colimits, cospans, circuits, and hypergraph categories.

6.4.1 Symmetric Monoidal Functors

Rough Definition 6.58. Let $(\mathcal{C}, I_{\mathcal{C}}, \otimes_{\mathcal{C}})$ and $(\mathcal{D}, I_{\mathcal{D}}, \otimes_{\mathcal{D}})$ be symmetric monoidal categories. To specify a *symmetric monoidal functor* (F, φ) between them,

(i) one specifies a functor $F \colon \mathcal{C} \to \mathcal{D}$,
(ii) one specifies a morphism $\varphi_I \colon I_{\mathcal{D}} \to F(I_{\mathcal{C}})$,
(iii) for each $c_1, c_2 \in \mathrm{Ob}(\mathcal{C})$, one specifies a morphism

$$\varphi_{c_1, c_2} \colon F(c_1) \otimes_{\mathcal{D}} F(c_2) \to F(c_1 \otimes_{\mathcal{C}} c_2),$$

natural in c_1 and c_2.

We call the various maps φ *coherence maps*. We require the coherence maps to obey bookkeeping axioms that ensure they are well behaved with respect to the symmetric monoidal structures on $\mathcal{C}$ and $\mathcal{D}$. If φ_I and φ_{c_1, c_2} are isomorphisms for all c_1, c_2, we say that (F, φ) is *strong*.

Example 6.59. Consider the power set functor P: **Set** $\to$ **Set**. It acts on objects by sending a set $S \in$ **Set** to its set of subsets $\mathrm{P}(S) := \{R \subseteq S\}$. It acts on morphisms by sending a function $f \colon S \to T$ to the image map $\mathrm{im}_f \colon \mathrm{P}(S) \to \mathrm{P}(T)$, which maps $R \subseteq S$ to $\{f(r) \mid r \in R\} \subseteq T$.

Now consider the symmetric monoidal structure $(\{1\}, \times)$ on **Set** from Example 4.38. To make P a symmetric monoidal functor, we need to specify a function $\varphi_I \colon \{1\} \to \mathrm{P}(\{1\})$ and for all sets S and T, a function $\varphi_{S,T} \colon \mathrm{P}(S) \times \mathrm{P}(T) \to \mathrm{P}(S \times T)$. One possibility is to define $\varphi_I(1)$ to be the maximal subset $\{1\} \subseteq \{1\}$, and given subsets $A \subseteq S$ and $B \subseteq T$, to define $\varphi_{S,T}(A, B)$ to be the product subset $A \times B \subseteq S \times T$. With these definitions, (P, φ) is a symmetric monoidal functor.

Exercise 6.60. Check that the maps $\varphi_{S,T}$ defined in Example 6.59 are natural in S and T. In other words, given $f \colon S \to S'$ and $g \colon T \to T'$, show that the diagram below commutes:

$$
\begin{array}{ccc}
\mathrm{P}(S) \times \mathrm{P}(T) & \xrightarrow{\ \varphi_{S,T}\ } & \mathrm{P}(S \times T) \\
{\scriptstyle \mathrm{im}_f \times \mathrm{im}_g}\big\downarrow & & \big\downarrow{\scriptstyle \mathrm{im}_{f \times g}} \\
\mathrm{P}(S') \times \mathrm{P}(T') & \xrightarrow[\ \varphi_{S',T'}\ ]{} & \mathrm{P}(S' \times T')
\end{array}
$$

$\diamond$

6.4.2 Decorated Cospans

Now that we have briefly introduced symmetric monoidal functors, we return to the task at hand: constructing a hypergraph category of circuits. To do so, we introduce the method of decorated cospans.

Circuits have lots of internal structure, but they also have some external ports – also called "terminals" – by which to interconnect them with others. Decorated cospans are ways of discussing exactly that: things with external ports and internal structure.

To see how this works, let us start with the following example circuit:

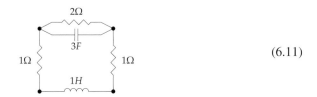

$$(6.11)$$

We might formally consider this as a graph on the set of four ports, where each edge is labeled by a type of circuit component (for example, the top edge would be labeled as a resistor of resistance 2Ω). For this circuit to be a morphism in some category, i.e. in order to allow for interconnection, we must equip the circuit with some notion of interface. We do this by marking the ports in the interface using functions from finite sets:

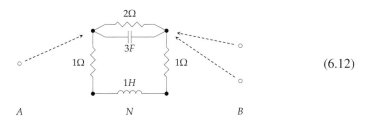

$$(6.12)$$

Let N be the set of nodes of the circuit. Here the finite sets A, B, and N are sets consisting of one, two, and four elements, respectively, drawn as points, and the values of the functions $A \to N$ and $B \to N$ are indicated by the gray arrows. This forms a cospan in the category of finite sets, for which the apex set N has been *decorated* by our given circuit.

Suppose we are given another such decorated cospan with input B

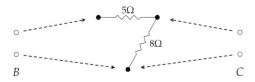

Since the output of the first equals the input of the second (both are B), we can stick them together in a single diagram:

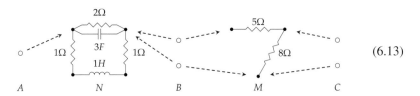

$$(6.13)$$

The composition is given by gluing the circuits along the identifications specified by B, resulting in the decorated cospan

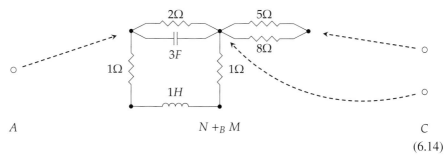

$$A \qquad\qquad\qquad N +_B M \qquad\qquad\qquad C$$

(6.14)

We've seen this sort of gluing before when we defined composition of cospans in Definition 6.39. But now there's this whole "decoration" thing; our goal is to formalize it.

Definition 6.61. Let $\mathcal{C}$ be a category with finite colimits, and $(F, \varphi) \colon (\mathcal{C}, +) \longrightarrow (\mathbf{Set}, \times)$ be a symmetric monoidal functor. An F-*decorated cospan* is a pair consisting of a cospan $A \xrightarrow{i} N \xleftarrow{o} B$ in $\mathcal{C}$ together with an element $s \in F(N)$.[6] We call (F, φ) the *decoration functor* and s the *decoration*.

The intuition here is to use $\mathcal{C} = \mathbf{FinSet}$, and, for each object $N \in \mathbf{FinSet}$, the functor F assigns the set of all legal decorations on a set N of nodes. When you choose an F-decorated cospan, you choose a set A of left-hand external ports, a set B of right-hand external ports, each of which maps to a set N of nodes, and you choose one of the available decorations on N nodes, taken from the set $F(N)$.

So, in our electrical circuit case, the decoration functor F sends a finite set N to the set of circuit diagrams – graphs whose edges are labeled by resistors, capacitors, etc. – that have N vertices.

Our goal is still to be able to compose such diagrams; so how does that work exactly? Basically one combines the way cospans are composed with the structures defining our decoration functor: namely F and φ.

Let $(A \xrightarrow{f} N \xleftarrow{g} B, s)$ and $(B \xrightarrow{h} P \xleftarrow{k} C, t)$ represent decorated cospans. Their composite is represented by the composite of the cospans $A \xrightarrow{f} N \xleftarrow{g} B$ and $B \xrightarrow{h} P \xleftarrow{k} C$, paired with the following element of $F(N +_B P)$:

$$F([\iota_N, \iota_P])(\varphi_{N,P}(s, t)). \tag{6.15}$$

That's rather compact! We'll unpack it, in a concrete case, in just a second. But let's record a theorem first.

---

[6] Just like in Definition 6.39, we should technically use equivalence classes of cospans. We will elide this point to get the bigger idea across. The interested reader should consult Section 6.6.

Theorem 6.62. Given a category $\mathcal{C}$ with finite colimits and a symmetric monoidal functor $(F, \varphi) \colon (\mathcal{C}, +) \longrightarrow (\mathbf{Set}, \times)$, there is a hypergraph category $\mathbf{Cospan}_F$ whose objects are the objects of $\mathcal{C}$, and whose morphisms are equivalence classes of F-decorated cospans.

The symmetric monoidal and hypergraph structures are derived from those on $\mathbf{Cospan}_{\mathcal{C}}$.

Exercise 6.63. Suppose you're worried that the notation $\mathbf{Cospan}_{\mathcal{C}}$ looks like the notation $\mathbf{Cospan}_F$, even though they're very different. An expert tells you "they're not so different; one is a special case of the other. Just use the constant functor $F(c) := \{*\}$." What does the expert mean? ◇

6.4.3 Electric Circuits

In order to work with the above abstractions, we will get a bit more precise about the circuits example and then have a detailed look at how composition works in decorated cospan categories.

Let's build some circuits

To begin, we'll need to choose which components we want in our circuit. This is simply a matter of what's in our electrical toolbox. Let's say we're carrying some lightbulbs, switches, batteries, and resistors of every possible resistance. That is, define a set

$$C :- \{\texttt{light}, \texttt{switch}, \texttt{battery}\} \sqcup \{x\Omega \mid x \in \mathbb{R}^+\}.$$

To be clear, the Ω are just labels; the above set is isomorphic to $\{\texttt{light}, \texttt{switch}, \texttt{battery}\} \sqcup \mathbb{R}^+$. But we write C this way to remind us that it consists of circuit components. If we wanted, we could also add inductors, capacitors, and even elements connecting more than two ports, like transistors, but let's keep things simple for now.

Given our set C, a C-circuit is just a graph (V, A, s, t), where $s, t \colon A \to V$ are the source and target functions, together with a function $\ell \colon A \to C$ labeling each edge with a certain circuit component from C.

For example, we might have the simple case of $V = \{1, 2\}$, $A = \{e\}$, $s(e) = 1$, $t(e) = 2$ – so e is an edge from 1 to 2 – and $\ell(e) = 3\Omega$. This represents a resistor with resistance 3Ω:

Note that in the formalism we have chosen, we have multiple ways to represent any circuit, as our representations explicitly choose directions for the edges. The above resistor could also be represented by the 'reversed graph', with data $V = \{1, 2\}$, $A = \{e\}$, $s(e) = 2$, $t(e) = 1$, and $\ell(e) = 3\Omega$.

Exercise 6.64. Write a tuple (V, A, s, t, ℓ) that represents the circuit in Eq. (6.11). ◇

A decoration functor for circuits

We want C-circuits to be our decorations, so let's use them to define a decoration functor as in Definition 6.61. We'll call the functor (Circ, ψ). We start by defining the functor part

$$\text{Circ}: (\mathbf{FinSet}, +) \longrightarrow (\mathbf{Set}, \times)$$

as follows. On objects, simply send a finite set V to the set of C-circuits:

$$\text{Circ}(V) := \{(V, A, s, t, \ell) \mid \text{where } s, t\colon A \to V, \ell\colon E \to C\}.$$

On morphisms, Circ sends a function $f\colon V \to V'$ to the function

$$\text{Circ}(f)\colon \text{Circ}(V) \longrightarrow \text{Circ}(V');$$
$$(V, A, s, t, \ell) \longmapsto \left(V', A, (s \fatsemi f), (t \fatsemi f), \ell\right).$$

This defines a functor; let's explore it a bit in an exercise.

Exercise 6.65. To understand this functor better, let $c \in \text{Circ}(\underline{4})$ be the circuit

$$3\Omega$$

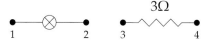

and let $f\colon \underline{4} \to \underline{3}$ be the function

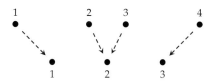

Draw a picture of the circuit $\text{Circ}(f)(c)$. ◇

We're trying to get a decoration functor (Circ, ψ) and so far we have Circ. For the coherence maps $\psi_{V,V'}$ for finite sets V, V', we define

$$\psi_{V,V'}\colon \text{Circ}(V) \times \text{Circ}(V') \longrightarrow \text{Circ}(V + V');$$
$$\left((V, A, s, t, \ell), (V', A', s', t', \ell')\right) \longmapsto (V + V', A + A', s + s', t + t', [\ell, \ell']).$$
$$(6.16)$$

This is simpler than it may look: it takes a circuit on V and a circuit on V', and just considers them together as a circuit on the disjoint union of vertices $V + V'$.

Exercise 6.66. Suppose we have circuits

$$b := \bullet\!\!-\!\!|\!\!|\!\!-\!\!\bullet \qquad \text{and} \qquad s := \bullet\!\!\diagup\!\!-\!\!\bullet$$

in $\text{Circ}(\underline{2})$. Use the definition of $\psi_{V,V'}$ from (6.16) to figure out what 4-vertex circuit $\psi_{2,2}(b, s) \in \text{Circ}(\underline{2} + \underline{2}) = \text{Circ}(\underline{4})$ should be, and draw a picture. ◇

Open circuits using decorated cospans

From the above data, just a monoidal functor $(\text{Circ}, \psi): (\textbf{FinSet}, +) \to (\textbf{Set}, \times)$, we can construct our promised hypergraph category of circuits!

Our notation for this category is $\textbf{Cospan}_{\text{Circ}}$. Following Theorem 6.62, the objects of this category are the same as the objects of **FinSet**, just finite sets. We'll reprise our notation from the introduction and Example 6.37, and draw these finite sets as collections of white circles ∘. For example, we'll represent the object $\underline{2}$ of $\textbf{Cospan}_{\text{Circ}}$ as two white circles:

∘   ∘

These white circles mark interface points of an open circuit.

More interesting than the objects, however, are the morphisms in $\textbf{Cospan}_{\text{Circ}}$. These are open circuits. By Theorem 6.62, a morphism $\underline{m} \to \underline{n}$ is a Circ-decorated cospan: that is, cospan $\underline{m} \to \underline{p} \leftarrow \underline{n}$ together with an element c of $\text{Circ}(\underline{p})$. As an example, consider the cospan $\underline{1} \overset{i_1}{\to} \underline{2} \overset{i_2}{\leftarrow} \underline{1}$ where $i_1(1) = 1$ and $i_2(1) = 2$, equipped with the battery element of $\text{Circ}(\underline{2})$ connecting node 1 and node 2. We'll depict this as follows:

$$\circ \dashrightarrow \bullet \!-\!\!|\!\!-\! \bullet \dashleftarrow \circ \tag{6.17}$$

Exercise 6.67. Morphisms of $\textbf{Cospan}_{\text{Circ}}$ are Circ-decorated cospans, as defined in Definition 6.61. This means (6.17) depicts a cospan together with a *decoration*, which is some C-circuit $(V, A, s, t, \ell) \in \text{Circ}(\underline{2})$. What is it?   ◇

Let's now see how the hypergraph operations in $\textbf{Cospan}_{\text{Circ}}$ can be used to construct electric circuits.

Composition in $\textbf{Cospan}_{\text{Circ}}$

First we'll consider composition. Consider the following decorated cospan from $\underline{1}$ to $\underline{1}$:

Since this and the circuit in (6.17) are both morphisms $\underline{1} \to \underline{1}$, we may compose them to get another morphism $\underline{1} \to \underline{1}$. How do we do this? There are two parts: to get the new cospan, we simply compose the cospans of our two circuits, and to get the new decoration, we use the formula $\text{Circ}([\iota_N, \iota_P])(\psi_{N,P}(s, t))$ from (6.15). Again, this is rather compact! Let's unpack it together.

We'll start with the cospans. The cospans we wish to compose are

(We simply ignore the decorations for now.) If we pushout over the common set $\underline{1} = \{\circ\}$, we obtain the pushout square

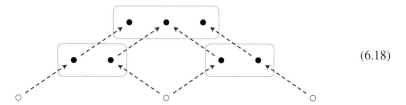

(6.18)

This means the composite cospan is

In the meantime, we already had you start unpacking the formula for the new decoration: you told us what the map $\psi_{2,2}$ does in Exercise 6.66. It takes the two decorations, both circuits in $\mathrm{Circ}(\underline{2})$, and turns them into the single disjoint circuit

in $\mathrm{Circ}(\underline{4})$. So this is what the $\psi_{N,P}(s,t)$ part means. What does $[\iota_N, \iota_P]$ mean? Recall that this is the copairing of the pushout maps, as described in Examples 6.13 and 6.24. In our case, the relevant pushout square is given by (6.18), and $[\iota_N, \iota_P]$ is in fact the function f from Exercise 6.65! This means the decoration on the composite cospan is

Putting this all together, the composite circuit is

Exercise 6.68. Refer back to the example at the beginning of Section 6.4.2. In particular, consider the composition of circuits in Eq. (6.13). Express the two circuits in this diagram as morphisms in $\mathbf{Cospan}_{\mathrm{Circ}}$, and compute their composite. Does it match the picture in Eq. (6.14)? ◇

Monoidal products in Cospan$_{\mathrm{Circ}}$

Monoidal products in $\mathbf{Cospan}_{\mathrm{Circ}}$ are much simpler than composition. On objects, we again just work as in **FinSet**: we take the disjoint union of finite sets. Morphisms again have a cospan, and a decoration. For cospans, we again just work in $\mathbf{Cospan}_{\mathrm{FinSet}}$: given two cospans $A \to M \leftarrow B$ and $C \to N \leftarrow D$, we take their coproduct cospan $A + C \to M + N \leftarrow B + D$. And for decorations, we use the map $\psi_{M,N} : \mathrm{Circ}(M) \times$

$\mathrm{Circ}(N) \to \mathrm{Circ}(M + N)$. So, for example, suppose we want to take the monoidal product of the open circuits

and

The result is given by stacking them. In other words, their monoidal product is:

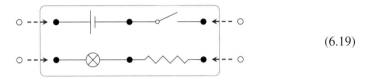

(6.19)

Easy, right?

We leave you to do two compositions of your own.

Exercise 6.69. Write x for the open circuit in (6.19). Also define cospans $\eta\colon 0 \to 2$ and $\eta\colon 2 \to 0$ as follows:

$$\eta := \quad \varnothing \begin{array}{|c|} \hline \bullet \\ \hline \end{array} \qquad \qquad \begin{array}{|c|} \hline \bullet \\ \hline \end{array} \varnothing \quad =: \epsilon$$

where each of these are decorated by the empty circuit $(\underline{1}, \varnothing, !, !, !) \in \mathrm{Circ}(\underline{1})$.[7]

Compute the composite $\eta \mathbin{\fatsemi} x \mathbin{\fatsemi} \epsilon$ in **Cospan**$_{\mathrm{Circ}}$. This is a morphism $\underline{0} \to \underline{0}$; we call such things *closed circuits*. ◇

6.5   Operads and Their Algebras

In Theorem 6.62 we described how decorating cospans builds a hypergraph category from a symmetric monoidal functor. We then explored how that works in the case that the decoration functor is somehow "all circuit graphs on a set of nodes."

In this book, we have devoted a great deal of attention to different sorts of compositional theories, from monoidal preorders to compact closed categories to hypergraph categories. Yet for an application you someday have in mind, it may be the case that none of these theories suffice. You need a different structure, customized to a particular situation. For example in [VSL15] the authors wanted to compose continuous dynamical systems with control-theoretic properties and realized that, in order for feedback to make sense, the wiring diagrams could not involve what they called "passing wires."

---

[7] As usual ! denotes the unique function, in this case from the empty set to the relevant codomain.

So to close our discussion of compositional structures, we want to quickly sketch something we can use as a sort of meta-compositional structure, known as an operad. We saw in Section 6.4.3 that we can build electric circuits from a symmetric monoidal functor **FinSet** → **Set**. Similarly we'll see that we can build examples of new algebraic structures from operad functors $\mathcal{O}$ → **Set**.

6.5.1 Operads Design Wiring Diagrams

Understanding that circuits are morphisms in a hypergraph category is useful: it means we can bring the machinery of category theory to bear on understanding electrical circuits. For example, we can build functors that express the compositionality of circuit semantics, i.e. how to derive the functionality of the whole from the functionality and interaction pattern of the parts. Or we can use the category-theoretic foundation to relate circuits to other sorts of network systems, such as signal flow graphs. Finally, the basic coherence theorems for monoidal categories and compact closed categories tell us that wiring diagrams give sound and complete reasoning in these settings.

However, one perhaps unsatisfying result is that the hypergraph category introduces artifacts like the domain and codomain of a circuit, which are not inherent to the structure of circuits or their composition. Circuits just have a single boundary interface, not "domains" and "codomains." This is not to say the above model is not useful: in many applications, a vector space does not have a preferred basis, but it is often useful to pick one so that we may use matrices (or signal flow graphs!). But it would be worthwhile to have a category-theoretic model that more directly represents the compositional structure of circuits. In general, we want the category-theoretic model to fit our desired application like a glove. Let us quickly sketch how this can be done.

Let's return to wiring diagrams for a second. We saw that wiring diagrams for hypergraph categories basically look like this:

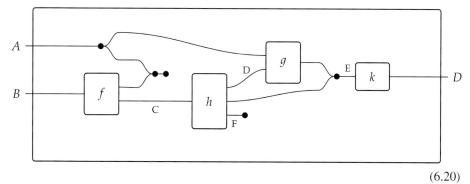

$$(6.20)$$

Note that if you had a box with A and B on the left and D on the right, you could plug the above diagram right inside it, and get a new open circuit. This is the basic move of operads.

But before we explain this, let's get where we said we wanted to go: to a model where there aren't ports on the left and ports on the right, there are just ports. We want a more succinct model of composition for circuit diagrams; something that looks more like this:

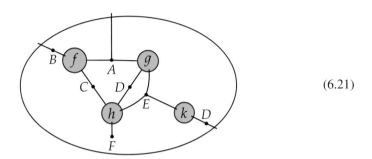

$$(6.21)$$

Do you see how diagrams Eq. (6.20) and Eq. (6.21) are actually exactly the same in terms of interconnection pattern? The only difference is that the latter does not have left/right distinction: we have lost exactly what we wanted to lose.

The cost is that the "boxes" f, g, h, k in Eq. (6.21) no longer have a left/right distinction; they're just circles now. That wouldn't be bad except that it means they can no longer represent morphisms in a category – like they used to above, in Eq. (6.20) – because morphisms in a category by definition have a domain and codomain. Our new circles have no such distinction. So now we need a whole new way to think about "boxes" categorically: if they're no longer morphisms in a category, what are they? The answer is found in the theory of operads.

In understanding operads, we will find we need to navigate one of the level shifts that we first discussed in Section 1.4.5. Notice that, for decorated cospans, we define a hypergraph *category* using a symmetric monoidal *functor*. This is reminiscent of our brief discussion of algebraic theories in Section 5.4.2, where we defined something called the theory of monoids as a prop $\mathcal{M}$, and defined monoids using functors $\mathcal{M} \to$ **Set**; see Remark 5.64. In the same way, we can view the category **Cospan**$_{\text{FinSet}}$ as some sort of "theory of hypergraph categories," and so define hypergraph categories as functors **Cospan**$_{\text{FinSet}} \to$ **Set**.

So that's the idea. An operad $\mathcal{O}$ will define a theory or grammar of composition, and operad functors $\mathcal{O} \to$ **Set**, known as $\mathcal{O}$-*algebras*, will describe particular applications that obey that grammar.

Rough Definition 6.70.   To specify an *operad* $\mathcal{O}$,

(i) one specifies a collection T, whose elements are called *types*;
(ii) for each tuple $(t_1, \ldots, t_n, t)$ of types, one specifies a set $\mathcal{O}(t_1, \ldots, t_n; t)$, whose elements are called *operations of arity* $(t_1, \ldots, t_n; t)$;
(iii) for each pair of tuples $(s_1, \ldots, s_m, t_i)$ and $(t_1, \ldots, t_n, t)$, one specifies a function

$$\circ_i : \mathcal{O}(s_1, \ldots, s_m; t_i) \times \mathcal{O}(t_1, \ldots, t_n; t)$$
$$\to \mathcal{O}(t_1, \ldots, t_{i-1}, s_1, \ldots, s_m, t_{i+1}, \ldots, t_n; t);$$

called *substitution*; and

(iv) for each type t, one specifies an operation $\mathrm{id}_t \in O(t;t)$ called the *identity operation*.

These must obey generalized identity and associativity laws.[8]

Let's ignore types for a moment and think about what this structure models. The intuition is that an operad consists of, for each n, a set of operations of arity n – i.e. all the operations that accept n arguments. If we take an operation f of arity m, and plug the output into the ith argument of an operation g of arity n, we should get an operation of arity $m + n - 1$: we have m arguments to fill in m, and the remaining $n - 1$ arguments to fill in g. Which operation of arity $m + n - 1$ do we get? This is described by the substitution function $\circ_i$, which says we obtain the operation $f \circ_i g \in \mathcal{O}(m+n-1)$. The coherence conditions say that these functions $\circ_i$ capture the following intuitive picture:

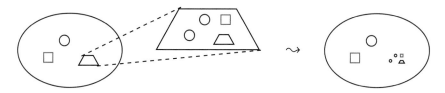

The types then allow us to specify the, well, types of the arguments – inputs – that each function takes. So making tea is a 2-ary operation, an operation with arity 2, because it takes in two things. To make tea, you need some warm water and you need some tea leaves.

Example 6.71. Context-free grammars are to operads as graphs are to categories. Let's sketch what this means. First, a context-free grammar is a way of describing a particular set of "syntactic categories" that can be formed from a set of symbols. For example, in English we have syntactic categories like nouns, determiners, adjectives, verbs, noun phrases, prepositional phrases, sentences, etc. The symbols are words, e.g. cat, dog, the, chases.

To define a context-free grammar on some alphabet, one specifies some *production rules*, which say how to form an entity in some syntactic category from a bunch of entities in other syntactic categories. For example, we can form a noun phrase from a determiner (the), an adjective (happy), and a noun (boy). Context-free grammars are important in both linguistics and computer science. In the former, they're a basic way to talk about the structure of sentences in natural languages. In the latter, they're crucial when designing parsers for programming languages.

So just as graphs present free categories, context-free grammars present free operads. This idea was first noticed in [HMP98].

---

[8] Often what we call types are called objects or colors, what we call operations are called morphisms, what we call substitution is called composition, and what we call operads are called multicategories. A formal definition can be found in [Lei04].

6.5.2   Operads from Symmetric Monoidal Categories

We shall see in Definition 6.75 that a large class of operads come from symmetric monoidal categories. Before we explain this, we give a couple of examples. Perhaps the most important operad is that of **Set**.

Example 6.72. The operad **Set** of sets has

(i) sets X as types,
(ii) functions $X_1 \times \cdots \times X_n \to Y$ as operations of arity $(X_1, \ldots, X_n; Y)$,
(iii) substitution defined by

$$(g \circ_i f)(x_1, \ldots, x_{i-1}, w_1, \ldots, w_m, x_{i+1}, \ldots, x_n)$$
$$= g\big(x_1, \ldots, x_{i-1}, f(w_1, \ldots, w_m), x_{i+1}, \ldots, x_n\big)$$

where $f \in \mathbf{Set}(W_1, \ldots, W_m; X_i)$, $g \in \mathbf{Set}(X_1, \ldots, X_n; Y)$, and hence $g \circ_i f$ is a function

$$(g \circ_i f): X_1 \times \cdots \times X_{i-1} \times W_1 \times \cdots \times W_m \times X_{i+1} \times \cdots \times X_n \longrightarrow Y,$$

(iv) identities $\mathrm{id}_X \in \mathbf{Set}(X; X)$ given by the identity function $\mathrm{id}_X : X \to X$.

Next, we give an example that reminds us what all this operad stuff was for: wiring diagrams.

Example 6.73. The operad **Cospan** of finite-set cospans has

(i) natural numbers $a \in \mathbb{N}$ as types,
(ii) cospans $a_1 + \cdots + a_n \to p \leftarrow b$ of finite sets as operations of arity $(a_1, \ldots, a_n; b)$,
(iii) substitution defined by pushout,
(iv) identities $\mathrm{id}_a \in \mathbf{Set}(a; a)$ just given by the identity cospan $a \xrightarrow{\mathrm{id}_a} a \xleftarrow{\mathrm{id}_a} a$.

This is the operadic analogue of the monoidal category $(\mathbf{Cospan}_{\mathbf{FinSet}}, 0, +)$.

We can depict operations in this operad using diagrams like those we drew above. For example, here's a picture of an operation:

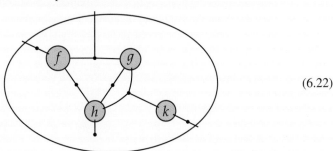

$$(6.22)$$

This is an operation of arity $(\underline{3}, \underline{3}, \underline{4}, \underline{2};\ \underline{3})$. Why? The circles marked f and g have 3 ports, h has 4 ports, k has 2 ports, and the outer circle has 3 ports: 3, 3, 4, 2; 3.

So how exactly is Eq. (6.22) a morphism in this operad? Well a morphism of this arity is, by (ii), a cospan $\underline{3} + \underline{3} + \underline{4} + \underline{2} \xrightarrow{a} p \xleftarrow{b} \underline{3}$. In the diagram above, the apex p is the set $\underline{7}$, because there are 7 nodes • in the diagram. The function a sends each port on one of the small circles to the node it connects to, and the function b sends each port of the outer circle to the node it connects to.

We are able to depict each operation in the operad **Cospan** as a wiring diagram. It is often helpful to think of operads as describing a wiring diagram grammar. The substitution operation of the operad signifies inserting one wiring diagram into a circle or box in another wiring diagram.

Exercise 6.74.

1. Consider the following cospan $f \in$ **Cospan**(2, 2; 2):

   Draw it as a wiring diagram with two inner circles, each with two ports, and one outer circle with two ports.
2. Draw the wiring diagram corresponding to the following cospan $g \in$ **Cospan**(2, 2, 2; 0):

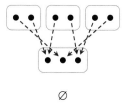

   $\varnothing$

3. Compute the cospan $g \circ_1 f$. What is its arity?
4. Draw the cospan $g \circ_1 f$. Do you see it as substitution?   ◇

We can turn any symmetric monoidal category into an operad in a way that generalizes the above two examples.

Definition 6.75. For any symmetric monoidal category $(\mathcal{C}, I, \otimes)$, there is an operad $\mathcal{O}_\mathcal{C}$, called the *operad underlying* $\mathcal{C}$, defined as having:

(i) $\mathrm{Ob}(\mathcal{C})$ as types,
(ii) morphisms $C_1 \otimes \cdots \otimes C_n \to D$ in $\mathcal{C}$ as the operations of arity $(C_1, \ldots, C_n; D)$,

(iii) substitution defined by

$$(f \circ_i g) := (\mathrm{id}\otimes, \dots, \otimes\mathrm{id} \otimes g \otimes \mathrm{id}\otimes, \dots, \otimes\mathrm{id}) \,\mathring{,}\, f,$$

(iv) identities $\mathrm{id}_a \in \mathcal{O}_{\mathcal{C}}(a; a)$ defined by id_a.

We can also turn any monoidal functor into what's called an operad functor.

6.5.3   The Operad for Hypergraph Props

An operad functor takes the types of one operad to the types of another, and then the operations of the first to the operations of the second in a way that respects this.

Rough Definition 6.76.   Suppose we are given two operads $\mathcal{O}$ and $\mathcal{P}$ with type collections T and U respectively. To specify an operad functor $F \colon \mathcal{O} \to \mathcal{P}$,

(i) one specifies a function $f \colon T \to U$,
(ii) for all arities $(t_1, \dots, t_n; t)$ in $\mathcal{O}$, one specifies a function

$$F \colon \mathcal{O}(t_1, \dots, t_n; t) \to \mathcal{P}(f(t_1), \dots, f(t_n); f(t))$$

such that composition and identities are preserved.

Just as **Set**-valued functors $\mathcal{C} \to$ **Set** from any category $\mathcal{C}$ are of particular interest – we saw them as database instances in Chapter 3 – so too are **Set**-valued functors $\mathcal{O} \to$ **Set** from any operad $\mathcal{O}$.

Definition 6.77.   An *algebra* for an operad $\mathcal{O}$ is an operad functor $F \colon \mathcal{O} \to$ **Set**.

We can think of functors $\mathcal{O} \to$ **Set** as defining a set of possible ways to fill the boxes in a wiring diagram. Indeed, each box in a wiring diagram represents a type t of the given operad $\mathcal{O}$ and an algebra $F \colon \mathcal{O} \to$ **Set** will take a type t and return a set $F(t)$ of fillers for box t. Moreover, given an operation (i.e. a wiring diagram) $f \in \mathcal{O}(t_1, \dots, t_n; t)$, we get a function $F(f)$ that takes an element of each set $F(t_i)$, and returns an element of $F(t)$. For example, it takes n circuits with interface $t_1, \dots, t_n$, respectively, and returns a circuit with boundary t.

Example 6.78. For electric circuits, the types are again finite sets, $T = \mathrm{Ob}(\textbf{FinSet})$, where each finite set $t \in T$ corresponds to a cell with t ports. Just as before, we have a set $\mathrm{Circ}(t)$ of fillers, namely the set of electric circuits with that t-marked terminals. As an operad algebra, $\mathrm{Circ} \colon \textbf{Cospan} \to$ **Set** transforms wiring diagrams like this one

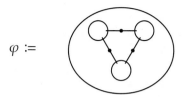

$$\varphi :=$$

into formulas that build a new circuit from a bunch of existing ones. In the above-drawn case, we would get a morphism $\mathrm{Circ}(\varphi) \in \mathbf{Set}(\mathrm{Circ}(2), \mathrm{Circ}(2), \mathrm{Circ}(2); \mathrm{Circ}(0))$, i.e. a function

$$\mathrm{Circ}(\varphi) \colon \mathrm{Circ}(2) \times \mathrm{Circ}(2) \times \mathrm{Circ}(2) \to \mathrm{Circ}(0).$$

We could apply this function to the three elements of $\mathrm{Circ}(2)$ shown here

and the result would be the closed circuit from the beginning of the chapter:

This is reminiscent of the story for decorated cospans: gluing fillers together to form hypergraph categories. An advantage of the decorated cospan construction is that one obtains an explicit category (where morphisms have domains and codomains and can hence be composed associatively), equipped with Frobenius structures that allow us to get around the strictures of domains and codomains. The operad perspective has other advantages. First, whereas decorated cospans can produce only some hypergraph categories, **Cospan**-algebras can produce any hypergraph category.

Proposition 6.79. There is an equivalence between **Cospan**-algebras and hypergraph props.

Another advantage of using operads is that one can vary the operad itself, from **Cospan** to something similar (like the operad of "cobordisms"), and get slightly different compositionality rules.

In fact, operads – with the additional complexity in their definition – can be customized even more than all compositional structures defined so far. For example, we can define operads of wiring diagrams where the wiring diagrams must obey precise conditions far more specific than the constraints of a category, such as requiring that the diagram itself has no wires that pass straight through it. In fact, operads are strong enough to define themselves: roughly speaking, there is an operad for operads. More precisely, the category of operads is equivalent to the category of algebras for a certain operad [Lei04, Example 2.2.23]. While operads can, of course, be generalized again, they conclude our march through an informal hierarchy of compositional structures, from preorders to categories to monoidal categories to operads.

6.6   Summary and Further Reading

This chapter began with a detailed exposition of colimits in the category of sets; as we saw, these colimits describe ways of joining or interconnecting sets. Our second way of talking about interconnection was the use of Frobenius monoids and hypergraph categories; we saw these two themes come together in the idea of a decorated cospan. The decorated cospan construction uses a certain type of structured functor to construct a certain type of structured category. More generally, we might be interested in other types of structured category, or other compositional structure. To address this, we briefly saw how these ideas fit into the theory of operads.

Colimits are a fundamental concept in category theory. For more on colimits, one might refer to any of the introductory category theory textbooks we mentioned in Section 3.6.

Special commutative Frobenius monoids and hypergraph categories were first defined, under the names "separable commutative Frobenius algebra" and "well-supported compact closed category," by Carboni and Walters [CW87; Car91]. The use of decorated cospans to construct them is detailed in [Fon15; Fon18; Fon16]. The application to networks of passive linear systems, such as certain electrical circuits, is discussed in [BF15], while further applications, such as to Markov processes and chemistry, can be found in [BFP16; BP17]. We recently learned from Grandis [Gra18] that a similar approach dates back to Darbo in the late 1960s, before operads or even symmetric monoidal functors were defined [Dar70]! For another interesting application of hypergraph categories, we recommend the pixel array method for approximating solutions to nonlinear equations [Spi+16]. The story of this chapter is fleshed out in a couple of recent, more technical papers [FS18b; FS18a].

Operads were introduced by May to describe compositional structures arising in algebraic topology [May72]; Leinster has written a great book on the subject [Lei04]. More recently, with collaborators author-David has discussed using operads in applied mathematics to model composition of structures in logic, databases, and dynamical systems [RS13; Spi13; VSL15].

7 Logic of Behavior: Sheaves, Toposes, and Internal Languages

7.1 How Can We Prove Our Machine is Safe?

Imagine you are trying to design a system of interacting components. You wouldn't be doing this if you didn't have a goal in mind: you want the system to do something, to behave in a certain way. In other words, you want to restrict its possibilities to a smaller set: you want the car to remain on the road, you want the temperature to remain in a particular range, you want the bridge to be safe for trucks to pass. Out of all the possibilities, your system should only permit some.

Since your system is made of components that interact in specified ways, the possible behavior of the whole – in any environment – is determined by the possible behaviors of each of its components in their local environments, together with the precise way in which they interact.[1] In this chapter, we will discuss a logic wherein one can describe general types of behavior that occur over time, and prove properties of a larger-scale system from the properties and interaction patterns of its components.

For example, suppose we want an autonomous vehicle to maintain a distance of some `safe` $\in \mathbb{R}$ from other objects. To do so, several components must interact: a sensor that approximates the real distance by an internal variable S', a controller that uses S' to decide what action A to take, and a motor that moves the vehicle with an acceleration based on A. This in turn affects the real distance S, so there is a feedback loop.

Consider the following model diagram:

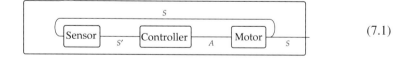

$$(7.1)$$

In the diagram shown, the distance S is exposed by the exterior interface. This just means we imagine S as being a variable that other components of a larger system may

---

[1] The well-known concept of emergence is not about possibilities, it is about prediction. Predicting the behavior of a system given predictions of its components is notoriously hard. The behavior of a double pendulum is chaotic – meaning extremely sensitive to initial conditions – whereas those of the two component pendulums are not. However, the set of possibilities for the double pendulum is completely understood: it is the set of possible angular positions and velocities of both arms. When we speak of a machine's properties in this chapter, we always mean the guarantees on its behaviors, not the probabilities involved, though the latter would certainly be an interesting thing to contemplate.

want to interact with. We could have exposed no variables (making it a closed system) or we could have exposed A and/or S' as well.

In order for the system to ensure $S \geq$ safe, we need each of the components to ensure a property of its own. But what are these components, "Sensor, Controller, Motor," and what do they do?

One way to think about any of the components is to open it up and see how it is put together; with a detailed study we may be able to say what it will do. For example, just as S was exposed in the diagram above, one could imagine opening up the "sensor" component box in Eq. (7.1) and seeing an interaction between subcomponents:

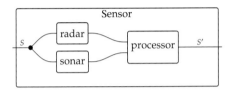

This ability to zoom in and see a single unit as being composed of others is important for design. But, at the end of the day, you eventually need to stop diving down and simply use the properties of the components in front of you to prove properties of the composed system. Have no fear: everything we do in this chapter will be fully compositional, i.e. compatible with opening up lower-level subsystems and using the fractal-like nature of composition. However, at a given time, your job is to design the system at a given level, taking the component properties of lower-level systems as given.

We will think of each component in terms of the relationship it maintains (through time) between the changing values on its ports. "Whenever I see a flash, I will increase pressure on the button": this is a relationship I maintain through time between the changing values on my eye port and my finger port. We will make this more precise soon, but fleshing out the situation in Eq. (7.1) should help. The sensor maintains a relationship between S and S', e.g. that the real distance S and its internal representation S' differ by no more than 5cm. The controller maintains a relationship between S' and the action signal A, e.g. that if at any time $S <$ safe, then within one second it will emit the signal $A =$ go. The motor maintains a relationship between A and S, e.g. that A dictates the second derivative of S by the formula

$$\left((A = \text{go}) \Rightarrow \ddot{S} > 1 \right) \wedge \left((A = \text{stop}) \Rightarrow \ddot{S} = 0 \right). \tag{7.2}$$

If we want to prove properties of the whole interacting system, then the relationships maintained by each component need to be written in a formal logical language, something like what we saw in Eq. (7.2). From that basis, we can use standard proof techniques to combine properties of subsystems into properties of the whole. This is our objective in the present chapter.

We have said how component systems, wired together in some arrangement, create larger-scale systems. We have also said that, given the wiring arrangement, the behavioral properties of the component systems dictate the behavioral properties of the whole. But what exactly are behavioral properties?

In this chapter, we want to give a formal language and semantics for a very general notion of behavior. Mathematics is itself a formal language; the usual style of mathematical modeling is to use any piece of this vast language at any time and for any reason. One uses "human understanding" to ensure that the different models are fitting together in an appropriate way when different systems are combined. The present work differs in that we want to find a domain-specific language for modeling behavior, any sort of behavior, and nothing but behavior. Unlike in the wide world of math, we want a setting where the only things that can be discussed are behaviors.

For this, we will construct what is called a *topos*, which is a special kind of category. Our topos, let's call it **BT**, will have behavior types – roughly speaking, sets whose elements can change through time – as its objects. An amazing fact about toposes[2] is that they come with an *internal language* that looks very much like the usual formal language of mathematics itself. Thus one can define graphs, groups, topological spaces, etc. in any topos. But in **BT**, what we call graphs will actually be graphs that change through time, and similarly what we call groups and spaces will actually be groups and spaces that change through time.

The topos **BT** has not only an internal language, but also a mathematical semantics using the notion of sheaves. Technically, a sheaf is a certain sort of functor, but one can imagine it as a space of possibilities, varying in a controlled way; in our case it will be a space of possible behaviors varying in a certain notion of time. Every property we prove in our logic of behavior types will have meaning in this category of sheaves.

When discussing systems and components – such as sensors, controllers, motors, etc. – we mentioned behavior types; these will be the objects in the topos **BT**. Every wire in the picture below will stand for a behavior type, and every box X will stand for a behavioral property, a relation that X maintains between the changing values on its ports.

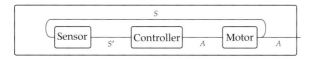

For example, we could imagine that the following hold.

- S (wire): The behavior of S over a time-interval $[a, b]$ is that of all continuous real-valued functions $[a, b] \to \mathbb{R}$.
- A (wire): The behavior of A over a time-interval $[a, b]$ is all piecewise constant functions, taking values in the finite set such as $\{\texttt{go}, \texttt{stop}\}$.
- controller (box): The relation $\{(S', A) \mid$ Eq. (7.2)$\}$, i.e. all behavioral pairs (S', A) that conform to what we said our controller is supposed to do in Eq. (7.2).

---

[2] The plural of topos is often written *topoi*, rather than toposes. This seems a bit fancy for our taste. As Johnstone suggests in [Joh77], we might ask those who "persist in talking about topoi whether, when they go out for a ramble on a cold day, they carry supplies of hot tea with them in thermoi." It's all in good fun; either term is perfectly reasonable and well accepted.

7.2   The Category Set as an Exemplar Topos

We want to think about a very abstract sort of thing, called a topos, because we shall see that behavior types form a topos. To get started, we begin with one of the easiest toposes to think about, namely the topos **Set** of sets. In this section we will discuss commonalities between sets and every other topos. We will go into some details about the category of sets, so as to give intuition for other toposes. In particular, we'll pay careful attention to the logic of sets, because we eventually want to understand the logic of behaviors.

Indeed, logic and sets are closely related. For example, the logical statement – more formally known as a predicate – `likes_cats` defines a function from the set P of people to the set $\mathbb{B} = \{\texttt{false}, \texttt{true}\}$ of truth values, where `Brendan` $\in P$ maps to `true` because he likes cats whereas `Ursula` $\in P$ maps to `false` because she does not. Alternatively, `likes_cats` also defines a subset of P, consisting of exactly the people that do like cats

$$\{p \in P \mid \texttt{likes\_cats}(p)\}.$$

In terms of these subsets, logical operations correspond to set operations, e.g. AND corresponds to intersection: indeed, the set of people for (i.e. mapped to `true` by) the predicate `likes_cats_AND_likes_dogs` is equal to the intersection of the set for `likes_cats` and the set for `likes_dogs`.

We saw in Chapter 3 that such operations, which are examples of database queries, can be described in terms of limits and colimits in **Set**. Indeed, the category **Set** has many such structures and properties, which together make logic possible in that setting. In this section we want to identify these properties, and show how logical operations can be defined using them.

Why would we want to abstractly find such structures and properties? In the next section, we'll start our search for other categories that also have them. Such categories, called toposes, will be **Set**-like enough to do logic, but have much more complex and interesting semantics. Indeed, we will discuss one whose logic allows us to reason not about properties of sets, but about behavioral properties of very general machines.

7.2.1   Set-like Properties Enjoyed by any Topos

Although we will not prove it in this book, toposes are categories that are similar to **Set** in many ways. Here are some facts that are true of any topos $\mathcal{E}$:

1. $\mathcal{E}$ has all limits,
2. $\mathcal{E}$ has all colimits,
3. $\mathcal{E}$ is cartesian closed,
4. $\mathcal{E}$ has epi-mono factorizations,
5. $\mathcal{E}$ has a subobject classifier $1 \xrightarrow{\texttt{true}} \Omega$.

In particular, since **Set** is a topos, all of the above facts are true for $\mathcal{E} = $ **Set**. Our first goal is to briefly review these concepts, focusing most on the subobject classifier.

Limits and colimits

We discussed limits and colimits briefly in Section 3.4.2, but the basic idea is that one can make new objects from old by taking products, using equations to define subobjects, forming disjoint unions, and taking quotient objects. One of the most important types of limit (resp. colimit) is that of pullbacks (resp. pushouts); see Example 3.84 and Definition 6.17. For our work below, we'll need to know a touch more about pullbacks than we have discussed so far, so let's begin there.

Suppose that $\mathcal{C}$ is a category and consider the diagrams below:

$$
\begin{array}{ccc}
A \longrightarrow B \longrightarrow C \\
\downarrow \quad \downarrow \;\lrcorner \quad \downarrow \\
D \longrightarrow E \longrightarrow F
\end{array}
\qquad\qquad
\begin{array}{ccc}
A \longrightarrow B \longrightarrow C \\
\downarrow \;\lrcorner \quad \downarrow \quad \downarrow \\
D \longrightarrow E \longrightarrow F
\end{array}
$$

In the left-hand square, the corner symbol $\lrcorner$ unambiguously means that the square (B, C, E, F) is a pullback. But in the right-hand square, does the corner symbol mean that (A, B, D, E) is a pullback or that (A, C, D, F) is a pullback? It's ambiguous but, as we next show, it becomes unambiguous if the right-hand square is a pullback.

Proposition 7.1. In the commutative diagram below, suppose that the (B, C, B', C') square is a pullback:

$$
\begin{array}{ccc}
A \longrightarrow B \longrightarrow C \\
\downarrow \;\lrcorner \quad \downarrow \;\lrcorner \quad \downarrow \\
A' \longrightarrow B' \longrightarrow C'
\end{array}
$$

Then the (A, B, A', B') square is a pullback iff the (A, C, A', C') rectangle is a pullback.

Exercise 7.2. Prove Proposition 7.1 using the definition of limit from Section 3.4.2. ◇

Epi-mono factorizations

The abbreviation "epi" stands for *epimorphism*, and the abbreviation "mono" stands for monomorphism. Epimorphisms are maps that act like surjections, and monomorphisms are maps that act like injections.[3] We can define them formally in terms of pushouts and pullbacks.

Definition 7.3. Let $\mathcal{C}$ be a category, and let $f \colon A \to B$ be a morphism. It is called a *monomorphism* (resp. *epimorphism*) if the square to the left is a pullback (resp. the square to the right is a pushout):

$$
\begin{array}{ccc}
A \xrightarrow{\;\mathrm{id}_A\;} A \\
{\scriptstyle \mathrm{id}_A}\downarrow \;\lrcorner \quad \downarrow{\scriptstyle f} \\
A \xrightarrow{\;f\;} B
\end{array}
\qquad\qquad
\begin{array}{ccc}
A \xrightarrow{\;f\;} B \\
{\scriptstyle f}\downarrow \quad \downarrow{\scriptstyle \mathrm{id}_B} \\
B \xrightarrow{\;\mathrm{id}_B\;} B
\end{array}
$$

---

[3] Surjections are sometimes called "onto" and injections are sometimes called "one-to-one," hence the Greek prefixes *epi* and *mono*.

Exercise 7.4.   Show that, in **Set**, monomorphisms are just injections:

1. Show that if f is a monomorphism then it is injective.
2. Show that if $f \colon A \to B$ is injective then it is a monomorphism.   ◇

Exercise 7.5.

1. Show that the pullback of an isomorphism along any morphism is an isomorphism. That is, suppose that $i \colon B' \to B$ is an isomorphism and $f \colon A \to B$ is any morphism. Show that i' is an isomorphism, in the following diagram:

$$
\begin{array}{ccc}
A' & \xrightarrow{\;f'\;} & B' \\
{\scriptstyle i'}\downarrow{\scriptstyle \cong} & \lrcorner & \cong\downarrow{\scriptstyle i} \\
A & \xrightarrow{\;f\;} & B
\end{array}
$$

2. Show that for any map $f \colon A \to B$, the square shown is a pullback:

$$
\begin{array}{ccc}
A & \xrightarrow{\;f\;} & B \\
\| & \lrcorner & \| \\
A & \xrightarrow{\;f\;} & B
\end{array}
$$

◇

Exercise 7.6.   Suppose the following diagram is a pullback in a category $\mathcal{C}$:

$$
\begin{array}{ccc}
A' & \xrightarrow{\;g\;} & A \\
{\scriptstyle f'}\downarrow & \lrcorner & \downarrow{\scriptstyle f} \\
B' & \xrightarrow{\;h\;} & B
\end{array}
$$

Use Proposition 7.1 and Exercise 7.5 to show that if f is a monomorphism, then so is f'.   ◇

Now that we have defined epimorphisms and monomorphisms, we can say what epi-mono factorizations are. We say that a morphism $f \colon C \to D$ in $\mathcal{E}$ has an epi-mono factorization if it has an "image"; that is, there is an object $\mathrm{im}(f)$, an epimorphism $C \twoheadrightarrow \mathrm{im}(f)$, and a monomorphism $\mathrm{im}(f) \rightarrowtail D$, whose composite is f.

In **Set**, epimorphisms are surjections and monomorphisms are injections. Every function $f \colon C \to D$ may be factored as a surjective function onto its image $\mathrm{im}(f) = \{f(c) \mid c \in C\}$, followed by the inclusion of this image into the codomain D. Moreover, this factorization is unique up to isomorphism.

Exercise 7.7. Factor the following function $f : \underline{3} \to \underline{3}$ as an epimorphism followed by a monomorphism.

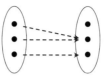

$\diamond$

This is the case in any topos $\mathcal{E}$: for any morphism $f : c \to d$, there exists an epimorphism e and a monomorphism m such that $f = (e \, \mathring{\scriptstyle 9} \, m)$ is their composite.

Cartesian closure

A category $\mathcal{C}$ being cartesian closed means that $\mathcal{C}$ has a symmetric monoidal structure given by products, and it is monoidal closed with respect to this. (We previously saw monoidal closure in Definition 2.57 (for preorders) and Proposition 4.45, as a corollary of compact closure.) Slightly more down-to-earth, cartesian closure means that, for any two objects $C, D \in \mathcal{C}$, there is a "hom-object" $D^C \in \mathcal{C}$ and a natural isomorphism for any $A \in \mathcal{C}$:

$$\mathcal{C}(A \times C, D) \cong \mathcal{C}(A, D^C). \tag{7.3}$$

Think of it this way. Suppose you're A and I'm C, and we're interacting through some game $f(-, -) : A \times C \to D$: for whatever action $a \in A$ that you take and action $c \in C$ that I take, $f(a, c)$ is some value in D. Since you're self-centered but loving, you think of this situation as though you're creating a game experience for me. When you do a, you make a game $f(a, -) : C \to D$ for me alone. In the formalism, D^C represents the set of games for me. So now you've transformed a two-player game, valued in D, into a one-player game, you're the player, valued in ... one-player games valued in D. This transformation is invertible – you can switch your point of view at will – and it's called *currying*. This is the content of Example 3.59.

Exercise 7.8. Let $\mathcal{V} = (V, \leq, I, \otimes)$ be a (unital, commutative) quantale – see Definition 2.66 – and suppose it satisfies the following for all $v, w, x \in V$:

- $v \leq I$,
- $v \otimes w \leq v$ and $v \otimes w \leq w$,
- if $x \leq v$ and $x \leq w$ then $x \leq v \otimes w$.

1. Show that $\mathcal{V}$ is a cartesian closed category, in fact a cartesian closed preorder.
2. Can every cartesian closed preorder be obtained in this way? $\diamond$

Subobject classifier

The concept of a subobject classifier requires more attention, because its existence has huge consequences for a category $\mathcal{C}$. In particular, it creates the setting for a rich system of *higher-order logic* to exist inside $\mathcal{C}$; it does so by providing some things called "truth values." The higher-order logic manifests in its fully glory when $\mathcal{C}$ has finite limits and is cartesian closed, because these facts give rise to the logical operations on truth values.[4] In particular, the higher-order logic exists in any topos.

We will explain subobject classifiers in as much detail as we can; in fact, it will be our subject for the rest of Section 7.2.

7.2.2   The Subobject Classifier

Before giving the definition of subobject classifiers, recall that monomorphisms in **Set** are injections, and any injection $X \rightarrowtail Y$ is isomorphic to a subset of Y. This gives a simple and useful way to conceptualize monomorphisms into Y when reading the following definition: it will do no harm to think of them as subobjects of Y.

Definition 7.9. Let $\mathcal{E}$ be a category with finite limits, i.e. with pullbacks and a terminal object 1. A *subobject classifier* in $\mathcal{E}$ consists of an object $\Omega \in \mathcal{E}$, together with a monomorphism $\mathtt{true}: 1 \to \Omega$, satisfying the following property: for any objects X and Y and monomorphism $m: X \rightarrowtail Y$ in $\mathcal{E}$, there is a unique morphism $\ulcorner m \urcorner: Y \to \Omega$ such that the diagram on the left of Eq. (7.4) is a pullback in $\mathcal{E}$:

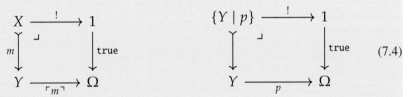

$$\begin{array}{ccc} X & \xrightarrow{\ !\ } & 1 \\ {\scriptstyle m}\downarrow & \lrcorner & \downarrow{\scriptstyle \mathtt{true}} \\ Y & \xrightarrow{\ \ulcorner m \urcorner\ } & \Omega \end{array} \qquad \begin{array}{ccc} \{Y \mid p\} & \xrightarrow{\ !\ } & 1 \\ \downarrow & \lrcorner & \downarrow{\scriptstyle \mathtt{true}} \\ Y & \xrightarrow{\ p\ } & \Omega \end{array} \qquad (7.4)$$

We refer to $\ulcorner m \urcorner$ as the *characteristic map* of m, or we say that $\ulcorner m \urcorner$ *classifies* m. Conversely, given any map $p: Y \to \Omega$, we denote the pullback of $\mathtt{true}$ as on the right of Eq. (7.4).
A *predicate* on Y is a morphism $Y \to \Omega$.

Definition 7.9 is a bit difficult to get one's mind around, partly because it is hard to imagine its consequences. It is like a superdense nugget from outer space, and through scientific explorations in the latter half of the twentieth century, we have found that it brings super powers to whichever categories possess it. We will explain some of the consequences below but, very quickly, the idea is the following.

When a category has a subobject classifier, it provides a translator, turning subobjects of any object Y into maps from that Y to the particular object Ω. Pullback of the monomorphism $\mathtt{true}: 1 \to \Omega$ provides a translator going back, turning maps $Y \to \Omega$ into subobjects of Y. We can replace our fantasy of the superdense nugget with a slightly more refined story: "any object Y understands itself – its parts and the logic of how they fit together – by asking questions of the oracle Ω, looking for what's true." Or to be fully precise but dry, "subobjects of Y are classified by predicates on Y."

---

[4] A category that has finite limits, is cartesian closed, and has a subobject classifier is called an *elementary topos*. We will not discuss these further, but they are the most general notion of topos in ordinary category theory. When someone says topos, you might ask "Grothendieck topos or elementary topos?," because there does not seem to be widespread agreement on which is the default.

Let's move from stories and slogans to concrete facts.

The subobject classifier in Set

Since **Set** is a topos, it has a subobject classifier. It will be a set with supposedly wonderful properties; what set is it?

The subobject classifier in **Set** is the set of booleans,

$$\Omega_{\text{Set}} := \mathbb{B} = \{\texttt{true}, \texttt{false}\}. \tag{7.5}$$

So, in **Set**, the truth values are true and false.

By definition (Def. 7.9), the subobject classifier comes equipped with a morphism, generically called $\texttt{true} \colon 1 \to \Omega$; in the case of **Set** it is played by the function $1 \to \{\texttt{true}, \texttt{false}\}$ that sends 1 to $\texttt{true}$. In other words, the morphism $\texttt{true}$ is aptly named in this case.

For sets, monomorphism just means injection, as we mentioned above. So Definition 7.9 says that, for any injective function $m \colon X \rightarrowtail Y$ between sets, we are supposed to be able to find a characteristic function $\ulcorner m \urcorner \colon Y \to \{\texttt{true}, \texttt{false}\}$ with some sort of pullback property. We propose the following definition of $\ulcorner m \urcorner$:

$$\ulcorner m \urcorner (y) := \begin{cases} \texttt{true} & \text{if } m(x) = y \text{ for some } x \in X, \\ \texttt{false} & \text{otherwise.} \end{cases}$$

In other words, if we think of X as a subset of Y, then we make $\ulcorner m \urcorner(y)$ equal to $\texttt{true}$ iff $y \in X$.

In particular, the subobject classifier property turns subsets $X \subseteq Y$ into functions $p \colon Y \to \mathbb{B}$, and vice versa. How it works is encoded in Definition 7.9, but the basic idea is that X will be the set of all things in Y that p sends to $\texttt{true}$:

$$X = \{y \in Y \mid p(y) = \texttt{true}\}. \tag{7.6}$$

This might help explain our abstract notation $\{Y \mid p\}$ in Eq. (7.4).

Exercise 7.10. Let $X = \mathbb{N} = \{0, 1, 2, \ldots\}$ and $Y = \mathbb{Z} = \{\ldots, -1, 0, 1, 2, \ldots\}$; we have $X \subseteq Y$, so consider it as a monomorphism $m \colon X \rightarrowtail Y$. It has a characteristic function $\ulcorner m \urcorner \colon Y \to \mathbb{B}$, as in Definition 7.9.

1. What is $\ulcorner m \urcorner(-5) \in \mathbb{B}$?
2. What is $\ulcorner m \urcorner(0) \in \mathbb{B}$?   ◇

Exercise 7.11.

1. Consider the identity function $\text{id}_\mathbb{N} \colon \mathbb{N} \to \mathbb{N}$. It is an injection, so it has a characteristic function $\ulcorner \text{id}_\mathbb{N} \urcorner \colon \mathbb{N} \to \mathbb{B}$. Give a concrete description of $\ulcorner \text{id}_\mathbb{N} \urcorner$, i.e. its exact value for each natural number $n \in \mathbb{N}$.
2. Consider the unique function $!_\mathbb{N} \colon \varnothing \to \mathbb{N}$ from the empty set. Give a concrete description of $\ulcorner !_\mathbb{N} \urcorner \colon \mathbb{N} \to \mathbb{B}$.   ◇

7.2.3   Logic in the Topos Set

As we said above, the subobject classifier of any topos $\mathcal{E}$ gives the setting in which to do logic. Before we explain a bit about how topos logic works in general, we continue to work concretely by focusing on logic in the topos **Set**.

Obtaining the AND operation

Consider the function $1 \to \mathbb{B} \times \mathbb{B}$ picking out the element $(\mathtt{true}, \mathtt{true})$. This is a monomorphism, so it defines a characteristic function $\ulcorner(\mathtt{true}, \mathtt{true})\urcorner: \mathbb{B} \times \mathbb{B} \to \mathbb{B}$. What function is it? By Eq. (7.6) the only element of $\mathbb{B} \times \mathbb{B}$ that can be sent to $\mathtt{true}$ is $(\mathtt{true}, \mathtt{true})$. Thus $\ulcorner(\mathtt{true}, \mathtt{true})\urcorner(P, Q) \in \mathbb{B}$ must be given by the following truth table

P	Q	$\ulcorner(\mathtt{true}, \mathtt{true})\urcorner(P, Q)$
true	true	true
true	false	false
false	true	false
false	false	false

This is exactly the truth table for the AND of P and Q, i.e. for $P \wedge Q$. In other words, $\ulcorner(\mathtt{true}, \mathtt{true})\urcorner = \wedge$. Note that this defines $\wedge$ as a function $\wedge: \mathbb{B} \times \mathbb{B} \to \mathbb{B}$, and we use the usual infix notation $x \wedge y := \wedge(x, y)$.

Obtaining the OR operation

Let's go backwards this time. The truth table for the OR of P and Q, i.e. that of the function $\vee: \mathbb{B} \times \mathbb{B} \to \mathbb{B}$ defining OR, is:

P	Q	$P \vee Q$
true	true	true
true	false	true
false	true	true
false	false	false

(7.7)

If we wanted to obtain this function as the characteristic function $\ulcorner m \urcorner$ of some subset $m: X \subseteq \mathbb{B} \times \mathbb{B}$, what subset would X be? By Eq. (7.6), X should be the set of $y \in Y$ that are sent to $\mathtt{true}$. Thus m is the characteristic map for the three element subset

$$X = \{(\mathtt{true}, \mathtt{true}), (\mathtt{true}, \mathtt{false}), (\mathtt{false}, \mathtt{true})\} \subseteq \mathbb{B} \times \mathbb{B}.$$

To prepare for later generalization of this idea in any topos, we want a way of thinking of X only in terms of properties listed at the beginning of Section 7.2.1. In fact, one can think of X as the union of $\{\mathtt{true}\} \times \mathbb{B}$ and $\mathbb{B} \times \{\mathtt{true}\}$ – an epi-mono factorization of a colimit of limits involving the subobject classifier and terminal object. This description will construct an analogous subobject of $\Omega \times \Omega$, and hence classify a map $\Omega \times \Omega \to \Omega$, in any topos $\mathcal{E}$.

Exercise 7.12. Every boolean has a negation, $\neg \texttt{false} = \texttt{true}$ and $\neg \texttt{true} = \texttt{false}$. The function $\neg \colon \mathbb{B} \to \mathbb{B}$ is the characteristic function of some thing, (*?*).

1. What sort of thing should (*?*) be? For example, should $\neg$ be the characteristic function of an object? A topos? A morphism? A subobject? A pullback diagram?
2. Now that you know the sort of thing (*?*) is, which thing of that sort is it?   ◇

Exercise 7.13. Given two booleans P, Q, define $P \Rightarrow Q$ to mean $P = (P \wedge Q)$.

1. Write down the truth table for the statement $P = (P \wedge Q)$:

P	Q	$P \wedge Q$	$P = (P \wedge Q)$
true	true	?	?
true	false	?	?
false	true	?	?
false	false	?	?

2. If you already have an idea what $P \Rightarrow Q$ should mean, does it agree with the last column of the table above?
3. What is the characteristic function $m \colon \mathbb{B} \times \mathbb{B} \to \mathbb{B}$ for $P \Rightarrow Q$?
4. What subobject does m classify?   ◇

Exercise 7.14. Consider the sets $E := \{n \in \mathbb{N} \mid n \text{ is even}\}$, $P := \{n \in \mathbb{N} \mid n \text{ is prime}\}$, and $T := \{n \in \mathbb{N} \mid n \geq 10\}$. Each is a subset of $\mathbb{N}$, so defines a function $\mathbb{N} \to \mathbb{B}$.

1. What is $\ulcorner E \urcorner(17)$?
2. What is $\ulcorner P \urcorner(17)$?
3. What is $\ulcorner T \urcorner(17)$?
4. Name the smallest three elements in the set classified by $(\ulcorner E \urcorner \wedge \ulcorner P \urcorner) \vee \ulcorner T \urcorner$.   ◇

Review

Let's take stock of where we are and where we're going. In Section 7.1, we set out our goal of proving properties about behavior, and we said that topos theory is a good mathematical setting for doing that. We are now at the end of Section 7.2, which was about **Set** as an exemplar topos. What happened?

In Section 7.2.1, we talked about properties of **Set** that are enjoyed by any topos: limits and colimits, cartesian closure, epi-mono factorizations, and subobject classifiers. Then in Section 7.2.2 we launched into thinking about the subobject classifier in general and in the specific topos **Set**, where it is the set $\mathbb{B}$ of booleans because any subset of Y is classified by a specific predicate $p \colon Y \to \mathbb{B}$. Finally, in Section 7.2.3 we discussed how to understand logic in terms of Ω: there are various maps $\wedge, \vee, \Rightarrow \colon \Omega \times \Omega \to \Omega$ and $\neg \colon \Omega \to \Omega$, etc., which serve as logical connectives. These are operations on truth values.

We have talked a lot about toposes, but we've only seen one so far: the category of sets. But we've actually seen more without knowing it: the category $\mathcal{C}\text{-}\mathbf{Inst}$ of instances on any database schema from Definition 3.51 is a topos. Such toposes are called *presheaf toposes* and are fundamental, but we will focus on *sheaf toposes*, because our topos of behavior types will be a sheaf topos.

Sheaves are fascinating, but highly abstract mathematical objects. They are not for the faint of mathematical heart (those who are faint of physical heart are welcome to proceed).

7.3   Sheaves

Sheaf theory began before category theory, e.g. in the form of something called "local coefficient systems for homology groups." However, its modern formulation in terms of functors and sites is due to Grothendieck, who also invented toposes.

The basic idea is that, rather than study spaces, we should study what happens *on* spaces. A space is merely the "site" at which things happen. For example, if we think of the plane $\mathbb{R}^2$ as a space, we might examine only points and regions in it. But if we think of $\mathbb{R}^2$ as a site where things happen, then we might think of things like weather systems throughout the plane, or sand dunes, or trajectories and flows of material. There are many sorts of things that can happen on a space, and these are the sheaves: a sheaf on a space is roughly "a sort of thing that can happen on the space." If we want to think about points or regions from the sheaf perspective, we would consider them as different points of view on what's happening. That is, it's all about what happens on a space: the parts of the space are just perspectives from which to watch the show.

This is reminiscent of databases. The schema of a database is not the interesting part; the data is what's interesting. To be clear, the schema of a database is a site – it's acting like the space – and the category of all instances on it is a topos. In general, we can think of any small category $\mathcal{C}$ as a site; the corresponding topos is the category of functors $\mathcal{C}^{op} \to$ **Set**.[5] Such functors are called *presheaves on* $\mathcal{C}$.

Did you notice that we just introduced a huge class of toposes? For any category $\mathcal{C}$, we said there is a topos of presheaves on it. So before we go on to sheaves, let's discuss this preliminary topic of presheaves. We will begin to develop some terminology and ways of thinking that will later generalize to sheaves.

7.3.1   Presheaves

Recall the definition of functor and natural transformation from Section 3.3. Presheaves are just functors, but they have special terminology that leads us to think about them in a certain geometric way.

Definition 7.15. Let $\mathcal{C}$ be a small category. A *presheaf* P on $\mathcal{C}$ is a functor $P \colon \mathcal{C}^{op} \to$ **Set**. To each object $c \in \mathcal{C}$, we refer to the set $P(c)$ as *the set of sections of P over c*. To each morphism $f \colon c' \to c$, we refer to the function $P(f) \colon P(c) \to P(c')$ as the *restriction map along f*. For any section $s \in P(c)$,

---

[5] The category of functors $\mathcal{C} \to$ **Set** is also a topos: use $\mathcal{C}^{op}$ as the defining site.

we may denote $P(f)(s) \in P(c')$, i.e. its restriction along f, by $s\big|_f$. If P and Q are presheaves, a *morphism* $\alpha\colon P \to Q$ between them is a natural transformation of functors

$$\mathcal{C}^{\mathrm{op}} \underset{Q}{\overset{P}{\Downarrow \alpha}} \mathbf{Set}.$$

Example 7.16. Let **ArShp** be the category shown below:

$$\mathbf{ArShp} := \boxed{\begin{array}{ccc} \text{Vertex} & \xrightarrow[\text{tgt}]{\text{src}} & \text{Pure Arrow} \\ \bullet & & \bullet \end{array}}$$

The reason we call our category **ArShp** is that we can imagine it as an "arrow shape."

$$\begin{array}{c} \text{Vertex} := \boxed{\bullet} \\[4pt] \text{src} \diagup \quad \diagdown \text{tgt} \\[4pt] \text{Pure Arrow} := \boxed{\longmapsto} \end{array} \qquad (7.8)$$

A presheaf on **ArShp** is a functor $I\colon \mathbf{ArShp}^{\mathrm{op}} \to \mathbf{Set}$, which is a database instance on **ArShp**^{op}. Note that **ArShp**^{op} is what we called **Gr** in Section 3.3.5; there we showed that database instances on **Gr** – i.e. presheaves on **ArShp** – are just directed graphs, e.g.

$$P := \boxed{\begin{array}{c} \bullet \longleftarrow \bullet \longrightarrow \bullet \qquad \bullet \nwarrow \\[6pt] \Downarrow \qquad\qquad \circlearrowleft \\[6pt] \bullet \underset{\longleftarrow}{\longrightarrow} \bullet \qquad \bullet \longrightarrow \bullet \end{array}} : \mathbf{ArShp}^{\mathrm{op}} \to \mathbf{Set}$$

Thinking of presheaves on any category $\mathcal{C}$, it often makes sense to imagine the objects of $\mathcal{C}$ as shapes of some sort, and the morphisms of $\mathcal{C}$ as continuous maps between shapes, just like we did for the arrow shape in Eq. (7.8). In that context, one can think of a presheaf P as a kind of lego construction: P is built out of the shapes in $\mathcal{C}$, connected together using the morphisms in $\mathcal{C}$. In the case where $\mathcal{C}$ is the arrow shape, a presheaf is a graph. So this would say that a graph is a sort of lego construction, built out of vertices and arrows connected together using the inclusion of a vertex as the source or target of an arrow. Can you see it?

This statement can be made pretty precise; though we cannot go through it here, the above lego idea is summarized by the formal statement that "the category of presheaves on $\mathcal{C}$ is the free colimit completion of $\mathcal{C}$." Ask a friendly neighborhood category theorist for details.

However one thinks of presheaves – in terms of lego assemblies or database instances – they're relatively straightforward. The difference between presheaves and sheaves is that sheaves take into account some sort of "covering information." The trivial notion of covering is to say that every object covers itself and nothing more; if one uses this trivial covering, presheaves and sheaves are the same thing. In our behavioral context we will need a nontrivial notion of covering, so sheaves and presheaves will be slightly different. Our next goal is to understand sheaves on a topological space.

7.3.2   Topological Spaces

We said in Section 7.3 that, rather than study spaces, we consider spaces as mere "sites" on which things happen. We also said the things that can happen on a space are called sheaves, and always form a type of category called a topos. To define a topos of sheaves, we must start with the site on which they exist.

Sites are very abstract mathematical objects, and we will not make them precise in this book. However, one of the easiest sorts of sites to think about are those coming from topological spaces: every topological space naturally has the structure of a site. We've talked about spaces for a while without making them precise; let's do so now.

Definition 7.17. Let X be a set, and let $\mathsf{P}(X) = \{U \subseteq X\}$ denote its set of subsets. A *topology* on X is a subset $\mathbf{Op} \subseteq \mathsf{P}(X)$, elements of which we call *open sets*,[6] satisfying the following conditions:

(a) Whole set: the subset $X \subseteq X$ is open, i.e. $X \in \mathbf{Op}$.
(b) Binary intersections: if $U, V \in \mathbf{Op}$ then $(U \cap V) \in \mathbf{Op}$.
(c) Arbitrary unions: if I is a set and if we are given an open set $U_i \in \mathbf{Op}$ for each i, then their union is also open, $\left(\bigcup_{i \in I} U_i \right) \in \mathbf{Op}$. We interpret the particular case where $I = \varnothing$ to mean that the empty set is open: $\varnothing \in \mathbf{Op}$.

If $U = \bigcup_{i \in I} U_i$, we say that $(U_i)_{i \in I}$ *covers* U.
A pair $(X, \mathbf{Op})$, where X is a set and $\mathbf{Op}$ is a topology on X, is called a *topological space*.
A *continuous function* between topological spaces $(X, \mathbf{Op}_X)$ and $(Y, \mathbf{Op}_Y)$ is a function $f \colon X \to Y$ such that, for every $U \in \mathbf{Op}_Y$, the preimage $f^{-1}(U)$ is in $\mathbf{Op}_X$.

At the very end of Section 7.3.1 we mentioned how sheaves differ from presheaves in that they take into account "covering information." The notion of covering an open set by a union of other open sets was defined in Definition 7.17, and it will come into play when we define sheaves in Definition 7.27.

---

[6] In other words, we refer to a subset $U \subseteq X$ as *open* if $U \in \mathbf{Op}$.

Example 7.18. The usual topology **Op** on $\mathbb{R}^2$ is based on "ϵ-balls." For any $\epsilon \in \mathbb{R}$ with $\epsilon > 0$, and any point $p = (x, y) \in \mathbb{R}^2$, define *the ϵ-ball centered at p* to be:

$$B(p; \epsilon) := \{p' \in \mathbb{R}^2 \mid d(p, p') < \epsilon\}.[7]$$

In other words, $B(x, y; \epsilon)$ is the set of all points within ϵ of (x, y).

For an arbitrary subset $U \subseteq \mathbb{R}^2$, we call it open and put it in **Op** if, for every $(x, y) \in U$, there exists a (small enough) $\epsilon > 0$ such that $B(x, y; \epsilon) \subseteq U$.

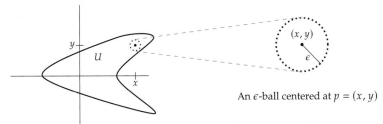

An ϵ-ball centered at $p = (x, y)$

An open set $U \subseteq \mathbb{R}^2$, a point $p = (x, y) \in U$, and an ϵ-ball $B(x, y; \epsilon) \subseteq U$

The same idea works if we replace $\mathbb{R}^2$ with any other metric space X (Definition 2.34): it can be considered as a topological space where the open sets are subsets U such that for any $p \in U$ there is an ϵ-ball centered at p and contained in U. So every metric space can be considered as a topological space.

Exercise 7.19. Consider the set $\mathbb{R}$. It is a metric space with $d(x_1, x_2) := |x_1 - x_2|$.

1. What is the one-dimensional analogue of ϵ-balls as found in Example 7.18? That is, for each $x \in \mathbb{R}$, define $B(x, \epsilon)$.
2. When is an arbitrary subset $U \subseteq \mathbb{R}$ called open, in analogy with Example 7.18?
3. Find three open sets U_1, U_2, and U in $\mathbb{R}$, such that $(U_i)_{i \in \{1,2\}}$ covers U.
4. Find an open set U and a collection $(U_i)_{i \in I}$ of opens sets where I is infinite, such that $(U_i)_{i \in I}$ covers U. ◇

Example 7.20. For any set X, there is a "coarsest" topology, having as few open sets as possible: $\mathbf{Op}_{\mathrm{crse}} = (\varnothing, X)$. There is also a "finest" topology, having as many open sets as possible: $\mathbf{Op}_{\mathrm{fine}} = P(X)$. The latter, $(X, P(X))$, is called the *discrete space on the set X*.

---

[7] Here, $d((x, y), (x', y')) := \sqrt{(x - x')^2 + (y - y')^2}$ is the usual "Euclidean distance" between two points. One can generalize d to any metric.

Exercise 7.21.

1. Verify that, for any set X, what we called $\mathbf{Op}_{\text{crse}}$ in Example 7.20 really is a topology, i.e. satisfies the conditions of Definition 7.17.
2. Verify also that $\mathbf{Op}_{\text{fine}}$ really is a topology.
3. Show that if $(X, \mathsf{P}(X))$ is discrete and $(Y, \mathbf{Op}_Y)$ is any topological space, then every function $X \to Y$ is continuous. ◇

Example 7.22. There are four topologies possible on $X = \{1, 2\}$. Two are $\mathbf{Op}_{\text{crse}}$ and $\mathbf{Op}_{\text{fine}}$ from Example 7.20. The other two are:

$$\mathbf{Op}_1 := \{\varnothing, \{1\}, X\} \quad \text{and} \quad \mathbf{Op}_2 := \{\varnothing, \{2\}, X\}.$$

The two topological spaces $(\{1, 2\}, \mathbf{Op}_1)$ and $(\{1, 2\}, \mathbf{Op}_2)$ are isomorphic; either one can be called *the Sierpinski space*.

The open sets of a topological space form a preorder

Given a topological space $(X, \mathbf{Op})$, the set $\mathbf{Op}$ has the structure of a preorder using the subset relation, $(\mathbf{Op}, \subseteq)$. It is reflexive because $U \subseteq U$ for any $U \in \mathbf{Op}$, and it is transitive because if $U \subseteq V$ and $V \subseteq W$ then $U \subseteq W$.

Recall from Section 3.2.3 that we can regard any preorder, and hence $\mathbf{Op}$, as a category: its objects are the open sets U and for any U, V the set of morphisms $\mathbf{Op}(U, V)$ is empty if $U \not\subseteq V$ and it has one element if $U \subseteq V$.

Exercise 7.23. Recall the Sierpinski space, say $(X, \mathbf{Op}_1)$ from Example 7.22.

1. Write down the Hasse diagram for its preorder of opens.
2. Write down all the covers. ◇

Exercise 7.24. Given any topological space $(X, \mathbf{Op})$, any subset $Y \subseteq X$ can be given the *subspace topology*, call it $\mathbf{Op}_{?\cap Y}$. This topology defines any $A \subseteq Y$ to be open, $A \in \mathbf{Op}_{?\cap Y}$, if there is an open set $B \in \mathbf{Op}$ such that $A = B \cap Y$.

1. Find a $B \in \mathbf{Op}$ that shows that the whole set Y is open, i.e. $Y \in \mathbf{Op}_{?\cap Y}$.
2. Show that $\mathbf{Op}_{?\cap Y}$ is a topology in the sense of Definition 7.17.[8]
3. Show that the inclusion function $Y \hookrightarrow X$ is a continuous function. ◇

Remark 7.25. Suppose $(X, \mathbf{Op})$ is a topological space, and consider the preorder $(\mathbf{Op}, \subseteq)$ of open sets. It turns out that $(\mathbf{Op}, \subseteq, X, \cap)$ is always a quantale in the sense of Definition 2.57. We will not need this fact, but we invite the reader to think about it a bit in Exercise 7.26.

---

[8] Hint 1: for any set I, collection of sets $(U_i)_{i \in I}$ with $U_i \subseteq X$, and set $V \subseteq X$, one has $\left(\bigcup_{i \in I} U_i\right) \cap V = \bigcup_{i \in I}(U_i \cap V)$. Hint 2: for any $U, V, W \subseteq X$, one has $(U \cap W) \cap (V \cap W) = (U \cap V) \cap W$.

Exercise 7.26. In Sections 2.3.2 and 2.3.3 we discussed how **Bool**-categories are preorders and **Cost**-categories are Lawvere metric spaces, and in Section 2.3.4 we imagined interpretations of $\mathcal{V}$-categories for other quantales $\mathcal{V}$.

If $(X, \mathbf{Op})$ is a topological space and $\mathcal{V}$ the corresponding quantale as in Remark 7.25, how might we imagine a $\mathcal{V}$-category?   ◇

7.3.3   Sheaves on Topological Spaces

To summarize where we are, a topological space $(X, \mathbf{Op})$ is a set X together with a bunch of subsets we call "open"; these open subsets form a preorder – and hence category – denoted **Op**. Sheaves on X will be presheaves on **Op** with a special property, aptly named the "sheaf condition."

Recall the terminology and notation for presheaves: a presheaf on **Op** is a functor $P\colon \mathbf{Op}^{\mathrm{op}} \to \mathbf{Set}$. Thus to every open set $U \in \mathbf{Op}$ we have a set $P(U)$, called the set of *sections over* U, and to every inclusion of open sets $V \subseteq U$ we have a function $P(U) \to P(V)$ called the *restriction*. If $s \in P(U)$ is a section over U, we may denote its restriction to V by $s|_V$. Recall that we say a collection of open sets $(U_i)_{i \in I}$ *covers* an open set U if $U = \bigcup_{i \in I} U_i$.

We are now ready to give the following definition, which comes in several waves: we first define matching families, then gluing, then sheaf condition, then sheaf, and finally the category of sheaves.

Definition 7.27.   Let $(X, \mathbf{Op})$ be a topological space, and let $P\colon \mathbf{Op}^{\mathrm{op}} \to \mathbf{Set}$ be a presheaf on **Op**.
Let $(U_i)_{i \in I}$ be a collection of open sets $U_i \in \mathbf{Op}$ covering U. A *matching family* $(s_i)_{i \in I}$ *of P-sections over* $(U_i)_{i \in I}$ consists of a section $s_i \in P(U_i)$ for each $i \in I$, such that for every $i, j \in I$, we have

$$s_i\big|_{U_i \cap U_j} = s_j\big|_{U_i \cap U_j}.$$

Given a matching family $(s_i)_{i \in I}$ for the cover $U = \bigcup_{i \in I} U_i$, we say that $s \in P(U)$ is a *gluing*, or *glued section*, of the matching family if $s|_{U_i} = s_i$ holds for all $i \in I$.
If there exists a unique gluing $s \in P(U)$ for every matching family $(s_i)_{i \in I}$, we say that P *satisfies the sheaf condition for the cover* $U = \bigcup_{i \in I} U_i$. If P satisfies the sheaf condition for every cover, we say that P is a *sheaf* on $(X, \mathbf{Op})$.

Thus a sheaf is just a presheaf satisfying the sheaf condition for every open cover. If P and Q are sheaves, then a *morphism* $f\colon P \to Q$ between these sheaves is just a morphism – that is, a natural transformation – between their underlying presheaves. We denote by $\mathbf{Shv}(X, \mathbf{Op})$, or more briefly $\mathbf{Shv}(X)$, the category of sheaves on $(X, \mathbf{Op})$.

The category of sheaves on X is a topos, but we'll get to that.

Example 7.28. Here is a funny – but very important – special case to which the notion of matching family applies. We do not give this example for intuition, but because (to emphasize) it's an important and easy-to-miss case. Just like the sum of no numbers is 0 and the product of no numbers is 1, the union of no sets is the empty set. Thus if we take $U = \varnothing \subseteq X$ and $I = \varnothing$, then the empty collection of subsets (one for each $i \in I$, of which there are none) covers U. In this case the empty tuple () counts a matching family of sections, and it is the only matching family for the empty cover of the empty set.

In other words, in order for a presheaf $P \colon \mathbf{Op}^{\mathrm{op}} \to \mathbf{Set}$ to be a sheaf, a necessary (but rarely sufficient) condition is that $P(\varnothing) \cong \{()\}$, i.e. $P(\varnothing)$ must be a set with one element.

Extended example: sections of a function

This example is for intuition, and gives a case where the "section" and "restriction" terminology are easy to visualize.

Consider the function $f \colon X \to Y$ shown below, where each element of X is sent to the element of Y immediately below it. For example, $f(a_1) = f(a_2) = a$, $f(b_1) = b$, and so on.

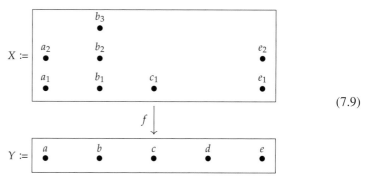

$$(7.9)$$

For each point $y \in Y$, the preimage set $f^{-1}(y) \subseteq X$ above it is often called the *fiber over y*. Note that different f's would arrange the eight elements of X differently over Y: elements of Y would have different fibers.

Exercise 7.29. Consider the function $f \colon X \to Y$ shown in Eq. (7.9).

1. What is the fiber of f over a?
2. What is the fiber of f over c?
3. What is the fiber of f over d?
4. Give an example of a function $f' \colon X \to Y$ for which every fiber has either one or two elements.   ◇

Let's consider X and Y as discrete topological spaces, so every subset is open, and f is automatically continuous (see Exercise 7.21). We will think of f as an arrangement of X over Y, in terms of fibers as above, and use it to build a sheaf on Y. To do this, we begin by building a presheaf – i.e. a functor $\operatorname{Sec}_f : \mathbf{Op}(Y)^{\mathrm{op}} \to \mathbf{Set}$ – and then we'll prove it's a sheaf.

Define the presheaf Sec_f on an arbitrary subset $U \subseteq Y$ by:

$$\operatorname{Sec}_f(U) := \{s : U \to X \mid (s \,\mathring{}\, f)(u) = u \text{ for all } u \in U\}.$$

One might describe $\operatorname{Sec}_f(U)$ as the set of all ways to pick a "cross-section" of the f arrangement over U. That is, an element $s \in \operatorname{Sec}_f(U)$ is a choice of one element per fiber over U.

As an example, let's say $U = \{a, b\}$. How many such s's are there in $\operatorname{Sec}_f(U)$? To answer this, let's clip the picture (7.9) and look only at the relevant part:

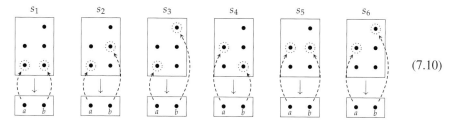

$$(7.10)$$

Looking at the picture (7.10), do you see how we get all cross-sections of f over U?

Exercise 7.30.   Refer to Eq. (7.9).

1. Let $V_1 = \{a, b, c\}$. Draw all the sections over it, i.e. all elements of $\operatorname{Sec}_f(V_1)$, as we did in Eq. (7.10).
2. Let $V_2 = \{a, b, c, d\}$. Again draw all the sections, $\operatorname{Sec}_f(V_2)$.
3. Let $V_3 = \{a, b, d, e\}$. How many sections (elements of $\operatorname{Sec}_f(V_3)$) are there?   ◇

By now you should understand the sections of $\operatorname{Sec}_f(U)$ for various $U \subseteq X$. This is Sec_f on objects, so you are half way to understanding Sec_f as a presheaf. That is, as a presheaf, Sec_f also includes a restriction map for every subset $V \subseteq U$. Luckily, the restriction maps are easy: if $V \subseteq U$, say $V = \{a\}$ and $U = \{a, b\}$, then given a section s as in Eq. (7.10), we get a section over V by "restricting" our attention to what s does on $\{a\}$.

$$s_1|_V = s_2|_V = s_3|_V \qquad\qquad s_4|_V = s_5|_V = s_6|_V$$

$$(7.11)$$

Exercise 7.31.

1. Write out the sets of sections $\operatorname{Sec}_f(\{a, b, c\})$ and $\operatorname{Sec}_f(\{a, c\})$.
2. Draw lines from the first to the second to indicate the restriction map.   ◇

Now we have understood Sec_f as a presheaf; we next explain how to see that it is a sheaf, i.e. that it satisfies the sheaf condition for every cover. To understand the sheaf condition, consider the set $U_1 = \{a, b\}$ and $U_2 = \{b, e\}$. These cover the set $U = \{a, b, e\} = U_1 \cup U_2$. By Definition 7.27, a matching family for this cover consists of a section over U_1 and a section over U_2 that agree on the overlap set, $U_1 \cap U_2 = \{b\}$.

So consider $s_1 \in \mathrm{Sec}_f(U_1)$ and $s_2 \in \mathrm{Sec}_f(U_2)$ shown below.

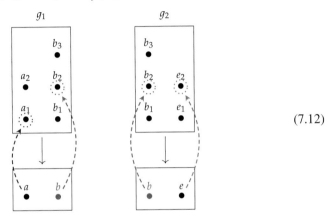

(7.12)

Since sections g_1 and g_2 agree on the overlap – they both send b to b_2 – the two sections shown in Eq. (7.12) can be glued to form a single section over $U = \{a, b, e\}$:

Glued section

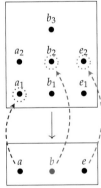

Exercise 7.32.   Again let $U_1 = \{a, b\}$ and $U_2 = \{b, e\}$, so the overlap is $U_1 \cap U_2 = \{b\}$.

1. Find a section $s_1 \in \mathrm{Sec}_f(U_1)$ and a section $s_2 \in \mathrm{Sec}_f(U_2)$ that *do not* agree on the overlap.
2. For your answer (s_1, s_2) in part 1, can you find a section $s \in \mathrm{Sec}_f(U_1 \cup U_2)$ such that $s|_{U_1} = s_1$ and $s|_{U_2} = s_2$?
3. Find a section $h_1 \in \mathrm{Sec}_f(U_1)$ and a section $h_2 \in \mathrm{Sec}_f(U_2)$ that *do* agree on the overlap, but which are different from our choice in Eq. (7.12).
4. Can you find a section $h \in \mathrm{Sec}_f(U_1 \cup U_2)$ such that $h|_{U_1} = h_1$ and $h|_{U_2} = h_2$?   ◇

Other examples of sheaves

The extended example above generalizes to any continuous function $f: X \to Y$ between topological spaces.

Example 7.33. Let $f: (X, \mathbf{Op}_X) \to (Y, \mathbf{Op}_Y)$ be a continuous function. Consider the functor $\mathrm{Sec}_f: \mathbf{Op}_Y^{\mathrm{op}} \to \mathbf{Set}$ given by

$$\mathrm{Sec}_f(U) := \{g: U \to X \mid g \text{ is continuous and } (g \,\overset{\circ}{,}\, f)(u) = u \text{ for all } u \in U\}.$$

The morphisms of $\mathbf{Op}_Y$ are inclusions $V \subseteq U$. Given $g: U \to X$ and $V \subseteq U$, what we call the restriction of g to V is the usual thing we mean by restriction, the same as it was in Eq. (7.11). One can again check that Sec_f is a sheaf.

Example 7.34. A nice example of a sheaf on a space M is that of vector fields on M. If you calculate the wind velocity at every point on Earth, you will have what's called a vector field on Earth. If you know the wind velocity at every point in Afghanistan and I know the wind velocity at every point in Pakistan, and our calculations agree around the border, then we can glue our information together to get the wind velocity over the union of the two countries. All possible wind velocity fields over all possible open sets of the Earth's surface together form the sheaf of vector fields.

Let's say this a bit more formally. A *manifold* M – you can just imagine a sphere such as the Earth's surface – always has something called a *tangent bundle*. It is a space TM whose points are pairs (m, v), where $m \in M$ is a point in the manifold and v is a tangent vector emanating from it. Here's a picture of one tangent plane – all the tangent vectors emanating from some fixed point – on a sphere:

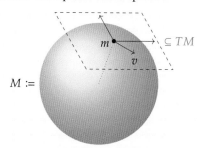

The tangent bundle TM includes the whole tangent plane shown above – including the three vectors drawn on it – as well as the tangent plane at every other point on the sphere.

The tangent bundle TM on a manifold M comes with a continuous map $\pi: TM \to M$ back down to the manifold, sending $(m, v) \mapsto m$. One might say that π "forgets the tangent vector and just remembers the point it emanated from." By Example 7.33, π defines a sheaf Sec_π. It could be called the sheaf of "tangent vector sections on M," but its usual name is the sheaf of *vector fields on M*. This is what we were describing when we spoke of the sheaf of wind velocities on Earth, above. Given an open subset $U \subseteq M$, an element $v \in \mathrm{Sec}_\pi(U)$ is called a vector field over U because it continuously assigns a tangent vector $v(u)$ to each point $u \in U$. The tangent vector at u tells us the velocity of the wind at that point.

Here's a fun digression: in the case of a spherical manifold M like the Earth, it's possible to prove that for every open set U, as long as $U \neq M$, there is a vector field $v \in \mathrm{Sec}_\pi(U)$ that is never 0: the wind could be blowing throughout U. However, a theorem of Poincaré says that if you look at the whole sphere, there is guaranteed to be a point $m \in M$ at which the wind is not blowing at all. It's like the eye of a hurricane or perhaps a cowlick. A cowlick in someone's hair occurs when the hair has no direction to go, so it sticks up! Hair sticking up would not count as a tangent vector: tangent vectors must start out lying flat along the head. Poincaré proved that if your head was covered completely with inch-long hair, there would be at least one cowlick. This difference between local sections (over arbitrary $U \subseteq X$) and global sections (over X) – namely that hair can be well combed whenever $U \neq X$ but cannot be well combed when $U = X$ – can be thought of as a generative effect, and can be measured by cohomology (see Section 1.5).

Exercise 7.35.   If M is a sphere as in Example 7.34, we know from Definition 7.27 that we can consider the category $\mathbf{Shv}(M)$ of sheaves on M; in fact, such categories are toposes and these are what we're getting to.

But are the sheaves on M the vector fields? That is, is there a one-to-one correspondence between sheaves on M and vector fields on M? If so, why? If not, how are sheaves on M and vector fields on M related?   ◇

Example 7.36. For every topological space $(X, \mathbf{Op})$, we have the topos of sheaves on it. The topos of sets, which one can regard as the story of set theory, is the category of sheaves on the one-point space $\{*\}$. In topos theory, we see the category of sets – a huge, amazing, and rich category – as corresponding to a single point. Imagine how much more complex arbitrary toposes are, when they can take place on much more interesting topological spaces (and in fact even more general "sites").

Exercise 7.37.   Consider the Sierpinski space $(\{1, 2\}, \mathbf{Op}_1)$ from Example 7.22.

1. What is the category $\mathbf{Op}$ for this space? (You may have already figured this out in Exercise 7.23; if not, do so now.)
2. What does a presheaf on $\mathbf{Op}$ consist of?
3. What is the sheaf condition for $\mathbf{Op}$?
4. How do we identify a sheaf on $\mathbf{Op}$ with a function?   ◇

7.4   Toposes

A *topos* is defined to be a category of sheaves.[9] So for any topological space $(X, \mathbf{Op})$, the category $\mathbf{Shv}(X, \mathbf{Op})$ defined in Definition 7.27 is a topos. In particular, taking the one-point space $X = \mathbf{1}$ with its unique topology, we find that the category $\mathbf{Set}$ is a topos, as we've been saying all along and saw again explicitly in Example 7.36. And for any database schema – i.e. finitely presented category – $\mathcal{C}$, the category $\mathcal{C}$-**Inst** of database instances on $\mathcal{C}$ is also a topos.[10] Toposes encompass both of these sources of examples, and many more.

Toposes are incredibly nice structures, for a variety of seemingly disparate reasons. In this sketch, the reason in focus is that every topos has many of the same structural properties that the category $\mathbf{Set}$ has. Indeed, we discussed in Section 7.2.1 that every topos has limits and colimits, is cartesian closed, has epi-mono factorizations, and has a subobject classifier (see Section 7.2.2). Using these properties, one can do logic with semantics in the topos $\mathcal{E}$. We explained this for sets, but now imagine it for sheaves on a topological space. There, the same logical symbols $\wedge, \vee, \neg, \Rightarrow, \exists, \forall$ become operations that mean something about sub-sheaves – e.g. vector fields, sections of continuous functions, etc. – not just subsets.

To understand this more deeply, we should say what the subobject classifier $\mathtt{true} \colon 1 \to \Omega$ is in more generality. We said that, in the topos $\mathbf{Set}$, the subobject classifier is the set of booleans $\Omega = \mathbb{B}$. In a sheaf topos $\mathcal{E} = \mathbf{Shv}(X, \mathbf{Op})$, the object $\Omega \in \mathcal{E}$ is a sheaf, not just a set. What sheaf is it?

7.4.1   The Subobject Classifier Ω in a Sheaf Topos

In this subsection we aim to understand the subobject classifier Ω, i.e. the object of truth values, in the sheaf topos $\mathbf{Shv}(X, \mathbf{Op})$. Since Ω is a sheaf, let's understand it by going through the definition of sheaf (Definition 7.27) slowly in this case. A sheaf Ω is a presheaf that satisfies the sheaf condition. As a presheaf it is just a functor $\Omega \colon \mathbf{Op}^{\mathrm{op}} \to \mathbf{Set}$; it assigns a set $\Omega(U)$ to each open $U \subseteq X$ and comes with a restriction map $\Omega(U) \to \Omega(V)$ whenever $V \subseteq U$. So in our quest to understand Ω, we first ask the question: what presheaf is it?

The answer to our question is that Ω is the presheaf that assigns to $U \in \mathbf{Op}$ the set of open subsets of U:

$$\Omega(U) := \{U' \in \mathbf{Op} \mid U' \subseteq U\}. \tag{7.13}$$

That was easy, right? And given the restriction map for $V \subseteq U$ is given by

$$\Omega(U) \to \Omega(V), \tag{7.14}$$
$$U' \mapsto U' \cap V.$$

---

[9] This is sometimes called a *sheaf topos* or a *Grothendieck topos*. There is a more general sort of topos called an *elementary topos* due to Lawvere.

[10] We said that a topos is a category of sheaves, yet database instances are presheaves; so how is $\mathcal{C}$-**Inst** a topos? Well, presheaves in fact count as sheaves. We apologize that this couldn't be clearer. All of this could be made formal if we were to introduce *sites*. Unfortunately, that concept is simply too abstract for the scope of this chapter.

One can check that this is functorial – see Exercise 7.39 – and after doing so we will still need to see that it satisfies the sheaf condition. But at least we don't have to struggle to understand Ω: it's a lot like **Op** itself.

Exercise 7.38.   Let $X = \{1\}$ be the one-point space. We said above that the subobject classifier of **Shv**(1) is the set $\mathbb{B}$ of booleans, but how does that align with the definition of Ω given in Eq. (7.13)? $\diamond$

Exercise 7.39.

1. Show that the definition of Ω given above in Eqs. (7.13) and (7.14) is functorial, i.e. that whenever $W \subseteq V \subseteq U$, the restriction map $\Omega(U) \to \Omega(V)$ followed by the restriction map $\Omega(V) \to \Omega(W)$ is the same as the restriction map $\Omega(U) \to \Omega(W)$.
2. Is that all that's necessary to conclude that Ω is a presheaf? $\diamond$

To see that Ω as defined in Eq. (7.13) satisfies the sheaf condition (see Definition 7.27), suppose that we have a cover $U = \bigcup_{i \in I} U_i$, and suppose we are given an element $V_i \in \Omega(U_i)$, i.e. an open set $V_i \subseteq U_i$, for each $i \in I$. Suppose further that, for all $i, j \in I$, it is the case that $V_i \cap U_j = V_j \cap U_i$, i.e. that the elements form a matching family. Define $V := \bigcup_{i \in I} V_i$; it is an open subset of U, so we can consider V as an element of $\Omega(U)$. The following verifies that V is indeed a gluing for the $(V_i)_{i \in I}$:

$$V \cap U_j = \left(\bigcup_{i \in I} V_i\right) \cap U_j = \bigcup_{i \in I}(V_i \cap U_j) = \bigcup_{i \in I}(V_j \cap U_i) = \left(\bigcup_{i \in I} U_i\right) \cap V_j = V_j.$$

In other words $V \cap U_j = V_j$ for any $j \in I$. So our Ω has been upgraded from presheaf to sheaf!

The eagle-eyed reader will have noticed that we haven't yet given all the data needed to define a subobject classifier. To turn the object Ω into a subobject classifier in good standing, we also need to give a sheaf morphism $\mathtt{true}\colon \{1\} \to \Omega$. Here $\{1\}\colon \mathbf{Op}^{\mathrm{op}} \to \mathbf{Set}$ is the terminal sheaf; it maps every open set to the terminal, one-element set $\{1\}$. The correct morphism $\mathtt{true}\colon \{1\} \to \Omega$ for the subobject classifier is the sheaf morphism that assigns, for every $U \in \mathbf{Op}$, the function $\{1\} = \{1\}(U) \to \Omega(U)$ sending $1 \mapsto U$, the largest open set $U \subseteq U$. From now on we denote $\{1\}$ simply as 1.

Upshot: truth values are open sets

The point is that the truth values in the topos of sheaves on a space $(X, \mathbf{Op})$ are the open sets of that space. When someone says "is property P true?," the answer is not yes or no, but "it is true on the open subset U." If this U is everything, $U = X$, then P is really true; if U is nothing, $U = \varnothing$, then P is really false. But, in general, it's just true in some places and not others.

Example 7.40. The category **Grph** of graphs is a presheaf topos, and one can also think of it as the category of instances for a database schema, as we saw in Example 7.16. The subobject classifier Ω in the topos **Gr** is thus a graph, so we can draw it. Here's what it looks like:

$$\Omega_{\mathbf{Grph}} =$$
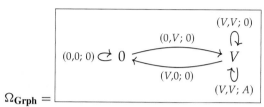

Finding Ω for oneself is easiest using something called the Yoneda lemma, but we have not introduced it. For a nice, easy introduction to the topos of graphs, see [Vig03]. The terminal graph is a single vertex with a single loop, and the graph homomorphism `true`: $1 \to \Omega$ sends that loop to $(V, V; A)$.

Given any graph G and subgraph $i\colon H \subseteq G$, we need to construct a graph homomorphism $\ulcorner H \urcorner\colon G \to \Omega$ classifying H. The idea is that, for each part of G, we decide how much of it is in H. A vertex in v in G is either in H or not; if so we send it to V and if not we send it to 0. But arrows a are more complicated. If a is in H, we send it $(V, V; A)$. But if it is not in H, the mathematics requires us to ask more questions: Is its source in H? Is its target in G? Both? Neither? Based on the answers to these questions we send a to $(V, 0;\ 0)$, $(0, V;\ 0)$, $(V, V;\ 0)$, or $(0, 0;\ 0)$, respectively.

Exercise 7.41. Consider the subgraph $H \subseteq G$ shown here:

$$
\begin{array}{ccc}
A & \xrightarrow{\quad} & B \\
\bullet & & \bullet \\
& & C \\
& & \bullet
\end{array}
\quad \subseteq \quad
\begin{array}{ccccccc}
A & \underset{g}{\overset{f}{\rightleftarrows}} & B & \xrightarrow{h} & C & \xrightarrow{i} & D \\
\bullet & & \bullet & & \bullet & & \bullet
\end{array}
$$

Find the graph homomorphism $\ulcorner H \urcorner\colon G \to \Omega$ classifying it. See Example 7.40. ◇

7.4.2 Logic in a Sheaf Topos

Let's consider the logical connectives, AND, OR, IMPLIES, and NOT. Suppose we have a topological space $X \in \mathbf{Op}$. Given two open sets U, V, considered as truth values $U, V \in \Omega(X)$, then their conjunction "U AND V" is their intersection, and their disjunction "U OR V" is their union;

$$(U \wedge V) := U \cap V \qquad \text{and} \qquad (U \vee V) := U \cup V. \tag{7.15}$$

These formulas are easy to remember, because $\wedge$ looks like $\cap$ and $\vee$ looks like $\cup$. The implication $U \Rightarrow V$ is the largest open set R such that $R \cap U \subseteq V$, i.e.

$$(U \Rightarrow V) := \bigcup_{\{R \in \mathbf{Op} \mid R \cap U \subseteq V\}} R. \tag{7.16}$$

In general, it is not easy to reduce Eq. (7.16) further, so implication is the hardest logical connective to think about topologically.

Finally, the negation of U is given by $\neg U := (U \Rightarrow \texttt{false})$, and this turns out to be relatively simple. By Eq. (7.16), it is the union of all R such that $R \cap U = \varnothing$, i.e. the union of all open sets in the complement of U. If you know topology, you might recognize that $\neg U$ is the "interior of the complement of U."

Example 7.42. Consider the real line $X = \mathbb{R}$ as a topological space (see Exercise 7.19). Let $U, V \in \Omega(X)$ be the open sets $U = \{x \in \mathbb{R} \mid x < 3\}$ and $V = \{x \in \mathbb{R} \mid -4 < x < 4\}$. Using interval notation, $U = (-\infty, 3)$ and $V = (-4, 4)$. Then

- $U \wedge V = (-4, 3)$,
- $U \vee V = (-\infty, 4)$,
- $\neg U = (3, \infty)$,
- $\neg V = (-\infty, -4) \cup (4, \infty)$,
- $(U \Rightarrow V) = (-4, \infty)$,
- $(V \Rightarrow U) = U$.

Exercise 7.43. Consider the real line $\mathbb{R}$ as a topological space, and consider the open subset $U = \mathbb{R} - \{0\}$.

1. What open subset is $\neg U$?
2. What open subset is $\neg\neg U$?
3. Is it true that $U \subseteq \neg\neg U$?
4. Is it true that $\neg\neg U \subseteq U$?   ◇

Above we explained operations on open sets, one corresponding to each logical connective; there are also open sets corresponding to the the symbols $\texttt{true}$ and $\texttt{false}$. We explore this in an exercise.

Exercise 7.44. Let $(X, \mathbf{Op})$ be a topological space.

1. Suppose the symbol $\texttt{true}$ corresponds to an open set such that, for any open set $U \in \mathbf{Op}$, we have $(\texttt{true} \wedge U) = U$. Which open set is it?
2. Other things we should expect from $\texttt{true}$ include $(\texttt{true} \vee U) = \texttt{true}$ and $(U \Rightarrow \texttt{true}) = \texttt{true}$ and $(\texttt{true} \Rightarrow U) = U$. Do these hold for your answer to 1?
3. The symbol $\texttt{false}$ corresponds to an open set $U \in \mathbf{Op}$ such that, for any open set $U \in \mathbf{Op}$, we have $(\texttt{false} \vee U) = U$. Which open set is it?
4. Other things we should expect from $\texttt{false}$ include $(\texttt{false} \wedge U) = \texttt{false}$ and $(\texttt{false} \Rightarrow U) = \texttt{true}$. Do these hold for your answer to 1?   ◇

Example 7.45. For a vector bundle $\pi : E \to X$ over a space X, the corresponding sheaf is Sec_π corresponding to its sections: to each open set $i_U : U \subseteq X$, we associate the set of functions $s : U \to E$ for which $s \mathbin{\raise.2ex\hbox{$\scriptstyle\circ$}} \pi = i_U$. For example, in the case of the tangent bundle $\pi : TM \to M$ (see Example 7.34), the corresponding sheaf, call it VF, associates to each U the set $\mathsf{VF}(U)$ of vector fields on U.

The internal logic of the topos can then be used to consider properties of vector fields. For example, one could have a predicate $\mathtt{Grad} : \mathsf{VF} \to \Omega$ that asks for the largest subspace $\mathtt{Grad}(v)$ on which a given vector field v comes from the gradient of some scalar function. One could also have a predicate that asks for the largest open set on which a vector field is nonzero. Logical operations like $\wedge$ and $\vee$ could then be applied to hone in on precise submanifolds throughout which various desired properties hold, and to reason logically about what other properties are forced to hold there.

7.4.3   Predicates

In English, a predicate is the part of the sentence that comes after the subject. For example "...is even" or "...likes the weather" are predicates. Not every subject makes sense for a given predicate; e.g. the sentence "7 is even" may be false, but it makes sense. In contrast, the sentence "2.7 is even" does not really make sense, and "2.7 likes the weather" certainly doesn't. In computer science, they might say "The expression '2.7 likes the weather' does not type check."

The point is that each predicate is associated to a type, namely the type of subject that makes sense for that predicate. When we apply a predicate to a subject of the appropriate type, the result has a truth value: "7 is even" is either true or false. Perhaps "Bob likes the weather" is true some days and false on others. In fact, this truth value might change by the year (bad weather this year), by the season, by the hour, etc. In English, we expect truth values of sentences to change over time, which is exactly the motivation for this chapter. We're working toward a logic where truth values change over time.

In a topos $\mathcal{E} = \mathbf{Shv}(X, \mathbf{Op})$, a predicate is a sheaf morphism $p : S \to \Omega$ where $S \in \mathcal{E}$ is a sheaf and $\Omega \in \mathcal{E}$ is the subobject classifier, the sheaf of truth values. By Definition 7.27 we get a function $p(U) : S(U) \to \Omega(U)$ for any open set $U \subseteq X$. In the above example – which we will discuss more carefully in Section 7.5 – if S is the sheaf of people (people come and go over time), Bob $\in S(U)$ is a person existing over a time U, and p is the predicate "likes the weather," then $p(\mathrm{Bob})$ is the set of times during which Bob likes the weather. So the answer to "Bob likes the weather" is something like "in summers yes, and also in April 2018 and May 2019 yes, but in all other times no." That's $p(\mathrm{Bob})$, the temporal truth value obtained by applying the predicate p to the subject Bob.

Exercise 7.46. Just now we described how a predicate $p : S \to \Omega$, such as "...likes the weather," acts on sections $s \in S(U)$, say $s = \mathrm{Bob}$. But, by Definition 7.9, any predicate $p \colon S \to \Omega$ also defines a subobject of $\{S \mid p\} \subseteq S$. Describe the sections of this subsheaf.   ◇

The poset of subobjects

For a topos $\mathcal{E} = \mathbf{Shv}(X, \mathbf{Op})$ and object (sheaf) $S \in \mathcal{E}$, the set of S-predicates $|\Omega^E| = \mathcal{E}(S, \Omega)$ is naturally given the structure of a poset, which we denote

$$(|\Omega^S|, \leq^S). \tag{7.17}$$

Given two predicates $p, q \colon S \to \Omega$, we say that $p \leq^S q$ if the first implies the second. More precisely, for any $U \in \mathbf{Op}$ and section $s \in S(U)$ we obtain two open subsets $p(s) \subseteq U$ and $q(s) \subseteq U$. We say that $p \leq^S q$ if $p(s) \subseteq q(s)$ for all $U \in \mathbf{Op}$ and $s \in S(U)$. We often drop the superscript from $\leq^S$ and simply write $\leq$. In formal logic notation, one might write $p \leq^S q$ using the $\vdash$ symbol, e.g. in one of the following ways:

$$s : S \mid p(s) \vdash q(s) \qquad \text{or} \qquad p(s) \vdash_{s:S} q(s).$$

In particular, if $S = 1$ is the terminal object, we denote $|\Omega^S|$ by $|\Omega|$, and refer to elements $p \in |\Omega|$ as *propositions*. They are just morphisms $p \colon 1 \to \Omega$.

This preorder is partially ordered – a poset – meaning that if $p \leq q$ and $q \leq p$ then $p = q$. The reason is that for any subsets $U, V \subseteq X$, if $U \subseteq V$ and $V \subseteq U$ then $U = V$.

Exercise 7.47.   Give an example of a space X, a sheaf $S \in \mathbf{Shv}(X)$, and two predicates $p, q \colon S \to \Omega$ for which $p(s) \vdash_{s:S} q(s)$ holds. You do not have to be formal.   ◇

All of the logical symbols ($\mathtt{true}, \mathtt{false}, \wedge, \vee, \Rightarrow, \neg$) from Section 7.4.2 make sense in any such poset $|\Omega^S|$. For any two predicates $p, q \colon S \to \Omega$, we define $(p \wedge q) \colon S \to \Omega$ by $(p \wedge q)(s) := p(s) \wedge q(s)$, and similarly for $\vee$. Thus one says that these operations are *computed pointwise* on S. With these definitions, the $\wedge$ symbol is the meet and the $\vee$ symbol is the join – in the sense of Definition 1.76 – for the poset $|\Omega^S|$.

With all of the logical structure we've defined so far, the poset $|\Omega^S|$ of predicates on S forms what's called a Heyting algebra. We will not define it here, but more information can be found in Section 7.6. We now move on to quantification.

7.4.4   Quantification

Quantification comes in two flavors: universal and existential, or "for all" and "there exists." Each takes in a predicate of $n+1$ variables and returns a predicate of n variables.

Example 7.48. Suppose we have two sheaves $S, T \in \mathbf{Shv}(X, \mathbf{Op})$ and a predicate $p \colon S \times T \to \Omega$. Let's say T represents what's considered newsworthy and S is again the set of people. So for a subset of time U, a section $t \in T(U)$ is something that's considered newsworthy throughout the whole of U, and a section $s \in S(U)$ is a person that lasts throughout the whole of U. Let's imagine the predicate p as "s is worried about t." Now recall from Section 7.4.3 that a predicate p does not simply return true or false; given a person s and a news-item t, it returns a truth value corresponding to the subset of times on which $p(s, t)$ is true.

"For all t in T, ... is worried about t" is itself a predicate on just one variable, S, which we denote

$$\forall(t : T).\, p(s, t).$$

Applying this predicate to a person s returns the times when that person is worried about everything in the news. Similarly, "there exists t in T such that s is worried about t" is also a predicate on S, which we denote $\exists(t : T).\, p(s, t)$. If we apply this predicate to a person s, we get the times when person s is worried about at least one thing in the news.

Exercise 7.49. In the topos **Set**, where $\Omega = \mathbb{B}$, consider the predicate $p \colon \mathbb{N} \times \mathbb{Z} \to \mathbb{B}$ given by

$$p(n, z) = \begin{cases} \texttt{true} & \text{if } n \leq |z|, \\ \texttt{false} & \text{if } n > |z|. \end{cases}$$

1. What is the set of $n \in \mathbb{N}$ for which the predicate $\forall(z : \mathbb{Z}).\, p(n, z)$ holds?
2. What is the set of $n \in \mathbb{N}$ for which the predicate $\exists(z : \mathbb{Z}).\, p(n, z)$ holds?
3. What is the set of $z \in \mathbb{Z}$ for which the predicate $\forall(n : \mathbb{N}).\, p(n, z)$ holds?
4. What is the set of $z \in \mathbb{Z}$ for which the predicate $\exists(n : \mathbb{N}).\, p(n, z)$ holds? $\diamond$

So, given p, we have a universally and an existentially quantified predicate $\forall(t : T).\, p(s, t)$ and $\exists(t : T).\, p(s, t)$ on S. How do we formally understand them as sheaf morphisms $S \to \Omega$ or, equivalently, as subsheaves of S?

Universal quantification

Given a predicate $p \colon S \times T \to \Omega$, the universally quantified predicate $\forall(t : T).\, p(s, t)$ takes a section $s \in S(U)$, for any open set U, and returns a certain open set $V \in \Omega(U)$. Namely, it returns the largest open set $V \subseteq U$ for which $p(s|_V, t) = V$ holds for all $t \in T(V)$.

Exercise 7.50. Suppose s is a person alive throughout the interval U. Apply the above definition to the example $p(s, t) = $ "person s is worried about news t" from Example 7.48. Here, $T(V)$ is the set of items that are in the news throughout the interval V.

1. What open subset of U is $\forall(t : T).\, p(s, t)$ for a person s?
2. Does it have the semantic meaning you'd expect, given the less formal description in Section 7.4.4? $\diamond$

Abstractly speaking, the universally quantified predicate corresponds to the subsheaf given by the following pullback:

$$
\begin{array}{ccc}
\forall_t p & \longrightarrow & 1 \\
\downarrow & \lrcorner & \downarrow {\scriptstyle \mathtt{true}^T} \\
S & \xrightarrow{\ p'\ } & \Omega^T
\end{array}
$$

where $p' : S \to \Omega^T$ is the currying of $S \times T \to \Omega$ and $\mathtt{true}^T$ is the currying of the composite $1 \times T \xrightarrow{\ !\ } 1 \xrightarrow{\ \mathtt{true}\ } \Omega$. See Eq. (7.3).

Existential quantification

Given a predicate $p \colon S \times T \to \Omega$, the existentially quantified predicate $\exists (t : T). \, p(s, t)$ takes a section $s \in S(U)$, for any open set U, and returns a certain open set $V \in \Omega(U)$, namely the union $V = \bigcup_i V_i$ of all the open sets V_i for which there exists some $t_i \in T(V_i)$ satisfying $p(s|_{V_i}, t_i) = V_i$. If the result is U itself, you might be tempted to think "ah, so there exists some $t \in T(U)$ satisfying $p(t)$," but that is not necessarily so. There is just a cover of $U = \bigcup U_i$ and local sections $t_i \in T(U_i)$, each satisfying p, as explained above. Thus the existential quantifier is doing a lot of work "under the hood," taking coverings into account without displaying that fact in the notation.

Exercise 7.51.   Apply the above definition to the "person s is worried about news t" predicate from Example 7.48.

1. What open set is $\exists (t : T). \, p(s, t)$ for a person s?
2. Does it have the semantic meaning you'd expect? $\diamond$

Abstractly speaking, the existentially quantified predicate is given as follows. Start with the subobject classified by p, namely $\{(s, t) \in S \times T \mid p(s, t)\} \subseteq S \times T$, compose with the projection $\pi_S \colon S \times T \to S$ as on the upper right; then take the epi-mono factorization of the composite as on the lower left:

$$
\begin{array}{ccc}
\{S \times T \mid p\} & \rightarrowtail & S \times T \\
\downarrow\!\!\!\!\downarrow & & \downarrow {\scriptstyle \pi_S} \\
\exists_t p & \rightarrowtail & S
\end{array}
$$

Then the bottom map is the desired subsheaf of S.

7.4.5   Modalities

Back in Example 1.115 we discussed modal operators – also known as modalities – saying they are closure operators on preorders which arise in logic. The preorders we were referring to are the ones discussed in Eq. (7.17): for any object $S \in \mathcal{E}$ there is the poset $(|\Omega^S|, \leq^S)$ of predicates on S, where $|\Omega^S| = \mathcal{E}(S, \Omega)$ is just the set of morphisms $S \to \Omega$ in the category $\mathcal{E}$.

Definition 7.52. A *modality* in **Shv**(X) is a sheaf morphism $j: \Omega \to \Omega$ satisfying three properties for all $U \subseteq X$ and $p, q \in \Omega(U)$:

(a) $p \leq j(p)$,
(b) $(j \, \begin{smallmatrix} \circ \\ \circ \end{smallmatrix} \, j)(p) \leq j(p)$,
(c) $j(p \wedge q) = j(p) \wedge j(q)$.

Exercise 7.53. Suppose $j: \Omega \to \Omega$ is a morphism of sheaves on X, such that $p \leq j(p)$ holds for all $U \subseteq X$ and $p \in \Omega(U)$. Show that for all $q \in \Omega(U)$ we have $j(j(q)) \leq j(q)$ iff $j(j(q)) = j(q)$. ◇

In Example 1.115 we informally said that for any proposition p, e.g. "Bob is in San Diego," there is a modal operator "assuming p, ..." Now we are in a position to make that formal.

Proposition 7.54. Fix a proposition $p \in |\Omega|$. Then

(a) the sheaf morphism $\Omega \to \Omega$ given by sending q to $p \Rightarrow q$ is a modality,
(b) the sheaf morphism $\Omega \to \Omega$ given by sending q to $p \vee q$ is a modality,
(c) the sheaf morphism $\Omega \to \Omega$ given by sending q to $(q \Rightarrow p) \Rightarrow p$ is a modality.

We cannot prove Proposition 7.54 here, but we give references in Section 7.6.

Exercise 7.55. Let S be the sheaf of people as in Section 7.4.3, and let $j: \Omega \to \Omega$ be "assuming Bob is in San Diego ..."

1. Name any predicate $p: S \to \Omega$, such as "likes the weather."
2. Choose a time-interval U. For an arbitrary person $s \in S(U)$, what sort of thing is $p(s)$, and what does it mean?
3. What sort of thing is $j(p(s))$ and what does it mean?
4. Is it true that $p(s) \leq j(p(s))$? Explain briefly.
5. Is it true that $j(j(p(s))) = j(p(s))$? Explain briefly.
6. Choose another predicate $q: S \to \Omega$. Is it true that $j(p \wedge q) = j(p) \wedge j(q)$? Explain briefly. ◇

7.4.6 Type Theories and Semantics

We have been talking about the logic of a topos in terms of open sets, but this is actually a conflation of two ideas that are really better left unconflated. The first is logic, or formal language, and the second is semantics, or meaning. The formal language looks like this:

$$\forall (t : T). \exists (s : S). \, f(s) = t \tag{7.18}$$

and semantic statements are like "the sheaf morphism $f: S \to T$ is an epimorphism." In the former, logical world, all statements are linguistic expressions formed according to strict rules, and all proofs are deductions that also follow strict rules. In the latter, semantic world, statements and proofs are about the sheaves themselves, as mathematical objects. We admit these are rough statements; again, our aim here is only to give a taste, an invitation to further reading.

To *provide semantics* for a logical system means to provide a compiler that converts each logical statement in the formal language into a mathematical statement about particular sheaves and their relationships. A computer can carry out logical deductions without knowing what any of them "mean" about sheaves. We say that semantics is *sound* if every formal proof is converted into a true fact about the relevant sheaves.

Every topos can be assigned a formal language, often called its *internal language*, in which to carry out constructions and formal proofs. This language has a sound semantics – a sort of logic-to-sheaf compiler – which goes under the name *categorical semantics* or *Kripke–Joyal semantics*. In this section we have sketched the basic ideas; we give references to the literature in Section 7.6.

Example 7.56. In every topos $\mathcal{E}$, and for every $f \colon S \to T$ in $\mathcal{E}$, the morphism f is an epimorphism if and only if Eq. (7.18) holds. For example, consider the case of database instances on a schema $\mathcal{C}$, say with 100 tables (one of which might be denoted $c \in \mathrm{Ob}(\mathcal{C})$) and 500 foreign key columns (one of which might be denoted $f \colon c \to c'$ in $\mathcal{C}$); see Eq. (3.2).

If S and T are two instances and f is a natural transformation between them, then we can ask the question of whether or not Eq. (7.18) holds. This simple formula is compiled by the Kripke–Joyal semantics into asking:

Is it true that for every table $c \in \mathrm{Ob}(\mathcal{C})$ and every row $t \in T(c)$ there exists a row $s \in S(c)$ such that $f(s) = t$?

This is exactly what it means for f to be surjective. Maybe this is not too impressive, but whether one is talking about databases or topological spaces, or complex ideas from algebraic geometry, Eq. (7.18) always compiles into the question of surjectivity. For topological spaces it would say something like:

Is it true that for every open set $U \subseteq X$ and every section $t \in T(u)$ of the bundle T, there exists an open covering of $(U_i \subseteq U)_{i \in I}$ of U and a section $s_i \in S(U_i)$ of the bundle S for each $i \in I$, such that $f(s_i) = t\big|_{u_i}$ is the restriction of s to U_i?

7.5   A Topos of Behavior Types

Now that we have discussed logic in a sheaf topos, we return to our motivating example, a topos of behavior types. We begin by discussing the topological space on which behavior types will be sheaves, a space called the *interval domain*.

Remark 7.57. Note that, above, we were thinking very intuitively about time, e.g. when we discussed people being worried about the news. Now we will be thinking about time in a different way, but there is no need to change your answers or reconsider the intuitive thinking done above.

7.5.1 The Interval Domain

The interval domain $\mathbb{IR}$ is a specific topological space, which we will use to model intervals of time. In other words, we will be interested in the category $\mathbf{Shv}(\mathbb{IR})$ of sheaves on the interval domain.

To give a topological space, one must give a pair $(X, \mathbf{Op})$, where X is a set of "points" and $\mathbf{Op}$ is a topology on X; see Definition 7.17. The set of points for $\mathbb{IR}$ is that of all finite closed intervals

$$\mathbb{IR} := \{[d, u] \subseteq \mathbb{R} \mid d \leq u\}.$$

For $a < b$ in $\mathbb{R}$, let $o_{[a,b]}$ denote the set $o_{[a,b]} := \{[d, u] \in \mathbb{IR} \mid a < d \leq u < b\}$; these are called *basic open sets*. The topology $\mathbf{Op}$ is determined by these basic open sets in that a subset U is open if it is the union of some collection of basic open sets.

Thus for example, $o_{[0,5]}$ is an open set: it contains every $[d, u]$ contained in the open interval $\{x \in \mathbb{R} \mid 0 < x < 5\}$. Similarly $o_{[4,8]}$ is an open set, but note that $o_{[0,5]} \cup o_{[4,8]} \neq o_{[0,8]}$. Indeed, the interval $[2, 6]$ is in the right-hand side but not the left.

Exercise 7.58.

1. Explain why $[2, 6] \in o_{[0,8]}$.
2. Explain why $[2, 6] \notin o_{[0,5]} \cup o_{[4,8]}$. ◇

Let $\mathbf{Op}$ denote the open sets of $\mathbb{IR}$, as described above, and let $\mathbf{BT} := \mathbf{Shv}(\mathbb{IR}, \mathbf{Op})$ denote the topos of sheaves on this space. We call it the topos of *behavior types*.

There is an important subspace of $\mathbb{IR}$, namely the usual space of real numbers $\mathbb{R}$. We see $\mathbb{R}$ as a subspace of $\mathbb{IR}$ via the isomorphism

$$\mathbb{R} \cong \{[d, u] \in \mathbb{IR} \mid d = u\}.$$

We discussed the usual topology on $\mathbb{R}$ in Example 7.18, but we also get a topology on $\mathbb{R}$ because it is a subset of $\mathbb{IR}$; i.e. we have the subspace topology as described in Exercise 7.24. These agree, as the reader can check.

Exercise 7.59. Show that a subset $U \subseteq \mathbb{R}$ is open in the subspace topology of $\mathbb{R} \subseteq \mathbb{IR}$ iff $U \cap \mathbb{R}$ is open in the usual topology on $\mathbb{R}$ defined in Example 7.18. ◇

7.5.2 Sheaves on $\mathbb{IR}$

We cannot go into much depth about the sheaf topos $\mathbf{BT} = \mathbf{Shv}(\mathbb{IR}, \mathbf{Op})$, for reasons of space; we refer the interested reader to Section 7.6. In this section we will briefly discuss what it means to be a sheaf on $\mathbb{IR}$, giving a few examples including that of the subobject classifier.

What is a sheaf on $\mathbb{IR}$?

A sheaf S on the interval domain ($\mathbb{IR}$, **Op**) is a functor $S\colon \mathbf{Op}^{\mathrm{op}} \to \mathbf{Set}$: it assigns to each open set U a set $S(U)$; how should we interpret this? An element $s \in S(U)$ is something that says it is an "event that takes place throughout the interval U." Given this U-event s together with an open subset of $V \subseteq U$, there is a V-event $s|_V$ that tells us what s is if we regard it as an event taking place throughout V. If $U = \bigcup_{i \in I} U_i$ and we can find matching U_i-events (s_i) for each $i \in I$, then the sheaf condition (Definition 7.27) says that they have a unique gluing, i.e. a U-event $s \in S(U)$ that encompasses all of them: $s|_{U_i} = s_i$ for each $i \in I$.

We said in Section 7.5.1 that every open set $U \subseteq \mathbb{IR}$ can be written as the union of basic open sets $o_{[a,b]}$. This implies that any sheaf S is determined by its values $S(o_{[a,b]})$ on these basic open sets. The sheaf condition furthermore implies that these vary continuously in a certain sense, which we can express formally as

$$S(o_{[a,b]}) \cong \lim_{\epsilon > 0} S(o_{[a-\epsilon, b+\epsilon]}).$$

However, rather than get into the details, we describe a few sorts of sheaves that may be of interest.

Example 7.60. For any set A there is a sheaf $\mathsf{A} \in \mathbf{Shv}(\mathbb{IR})$ that assigns to each open set U the set $\mathsf{A}(U) := A$. This allows us to refer to integers, or real numbers, or letters of an alphabet, as though they were behaviors. What sort of behavior is $7 \in \mathbb{N}$? It is the sort of behavior that never changes: it's always seven. Thus A is called the *constant sheaf on A*.

Example 7.61. Fix any topological space $(X, \mathbf{Op}_X)$. Then there is a sheaf F_X of *local functions from $\mathbb{IR}$ to X*. That is, for any open set $U \in \mathbf{Op}_{\mathbb{IR}}$, we assign the set $F_X(U) := \{f\colon U \to X \mid f$ is continuous$\}$. There is also the sheaf G_X of local functions on the subspace $\mathbb{R} \subseteq \mathbb{IR}$. That is, for any open set $U \in \mathbf{Op}_{\mathbb{IR}}$, we assign the set $G_X(U) := \{f\colon U \cap \mathbb{R} \to X \mid f$ is continuous$\}$.

Exercise 7.62. Let's check that Example 7.61 makes sense. Fix any topological space $(X, \mathbf{Op}_X)$ and any subset $R \subseteq \mathbb{IR}$ of the interval domain. Define $H_X(U) := \{f\colon U \cap R \to X \mid f$ is continuous$\}$.

1. Is H_X a presheaf? If not, why not; if so, what are the restriction maps?
2. Is H_X a sheaf? Why or why not?   ◇

Example 7.63. Another source of examples comes from the world of open hybrid dynamical systems. These are machines whose behavior is a mixture of continuous movements – generally imagined as trajectories through a vector field – and discrete jumps. These jumps are imagined as being caused by signals that spontaneously arrive. Over any interval of time, a hybrid system has certain things that it can do and certain things that it cannot. Although we will not make this precise here, there is a construction for converting any hybrid system into a sheaf on $\mathbb{IR}$; we will give references in Section 7.6.

We refer to sheaves on $\mathbb{IR}$ as behavior types because almost any sort of behavior one can imagine is a behavior type. Of course, a complex behavior type – such as the way someone acts when they are in love – would be extremely hard to write down. But the idea is straightforward: for any interval of time, say a three-day interval $(d, d + 3)$, let $L(d, d+3)$ denote the set of all possible behaviors a person who is in love could possibly have. Obviously it's a big, unwieldy set, and no one would want to make precise. But to the extent that one can imagine that sort of behavior as occurring through time, they could imagine the corresponding sheaf.

The subobject classifier as a sheaf on $\mathbb{IR}$

In any sheaf topos, the subobject classifier Ω is itself a sheaf. It is responsible for the truth values in the topos. As we said in Section 7.4.1, when it comes to sheaves on a topological space $(X, \mathbf{Op})$, truth values are open subsets $U \in \mathbf{Op}$.

BT is the topos of sheaves on the space $(\mathbb{IR}, \mathbf{Op})$, as defined in Section 7.5.1. As always, the subobject classifier Ω assigns to any $U \in \mathbf{Op}$ the set of open subsets of U, so these are the truth values. But what do they mean? The idea is that every proposition, such as "Bob likes the weather," returns an open set U, as if to respond that Bob likes the weather "… throughout time period U." Let's explore this a bit more.

Suppose Bob likes the weather throughout the interval $(0, 5)$ and throughout the interval $(4, 8)$. We would probably conclude that Bob likes the weather throughout the interval $(0, 8)$. But what about the more ominous statement "a single pair of eyes has remained watching position p." Then just because it's true on $(0, 5)$ and on $(4, 8)$, does not imply that it has been true on $(0, 8)$: there may have been a change of shift, where one watcher was relieved from their post by another watcher. As another example, consider the statement "the stock market did not go down by more than 10 points." This might be true on $(0, 5)$ and true on $(4, 8)$ but not on $(0, 8)$. In order to capture the semantics of statements like these – statements that take time to evaluate – we must use the space $\mathbb{IR}$ rather than the space $\mathbb{R}$.

7.5.3   Safety Proofs in Temporal Logic

We now have at least a basic idea of what goes into a proof of safety, say for autonomous vehicles, or airplanes in the national airspace system. In fact, the underlying ideas of this chapter came out of a project between MIT, Honeywell Inc., and NASA [SSV18]. The background for the project was that the National Airspace System consists of many different systems interacting: interactions between airplanes, each of which is an interaction between physics, humans, sensors, and actuators, each of which is an inter-action between still more basic parts. The same sort of story would hold for a fleet of autonomous vehicles, as in the introduction to this chapter.

Suppose that each of the systems – at any level – is guaranteed to satisfy some prop-erty. For example, perhaps we can assume that an engine is either out of gas, has a broken fuel line, or is following the orders of a human driver or pilot. If there is a rup-ture in the fuel line, the sensors will alert the human within three seconds, etc. Each of the components interact with a number of different variables. In the case of airplanes, a pilot interacts with the radio, the positions of the dials, the position of the thruster, and the visual data in front of her. The component – here the pilot – is guaranteed to keep these variables in some relation: "if I see something, I will say something" or "if the dials are in position `bad_pos`, I will engage the thruster within 1 second." We call these guarantees *behavior contracts*.

All of the above can be captured in the topos **BT** of behavior types. The variables are behavior types: the altimeter is a variable whose value $\theta \in \mathbb{R}_{\geq 0}$ is changing continuously with respect to time. The thruster is also a continuously changing variable whose value is in the range [0, 1], etc.

The guaranteed relationships – behavior contracts – are given by predicates on vari-ables. For example, if the pilot will always engage the thruster within one second of the display dials being in position `bad_pos`, this can be captured by a predicate $p:$ `dials` $\times$ `thrusters` $\to \Omega$. While we have not written out a formal language for p, one could imagine the predicate $p(D, T)$ for $D :$ `dials` and $T :$ `thrusters` as

$$
\forall (t : \mathbb{R}).\ @_t\big(\texttt{bad\_pos}(D)\big) \Rightarrow \\
\exists (r : \mathbb{R}).\ (0 < r < 1) \wedge \forall(r' : \mathbb{R}).\ 0 \leq r' \leq 5 \Rightarrow @_{t+r+r'}\big(\texttt{engaged}(T)\big). \tag{7.19}
$$

Here $@_t$ is a modality, as we discussed in Definition 7.52; in fact it turns out to be one of type 3 from Proposition 7.54, but we cannot go into that. For a proposition q, the statement $@_t(q)$ says that q is true in some small enough neighborhood around t. So (7.19) says "starting within one second of whenever the dials say that we are in a bad position, I'll engage the thrusters for five seconds."

Given an actual playing-out-of-events over a time period U, i.e. actual section $D \in$ `dials`(U) and section $T \in$ `thrusters`(U), the predicate Eq. (7.19) will hold on certain parts of U and not others, and this is the truth value of p. Hopefully the pilot upholds her behavior contract at all times she is flying, in which case the truth value will be `true` throughout that interval U. But if the pilot breaks her contract over certain intervals, then this fact is recorded in Ω.

The logic allows us to record axioms like that shown in Eq. (7.19) and then rea-son from them: e.g. if the pilot and the airplane, and at least one of the three radars, uphold their contracts then safe separation will be maintained. We cannot give fur-ther details here, but these matters have been worked out in detail in [SS18]; see Section 7.6.

7.6   Summary and Further Reading

This chapter was about modeling various sorts of behavior using sheaves on a space of time-intervals. Behavior may seem like it's something that occurs now in the present, but in fact our memory of past behavior informs what the current behavior means. In order to commit to anything, to plan or complete any sort of process, one needs to be able to reason over time-intervals. The nice thing about temporal sheaves – indeed sheaves on any site – is that they fit into a categorical structure called a topos, which has many useful formal properties. In particular, it comes equipped with a higher-order logic with which we can formally reason about how temporal sheaves work together when combined in larger systems. A much more detailed version of this story was presented in [SS18]. But it would have been impossible without the extensive edifice of topos theory and domain theory that has been developed over the past six decades.

Sheaf toposes were invented by Grothendieck and his school in the 1960s [AGV71] as an approach to proving conjectures at the intersection of algebraic geometry and number theory, called the Weil conjectures. Soon after, Lawvere and Tierney recognized that toposes had all the structure necessary to do logic, and with a whole host of other category theorists, the subject was developed to an impressive extent in many directions. For a much more complete history, see [McL90].

There are many sorts of references on topos theory. One that starts by introducing categories and then moves to toposes, focusing on logic, is [McL92]. Our favorite treatment is perhaps [MM92], where the geometric aspects play a central role. Finally, Johnstone has done the field a huge favor by collecting large amounts of the theory into a single two-volume set [Joh02]; it is very dense, but an essential reference for the serious student or researcher. For just categorical (Kripke–Joyal) semantics of logic in a topos, one should see [MM92], [Jac99], or [LS88].

We did not mention domain theory much in this chapter, aside from referring to the interval domain. But domains, in the sense of Dana Scott, play an important role in the deeper aspects of temporal type theory. A good reference is [Gie+03], but for an introduction we suggest [AJ94].

In some sense our application area has been a very general sort of dynamical system. Other categorical approaches to this subject include [JNW96], [HTP03], [AS05], and [Law86], though there are many others.

We hope you have enjoyed the seven sketches in this book. As a next step, consider running a reading course on applied category theory with some friends or colleagues. Simultaneously, we hope you begin to search out categorical ways of thinking about familiar subjects. Perhaps you'll find something you want to contribute to this growing field of applied category theory, or as we sometimes call it, the field of compositionality.

Appendix Exercise Solutions

A.1   Solutions for Chapter 1

Solution to Exercise 1.1
For each of the following properties, we need to find a function $f : \mathbb{R} \to \mathbb{R}$ that preserves it, and another function – call it g – that does not.

Order-preserving: Take $f(x) = x + 5$; if $x \leq y$ then $x + 5 \leq y + 5$, so f is order-preserving. Take $g(x) := -x$; even though $1 \leq 2$, the required inequality $-1 \overset{?}{\leq} -2$ does not hold, so g is not order-preserving.

Metric-preserving: Take $f(x) := x + 5$; for any x, y we have $|x - y| = |(x + 5) - (y + 5)|$ by the rules of arithmetic, so $|x - y| = |f(x) - f(y)|$, meaning f preserves metric. Take $g(x) := 2 * x$; then with $x = 1$ and $y = 2$ we have $|x - y| = 1$ but $|2x - 2y| = 2$, so g does not preserve the metric.

Addition-preserving: Take $f(x) := 3 * x$; for any x, y we have $3 * (x + y) = (3 * x) + (3 * y)$, so f preserves addition. Take $g(x) := x + 1$; then with $x = 0$ and $y = 0$, we have $g(x + y) = 1$, but $g(x) + g(y) = 2$, so g does not preserve addition.

Solution to Exercise 1.2
Here is the join of the two systems:

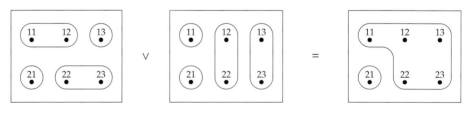

Solution to Exercise 1.3
1. Here is the Hasse diagram for partitions of the two-element set $\{\bullet, *\}$:

$$(12) = \boxed{\bullet \quad\quad *}$$

$\uparrow$

$$(1)(2) = \enclose{circle}{\bullet} \quad \enclose{circle}{*}$$

2. Here is a picture (using text, rather than circles) for partitions of the set $\{1, 2, 3, 4\}$:

For the remaining parts, we choose $A = (12)(3)(4)$ and $B = (13)(2)(4)$.

3. $A \vee B = (123)(4)$.
4. Yes, it is true that $A \leq (A \vee B)$ and that $B \leq (A \vee B)$.
5. The systems C with $A \leq C$ and $B \leq C$ are: $(123)(4)$ and (1234).
6. Yes, it is true that in each case $(A \vee B) \leq C$.

Solution to Exercise 1.4
1. $\texttt{true} \vee \texttt{false} = \texttt{true}$.
2. $\texttt{false} \vee \texttt{true} = \texttt{true}$.
3. $\texttt{true} \vee \texttt{true} = \texttt{true}$.
4. $\texttt{false} \vee \texttt{false} = \texttt{false}$.

Solution to Exercise 1.6
1. This is true: a natural number is exactly an integer that is at least 0.
2. This is false: $0 \in \mathbb{N}$ but $0 \notin \{n \in \mathbb{Z} \mid n \geq 1\}$.
3. This is true: no elements of $\mathbb{Z}$ are strictly between 1 and 2.

Solution to Exercise 1.7
1. The eight subsets of $B := \{1, 2, 3\}$ are

$$\varnothing, \quad \{1\}, \quad \{2\}, \quad \{3\}, \quad \{1, 2\}, \quad \{1, 3\}, \quad \{2, 3\}, \quad \{1, 2, 3\}.$$

2. The union of $\{1, 2, 3\}$ and $\{1\}$ is $\{1, 2, 3\} \cup \{1\} = \{1, 2, 3\}$.
3. The six elements of $\{h, 1\} \times \{1, 2, 3\}$ are

$$(h, 1), \quad (h, 2), \quad (h, 3), \quad (1, 1), \quad (1, 2), \quad (1, 3).$$

4. The five elements of $\{h, 1\} \sqcup \{1, 2, 3\}$ are

$$(h, 1), \quad (1, 1), \quad (1, 2), \quad (2, 2), \quad (3, 2).$$

5. The four elements of $\{h, 1\} \cup \{1, 2, 3\}$ are

$$h, \quad 1, \quad 2, \quad 3.$$

Solution to Exercise 1.11
Suppose that A is a set and $\{A_p\}_{p \in P}$ and $\{A'_{p'}\}_{p' \in P'}$ are two partitions of A such that for each $p \in P$ there exists a $p' \in P'$ with $A_p = A'_{p'}$.

1. Given $p \in P$, suppose we had $p'_1, p'_2 \in P'$ such that $A_p = A'_{p'_1}$ and $A_p = A'_{p'_2}$. Well then $A_{p'_1} = A_{p'_2}$, so in particular $A_{p'_1} \cap A_{p'_2} = A_{p'_1}$. By Definition 1.10 of partition, $A_{p'_1} \neq \varnothing$, and yet if $p_1 \neq p_2$ then $A_{p'_1} \cap A_{p'_2} = \varnothing$. This can't be, so we must have $p'_1 = p'_2$, as desired.

2. Suppose we are given $p' \in P'$; we want to show that there is a $p \in P$ such that $A_p = A'_{p'}$. Since $A'_{p'} \neq \varnothing$ is nonempty by definition, we can pick some $a \in A'_{p'}$; since $A'_{p'} \subseteq A$, we have $a \in A$. Finally, since $A = \bigcup_{p \in P} A_p$, there is some p with $a \in A_p$. This is our candidate p; now we show that $A_p = A'_{p'}$. By assumption there is some $p'' \in P'$ with $A_p = A'_{p''}$, so now $a \in A'_{p''}$ and $a \in A'_{p'}$, so $a \in A'_{p'} \cap A'_{p''}$. Again by definition, having a nonempty intersection means $p' = p''$. So we conclude that $A_p = A_{p'}$.

Solution to Exercise 1.12

The pairs (a, b) such that $a \sim b$ are:

$$(11, 11), \quad (11, 12), \quad (12, 11), \quad (12, 12), \quad (13, 13),$$
$$(21, 21), \quad (22, 22), \quad (22, 23), \quad (23, 22), \quad (23, 23).$$

Solution to Exercise 1.15

1. One aspect in the definition of the parts is that they are connected, and one aspect of that is that they are nonempty. So each part A_p is nonempty.

2. Suppose $p \neq q$, i.e. A_p and A_q are not exactly the same set. To prove $A_p \cap A_q = \varnothing$, we suppose otherwise and derive a contradiction. So suppose there exists $a \in A_p \cap A_q$; we will show that $A_p = A_q$, which contradicts an earlier hypothesis. To show that these two subsets are equal, it suffices to show that $a' \in A_p$ iff $a' \in A_q$ for all $a' \in A$. Suppose $a' \in A_p$; then because A_p is connected, we have $a \sim a'$. And because A_q is closed, $a' \in A_q$. In a similar way, if $a' \in A_q$ then because A_q is connected and A_p is closed, $a' \in A_p$, and we are done.

3. To show that $A = \bigcup_{p \in P} A_p$, it suffices to show that, for each $a \in A$, there is some $p \in P$ such that $a \in A_p$. We said that P was the set of closed and connected subsets of A, so it suffices to show that there is some closed and connected subset containing a. Let $X := \{a' \in A \mid a' \sim a\}$; we claim it is closed and connected and contains a. To see X is closed, suppose $a' \in X$ and $b \sim a'$; then $b \sim a$ by transitivity and symmetry of $\sim$, so $b \in X$. To see that X is connected, suppose $b, c \in X$; then $b \sim a$ and $c \sim a$ so $b \sim c$ by the transitivity and symmetry of $\sim$. Finally, $a \in X$ by the reflexivity of $\sim$.

Solution to Exercise 1.19

1. The unique function $\varnothing \to \{1\}$ is injective but not surjective.
2. The unique function $\{a, b\} \to \{1\}$ is surjective but not injective.
3. The second and third are not functions; the first and fourth are functions.

4. Neither the second nor third is "total." Moreover, the second one is not deterministic. The first one is a function which is not injective and not surjective. The fourth one is a function which is both injective and surjective.

Solution to Exercise 1.20

By Definition 1.17, a function $f \colon A \to \varnothing$ is a subset $F \subseteq A \times \varnothing$ such that, for all $a \in A$, there exists a unique $b \in \varnothing$ with $(a, b) \in F$. But there are no elements $b \overset{?}{\in} \varnothing$, so if F is to have the above property, there can be no $a \in A$ either; i.e. A must be empty.

Solution to Exercise 1.22

Below each partition, we draw a corresponding surjection out of $\{\bullet, \ast, \circ\}$:

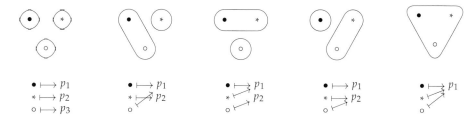

Solution to Exercise 1.33

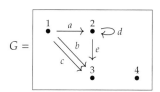

Arrow a	**Source** $s(a) \in V$	**Target** $t(a) \in V$
a	1	2
b	1	3
c	1	3
d	2	2
e	2	3

Solution to Exercise 1.35

The graph G from Exercise 1.33 is a strange Hasse diagram because it has two arrows $1 \to 3$ and a loop, both of which are "useless" from a preorder point-of-view. But that does not prevent our formula from working. The preorder $(P, \leq)$ is given by taking $P := V = \{1, 2, 3, 4\}$ and writing $p \leq q$ whenever there exists a path from p to q. So:

$$1 \leq 1, \quad 1 \leq 2, \quad 1 \leq 3, \quad 2 \leq 2, \quad 2 \leq 3, \quad 3 \leq 3, \quad 4 \leq 4.$$

Solution to Exercise 1.36

A collection of points, e.g. $\boxed{\bullet \quad \bullet \quad \bullet}$ is a Hasse diagram, namely for the discrete order, i.e. for the order where $x \leq y$ iff $x = y$.

Solution to Exercise 1.37

Let's write the five elements of X as

$$(\bullet)(\circ)(\ast), \quad (\bullet\circ)(\ast), \quad (\bullet\ast)(\circ), \quad (\bullet)(\circ\ast), \quad (\bullet \circ \ast).$$

Our job is to write down all 12 pairs of $x_1, x_2 \in X$ with $x_1 \leq x_2$. Here they are:

$$(\bullet)(\circ)(*) \leq (\bullet)(\circ)(*), \qquad (\bullet)(\circ)(*) \leq (\bullet\circ)(*), \qquad (\bullet)(\circ)(*) \leq (\bullet*)(\circ),$$

$$(\bullet)(\circ)(*) \leq (\bullet)(\circ*), \qquad (\bullet)(\circ)(*) \leq (\bullet \circ *), \qquad (\bullet\circ)(*) \leq (\bullet\circ)(*),$$

$$(\bullet\circ)(*) \leq (\bullet \circ *), \qquad (\bullet*)(\circ) \leq (\bullet*)(\circ), \qquad (\bullet*)(\circ) \leq (\bullet \circ *),$$

$$(\bullet)(\circ*) \leq (\bullet)(\circ*), \qquad (\bullet)(\circ*) \leq (\bullet \circ *), \qquad (\bullet \circ *) \leq (\bullet \circ *).$$

Solution to Exercise 1.39

The statement in the text is almost correct. It is correct to say that a discrete preorder is one where x and y are comparable if and only if $x = y$.

Solution to Exercise 1.41

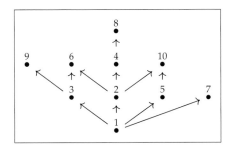

No, it is not a total order; for example $4 \not\leq 6$ and $6 \not\leq 4$.

Solution to Exercise 1.43

Yes, the usual $\leq$ ordering is a total order on $\mathbb{R}$: for every $a, b \in \mathbb{R}$ either $a \leq b$ or $b \leq a$.

Solution to Exercise 1.46

The Hasse diagrams for $\mathsf{P}(\varnothing)$, $\mathsf{P}\{1\}$, and $\mathsf{P}\{1, 2\}$ are

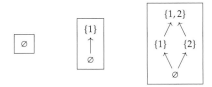

Solution to Exercise 1.48

The coarsest partition on S corresponds to the unique function $!: S \to \{1\}$. The finest partition on S corresponds to the identity function $\mathrm{id}_S : S \to S$.

Solution to Exercise 1.50

If X has the discrete preorder, then every subset U of X is an upper set: indeed, if $p \in U$, the only q such that $p \leq q$ is p itself, so q is definitely in U! This means that $\mathsf{U}(X)$ contains all subsets of X, so it's exactly the power set, $\mathsf{U}(X) = \mathsf{P}(X)$.

Solution to Exercise 1.52

The product preorder and its upper set preorder are:

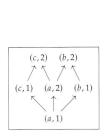

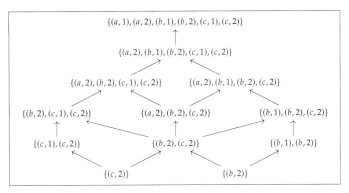

Solution to Exercise 1.58

With $X = \{0, 1, 2\}$, the Hasse diagram for $\mathsf{P}(X)$, the preorder $0 \le \cdots \le 3$, and the cardinality map between them are shown below:

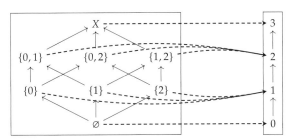

Solution to Exercise 1.60

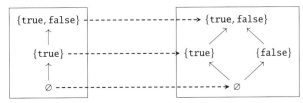

Solution to Exercise 1.61

1. Let $q \in \uparrow p$, and suppose $q \le q'$. Since $q \in \uparrow p$, we have $p \le q$. Thus by transitivity $p \le q'$, so $q' \in \uparrow p$. Thus $\uparrow p$ is an upper set.

2. Suppose $p \le q$ in P; this means that $q \le^{\mathrm{op}} p$ in P^{op}. We must show that $\uparrow q \subseteq \uparrow p$. Take any $q' \in \uparrow q$. Then $q \le q'$, so by transitivity $p \le q'$, and hence $q' \in \uparrow p$. Thus $\uparrow q \subseteq \uparrow p$.

3. Monotonicity of $\uparrow$ says that $p \le p'$ implies $\uparrow(p') \subseteq \uparrow(p)$. We must prove the other direction, that if $p \not\le p'$ then $\uparrow(p') \not\subseteq \uparrow(p)$. This is straightforward, since by reflexivity we always have $p' \in \uparrow(p')$, but if $p \not\le p'$, then $p' \notin \uparrow(p)$, so $\uparrow(p') \not\subseteq \uparrow(p)$.

4. The map $\uparrow\colon P^{\mathrm{op}} \to \mathsf{U}(P)$ can be depicted:

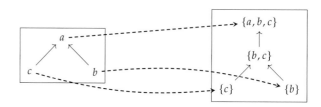

Solution to Exercise 1.62

Suppose $(P, \leq_P)$ is a discrete preorder and that $(Q, \leq_Q)$ is any preorder. We want to show that every function $f\colon P \to Q$ is monotone, i.e. that if $p_1 \leq_P p_2$ then $f(p_1) \leq_Q f(p_2)$. But in P we have $p_1 \leq_P p_2$ iff $p_1 = p_2$; that's what discrete means. If $p_1 \leq_P p_2$ then $p_1 = p_2$, so $f(p_1) = f(p_2)$, so $f(p_1) \leq f(p_2)$.

Solution to Exercise 1.64

Let $X = \mathbb{Z} = \{\ldots, -2, -1, 0, 1, \ldots\}$ be the set of all integers, and let $Y = \{n, z, p\}$; let $f\colon X \to Y$ send negative numbers to n, zero to z, and positive integers to p. This is surjective because all three elements of Y are hit.

We consider two partitions of Y, namely $P := (nz)(p)$ and $Q := (np)(z)$. Technically, these are notation for $\{\{n, z\}, \{p\}\}$ and $\{\{n, p\}, \{z\}\}$ as sets of disjoint subsets whose union is Y. Their pulled back partitions are $f^*P = (\ldots, -2, -1, 0)(1, 2, \ldots)$ and $f^*Q = (0)(\ldots, -2, -1, 1, 2, \ldots)$, or technically

$$f^*(P) = \{\{x \in \mathbb{Z} \mid x \leq 0\}, \{x \in \mathbb{Z} \mid x \geq 1\}\} \quad \text{and} \quad f^*(Q) = \{\{0\}, \{x \in \mathbb{Z} \mid x \neq 0\}\}.$$

Solution to Exercise 1.66

We have preorders $(P, \leq_P)$, $(Q, \leq_Q)$, and $(R, \leq_R)$, and we have monotone maps $f\colon P \to Q$ and $g\colon Q \to R$.

1. To see that id_P is monotone, we need to show that if $p_1 \leq_P p_2$ then $\mathrm{id}_P(p_1) \leq \mathrm{id}_P(p_2)$. But $\mathrm{id}_P(p) = p$ for all $p \in P$, so this is clear.
2. We have that $p_1 \leq_P p_2$ implies $f(p_1) \leq_Q f(p_2)$ and that $q_1 \leq_Q q_2$ implies $g(q_1) \leq_R g(q_2)$. By substitution, $p_1 \leq_P p_2$ implies $g(f(p_1)) \leq_R g(f(p_2))$ which is exactly what is required for $(f \,\mathring{\S}\, g)$ to be monotone.

Solution to Exercise 1.68

We need to show that if $(P, \leq_P)$ is both skeletal and dagger, then it is discrete. So suppose it is skeletal, i.e. $p_1 \leq p_2$ and $p_2 \leq p_1$ implies $p_1 = p_2$. And suppose it is dagger, i.e. $p_1 \leq p_2$ implies $p_2 \leq p_1$. Well then $p_1 \leq p_2$ implies $p_1 = p_2$, and this is exactly the definition of P being discrete.

Solution to Exercise 1.72

The map Φ from Section 1.1.1 took partitions of $\{\bullet, *, \circ\}$ and returned true or false based on whether or not $\bullet$ was in the same partition as $*$. We need to see that it's actually a monotone map

$\Phi \colon \mathsf{Prt}(\{\bullet, *, \circ\}) \to \mathbb{B}$. So suppose P, Q are partitions with $P \leq Q$; we need to show that if $\Phi(P) = \mathtt{true}$ then $\Phi(Q) = \mathtt{true}$.

By definition $P \leq Q$ means that P is finer than Q: i.e. P differentiates more stuff, and Q lumps more stuff together. Technically, $x \sim_P y$ implies $x \sim_Q y$ for all $x, y \in \{\bullet, *, \circ\}$. Applying this to $\bullet, *$ gives the result.

Solution to Exercise 1.74

Given a function $f \colon P \to Q$, we have $f^* \colon \mathsf{U}(Q) \to \mathsf{U}(P)$ given by $U \mapsto f^{-1}(U)$. But upper sets in Q are classified by monotone maps $u \colon Q \to \mathbb{B}$, and similarly for P; our job is to show that $f^*(U)$ is given by composing the classifier u with f.

Given an upper set $U \subseteq Q$, let $u \colon Q \to \mathbb{B}$ be the corresponding monotone map, which sends $q \mapsto \mathtt{true}$ iff $q \in U$. Then $(f \,\mathring{,}\, u) \colon P \to \mathbb{B}$ sends $p \mapsto \mathtt{true}$ iff $f(p) \in U$; it corresponds to the upper set $\{p \in P \mid f(p) \in U\}$ which is exactly $f^{-1}(U)$.

Solution to Exercise 1.75

1. 0 is a lower bound for $S = \{\frac{1}{n+1} \mid n \in \mathbb{N}\}$ because $0 \leq \frac{1}{n+1}$ for any $n \in \mathbb{N}$.
2. Suppose that b is a lower bound for S; we want to see that $b \leq 0$. If one believes to the contrary that $0 < b$, then consider $1/b$; it is a real number so we can find a natural number n that's bigger $1/b < n < n + 1$. This implies $1 < b(n + 1)$ and hence $\frac{1}{n+1} < b$, but that is a contradiction of b being a lower bound for S. The false believer is defeated!

Solution to Exercise 1.80

We have a preorder $(P, \leq)$, an element $p \in P$, and a subset $A = \{p\}$ with one element.

1. To see that $\bigwedge A \cong p$, we need to show that $p \leq a$ for all $a \in A$ and that if $q \leq a$ for all $a \in A$ then $q \leq p$. But the only $a \in A$ is $a = p$, so both are obvious.
2. We know p is a meet of A, so if q is also a meet of A then $q \leq a$ for all $a \in A$ so $q \leq p$; similarly $p \leq a$ for all $a \in A$, so $p \leq q$. Then by definition we have $p \cong q$, and since $(P, \leq)$ is a partial order, $p = q$.
3. The analogous facts are true when $\bigwedge$ is replaced by $\bigvee$; the only change in the argument is to replace $\leq$ by $\geq$ and "meet" by "join" everywhere.

Solution to Exercise 1.85

The meet of 4 and 6 is the highest number in the order that divides both of them; the numbers dividing both are 1 and 2, and 2 is higher, so $4 \wedge 6 = 2$. Similar reasoning shows that $4 \vee 6 = 12$. The meet is the "greatest common divisor" and the join is the "least common multiple," and this holds up for all pairs $m, n \in \mathbb{N}$ not just 4, 6. (Here $\gcd(0, a)$ is defined to be 0.)

Solution to Exercise 1.89

Since f is monotone, the facts that $a \leq a \vee b$ and $b \leq a \vee b$ imply that $f(a) \leq f(a \vee b)$ and $f(b) \leq f(a \vee b)$. But, by definition of join, $f(a) \vee f(b)$ is the largest element with that property, so $f(a) \vee f(b) \leq f(a \vee b)$, as desired.

Solution to Exercise 1.92

By analogy with Example 1.91, the right adjoint for $(3 \times -)$ should be $\lfloor -/3 \rfloor$. But to prove this is correct, we must show that for any any $r \in \mathbb{R}$ and $z \in \mathbb{Z}$ we have $z \leq \lfloor r/3 \rfloor$ iff $3 * z \leq r$.

Suppose the largest integer below $r/3$ is $z' := \lfloor r/3 \rfloor$. Then $z \le z'$ implies $3 * z \le 3 * z' \le 3 * r/3 = r$, giving one direction. For the other, suppose $3 * z \le r$. Then dividing both sides by 3, we have $z = 3 * z/3 \le r/3$. Since z is an integer below $r/3$ it is below $\lfloor r/3 \rfloor$ because $\lfloor r/3 \rfloor$ is the greatest integer below $r/3$, and we are done.

Solution to Exercise 1.93

1. We need to check that for all nine pairs $\{(p, q) \mid 1 \le p \le 3 \text{ and } 1 \le q \le 3\}$ we have $f(p) \le q$ iff $p \le g(q)$, where f and g are the functions shown here:

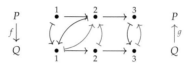

   When $p = q = 1$ we have $f(p) = 1$ and $g(q) = 2$, so both $f(p) = 1 \le 1 = q$ and $p = 1 \le g(q)$; it works! The same sort of story happens when (p, q) is $(1, 2)$, $(1, 3)$, $(2, 1)$, $(2, 2)$, $(2, 3)$, and $(3, 3)$. A different story happens for $p = 3, q = 1$ and $p = 3, q = 2$. In those cases $f(p) = 3$ and $g(q) = 2$, and neither inequality holds: $f(p) \not\le q$ and $p \not\le g(q)$. But that's fine, we still have $f(p) \le q$ iff $p \le g(q)$ in all nine cases, as desired.

2.

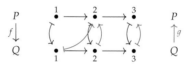

   Here f is *not* left adjoint to g because $f(2) \not\le 1$ but $2 \le g(1)$.

Solution to Exercise 1.95

1. Let's suppose we have a monotone map $L \colon \mathbb{Z} \to \mathbb{R}$ that's left adjoint to $\lceil -/3 \rceil$ and see what happens. Writing $C(r) := \lceil r/3 \rceil$, then for all $z \in \mathbb{Z}$ and $r \in \mathbb{R}$ we have $L(z) \le r$ iff $z \le C(r)$ by definition of adjunction. So take $z = 1$ and $r = .01$; then $\lceil r/3 \rceil = 1$ so $z \le C(r)$, and hence $L(z) \le r$, i.e. $L(1) \le 0.01$. In the same way $L(1) \le r$ for all $r > 0$, so $L(1) \le 0$. By definition of adjunction $1 \le C(0) = \lceil 0/3 \rceil = 0$, a contradiction.

2. There's no left adjoint, because starting with an arbitrary one, we derived a contradiction.

Solution to Exercise 1.97

We have $S = \{1, 2, 3, 4\}$, $T = \{12, 3, 4\}$, and $g \colon S \to T$ the "obvious" function between them; see Example 1.96. Take c_1, c_2, c_3, c_4 to be the following partitions:

$$c_1 = \quad c_2 = \quad c_3 = \quad c_4 =$$

Then the induced partitions $g_!(c_1)$, $g_!(c_2)$, $g_!(c_3)$, and $g_!(c_4)$ on T are:

$$g_!(c_1) = \quad g_!(c_2) = \quad g_!(c_3) = \quad g_!(c_4) =$$

Solution to Exercise 1.99

Here are the partitions on $S = \{1, 2, 3, 4\}$ induced via g^* by the five partitions on $T = \{12, 3, 4\}$:

Solution to Exercise 1.100

1. We choose the following partition c on S and compute its pushforward $g_!(c)$:

2. Let d be the partition as shown, which was chosen to be coarser than $g_!(c)$.

3. Let e be the partition as shown, which was chosen to *not* be coarser than $g_!(c)$.

4. Here are $g^*(d)$ and $g^*(e)$:

5. Comparing c, the left-hand partition in part 1., with $g^*(d)$ and $g^*(e)$, we indeed have $c \leq g^*(d)$ but $c \not\leq g^*(e)$, as desired.

Solution to Exercise 1.102

Suppose P and Q are preorders, and that $f \colon P \leftrightarrows Q \colon g$ are monotone maps.

1. Suppose f is left adjoint to g. By definition this means $f(p) \leq q$ iff $p \leq g(q)$, for all $p \in P$ and $q \in Q$. Then starting with the reflexivity fact $g(q) \leq g(q)$, the definition with $p := g(q)$ gives $f(g(q)) \leq q$ for all q.

2. Suppose that $p \leq g(f(p))$ and $f(g(q)) \leq q$ for all $p \in P$ and $q \in Q$. We first want to show that $p \leq g(q)$ implies $f(p) \leq q$, so assume $p \leq g(q)$. Then applying the monotone map f to both sides, we have $f(p) \leq f(g(q))$, and then by transitivity $f(g(q)) \leq q$ implies $f(p) \leq q$, as desired. The other direction is similar.

Solution to Exercise 1.103

1. Suppose that $f : P \to Q$ has two right adjoints, $g, g' : Q \to P$. We want to show that $g(q) \cong g'(q)$ for all $q \in Q$. We will prove $g(q) \le g'(q)$; the inequality $g'(q) \le g(q)$ is similar. To do this, we use the fact that $p \le g'(f(p))$ and $f(g(q)) \le q$ for all p, q by Eq. (1.7). Then the trick is to reason as follows:

$$g(q) \le g'(f(g(q))) \le g'(q).$$

2. It is the same for left adjoints.

Solution to Exercise 1.105

Suppose $f : P \to Q$ is left adjoint to $g : Q \to P$. Let $A \subseteq P$ be any subset and let $j := \bigvee A$ be its join. Then since f is monotone $f(a) \le f(j)$ for all $a \in A$, so $f(j)$ is an upper bound for the set $f(A)$. We want to show that it is the least upper bound, so take any other upper bound b for $f(A)$, meaning we have $f(a) \le b$ for all $a \in A$. Then by definition of adjunction, we also have $a \le g(b)$ for all $a \in A$. By definition of join, we have $j \le g(b)$. Again by definition of adjunction $f(j) \le b$, as desired.

Solution to Exercise 1.107

We want to show that in the following picture, g is really right adjoint to f:

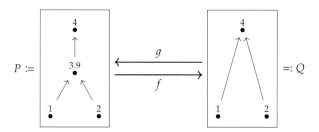

Here g preserves labels and f rounds 3.9 to 4.

There are 12 tiny things to check: for each $p \in P$ and $q \in Q$, we need to see that $f(p) \le q$ iff $p \le g(q)$.

p	q	$f(p)$	$g(q)$	$f(p) \le^? q$	$p \le^? g(q)$	**Same?**
1	1	1	1	yes	yes	yes!
1	2	1	2	no	no	yes!
1	4	1	4	yes	yes	yes!
2	1	2	1	no	no	yes!
2	2	2	2	yes	yes	yes!
2	4	2	4	yes	yes	yes!
3.9	1	4	1	no	no	yes!
3.9	2	4	2	no	no	yes!
3.9	4	4	4	yes	yes	yes!
4	1	4	1	no	no	yes!
4	2	4	2	no	no	yes!
4	4	4	4	yes	yes	yes!

Solution to Exercise 1.110

Consider the function shown below, which "projects straight down":

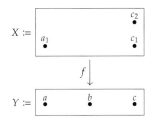

1. Let $B_1 := \{a, b\}$ and $B_2 := \{c\}$. Then $f^*(B_1) = \{a_1\}$ and $f^*(B_2) = \{c_1, c_2\}$.
2. Let $A_1 := \varnothing$ and $A_2 := \{a_1, c_1\}$. Then $f_!(A_1) = \varnothing$ and $f_!(A_2) = \{a, c\}$.
3. With the same A_1 and A_2, we compute $f_*(A_1) = \{b\}$ and $f_*(A_2) = \{a, b\}$.

Solution to Exercise 1.111

Assume $f : P \to Q$ is left adjoint to $g : Q \to P$.

1. By Proposition 1.101 we have that $p \le g(f(p))$, and of course $g(f(p))$ and $(f \,\mathring{\circ}\, g)(p)$ mean the same thing.
2. We want to show that $g(f(g(f(p)))) \le g(f(p))$ and $g(f(p)) \le g(f(g(f(p))))$ for all p. The latter is just the fact that $p' \le g(f(p'))$ for any p', applied with $g(f(p))$ in place of p'. The former uses that $f(g(q)) \le q$, with $f(p)$ substituted for q: this gives $f(g(f(p))) \le f(p)$, and then we apply g to both sides.

Solution to Exercise 1.116

We denote tuples (a, b) by ab for space reasons. So the relation $\{(1, 1), (1, 2), (2, 1)\}$ will be denoted $\{11, 12, 21\}$.

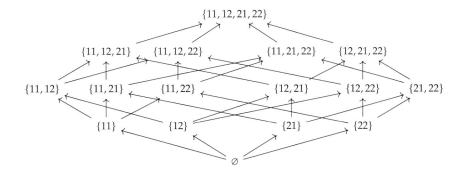

Solution to Exercise 1.117

Let $S := \{1, 2, 3\}$.

1. Let $\le$ be the preorder with $1 \le 2$, and of course $1 \le 1$, $2 \le 2$, and $3 \le 3$. Then $U(\le) = \{(1, 1), (1, 2), (2, 2), (3, 3)\}$.
2. Let $Q := \{(1, 1)\}$ and $Q' := \{(2, 1)\}$.

3. The closure $Cl(Q)$ of Q is the smallest preorder containing $(1, 1)$, which is $Cl(Q) = \{(1, 1), (2, 2), (3, 3)\}$. Similarly, $Cl(Q') = \{(1, 1), (2, 1), (2, 2), (3, 3)\}$. It is easy to see that $Cl(Q) \sqsubseteq \leq$ because every ordered pair in $Cl(Q)$ is also in $\leq$.

4. It is easy to see that $Cl(Q') \not\sqsubseteq \leq$ because the ordered pair $(2, 1)$ is in $Cl(Q')$ but is not in $\leq$.

A.2   Solutions for Chapter 2

Solution to Exercise 2.4

The expert is right! The proposal violates property (a) when $x_1 = -1$, $x_2 = 0$, $y_1 = -1$, and $y_2 = 1$. Indeed $-1 \leq -1$ and $0 \leq 1$, but $-1 * 0 = 0 \not\leq -1 = -1 * 1$.

Solution to Exercise 2.6

To check that $(\mathbf{Disc}(M), =, *, e)$ is a symmetric monoidal preorder, we need to check our proposed data obeys conditions (a)–(d) of Definition 2.1. Condition (a) just states the tautology that $x_1 \otimes x_2 = x_1 \otimes x_2$, conditions (b) and (c) are precisely the equations Eq. (2.2), and (d) is the commutativity condition. So we're done. We leave it to you to decide whether we were telling the truth when we said it was easy.

Solution to Exercise 2.10

1. Here is a line-by-line proof, where we write the reason for each step in parentheses on the right. Recall we call the properties (a) and (c) in Definition 2.1 *monotonicity* and *associativity*, respectively.

$$
\begin{aligned}
t + u &\leq (v + w) + u && \text{(monotonicity, } t \leq v + w, u \leq u) \\
&= v + (w + u) && \text{(associativity)} \\
&\leq v + (x + z) && \text{(monotonicity, } v \leq v, w + u \leq x + v) \\
&= (v + x) + z && \text{(associativity)} \\
&\leq y + z. && \text{(monotonicity, } v + x \leq y, z \leq z)
\end{aligned}
$$

2. We use reflexivity when we assert that $u \leq u$, $v \leq v$ and $z \leq z$, and use transitivity to assert that the above sequence of inequalities implies the single inequality $t + u \leq y + z$.

3. We know that the symmetry axiom is not necessary because no pair of wires cross.

Solution to Exercise 2.11

Condition (a), monotonicity, says that if $x \to y$ and $z \to w$ are reactions, then $x + z \to y + w$ is a reaction. Condition (b), unitality, holds as 0 represents having no material, and adding no material to some other material does not change it. Condition (c), associativity, says that when combining three collections x, y, and z of molecules it doesn't matter whether you combine x and y and then z, or combine x with y already combined with z. Condition (d), symmetry, says that combining x with y is the same as combining y with x. All these are true in our model of chemistry, so $(Mat, \to, 0, +)$ forms a symmetric monoidal preorder.

Solution to Exercise 2.13

The monoidal unit must be `false`. The symmetric monoidal preorder does satisfy the rest of the conditions; this can be verified just by checking all cases.

Solution to Exercise 2.15
The monoidal unit is the natural number 1. Since we know that $(\mathbb{N}, \leq)$ is a preorder, we just need to check that $*$ is monotonic, associative, unital with 1, and symmetric. These are all familiar facts from arithmetic.

Solution to Exercise 2.17
This proposed monoidal product is not monotonic: we have $1|1$ and $1|2$, but $(1+1) \nmid (1+2)$.

Solution to Exercise 2.18
1.

min	no	maybe	yes
no	no	no	no
maybe	no	maybe	maybe
yes	no	maybe	yes

2. We need to show that (a) if $x \leq y$ and $z \leq w$, then $\min(x, z) \leq \min(y, w)$, (b) $\min(x, \text{yes}) = x = \min(\text{yes}, x)$, (c) $\min(\min(x, y), z) = \min(x, \min(y, z))$, and (d) $\min(x, y) = \min(y, x)$. The most straightforward way is to just check all cases.

Solution to Exercise 2.19
Yes, $(\mathsf{P}(S), \leq, S, \cap)$ is a symmetric monoidal preorder.

Solution to Exercise 2.20
Depending on your mood, you might come up with either of the following. First, we could take the monoidal unit to be some statement `true` that is true for all natural numbers, such as "n is a natural number." We can pair this unit with the monoidal product $\wedge$, which takes statements P and Q and makes the statement $P \wedge Q$, where $(P \wedge Q)(n)$ is true if $P(n)$ and $Q(n)$ are true, and false otherwise. Then $(\mathrm{Prop}^{\mathbb{N}}, \leq, \texttt{true}, \wedge)$ forms a symmetric monoidal preorder.

Another option is to take define `false` to be some statement that is false for all natural numbers, such as "$n + 10 \leq 1$" or "n is made of cheese." We can also define $\vee$ such that $(P \vee Q)(n)$ is true if and only if at least one of $P(n)$ and $Q(n)$ is true. Then $(\mathrm{Prop}^{\mathbb{N}}, \leq, \texttt{false}, \vee)$ forms a symmetric monoidal preorder.

Solution to Exercise 2.23
These conditions have nothing to do with the order in $\mathcal{X}^{\mathrm{op}}$: they simply state that $I \otimes x = x = x \otimes I$, and $(x \otimes y) \otimes z = x \otimes (y \otimes z)$, and $x \otimes y = y \otimes x$. Since these are true in $\mathcal{X}$, they are true in $\mathcal{X}^{\mathrm{op}}$.

Solution to Exercise 2.24
1. The preorder **Cost**^{op} has underlying set $[0, \infty]$, and the usual increasing order on real numbers $\leq$ as its order.
2. Its monoidal unit is 0.
3. Its monoidal product is $+$.

Solution to Exercise 2.27

1. The map g is monotonic as $g(\texttt{false}) = \infty \geq 0 = g(\texttt{true})$, satisfies condition (a) since $0 \geq 0 = g(\texttt{true})$, and satisfies condition (b) since

$$
\begin{aligned}
g(\texttt{false}) + g(\texttt{false}) &= \infty + \infty \geq \infty = g(\texttt{false} \wedge \texttt{false}), \\
g(\texttt{false}) + g(\texttt{true}) &= \infty + 0 \geq \infty = g(\texttt{false} \wedge \texttt{true}), \\
g(\texttt{true}) + g(\texttt{false}) &= 0 + \infty \geq \infty = g(\texttt{true} \wedge \texttt{false}), \\
g(\texttt{true}) + g(\texttt{true}) &= 0 + 0 \geq 0 = g(\texttt{true} \wedge \texttt{true}).
\end{aligned}
$$

2. Since all the inequalities regarding (a) and (b) above are in fact equalities, g is a strict monoidal monotone.

Solution to Exercise 2.28

The answer to all these questions is yes: d and u are both strict monoidal monotones. Here is one way to interpret this. The function d asks "is $x = 0$?" This is monotonic, 0 is 0, and the sum of two elements of $[0, \infty]$ is 0 if and only if they are both 0. The function u asks "is x finite?" Similarly, this is monotonic, 0 is finite, and the sum of x and y is finite if and only if x and y are both finite.

Solution to Exercise 2.29

1. Yes, multiplication is monotonic in $\leq$, unital with respect to 1, associative, and symmetric, so $(\mathbb{N}, \leq, 1, *)$ is a monoidal preorder. We also met this preorder in Exercise 2.15.
2. The map $f(n) = 1$ for all $n \in \mathbb{N}$ defines a monoidal monotone $f : (\mathbb{N}, \leq, 0, +) \to (\mathbb{N}, \leq, 1, *)$.
3. $(\mathbb{Z}, \leq, *, 1)$ is not a monoidal preorder because $*$ is not monotone. Indeed $-1 \leq 0$ but $(-1 * -1) \not\leq (0 * 0)$.

Solution to Exercise 2.33

1. Let $(P, \leq)$ be a preorder. How is this a **Bool**-category? Following Example 2.31, we can construct a **Bool**-category $\mathcal{X}_P$ with P as its set of objects, and with $\mathcal{X}_P(p, q) := \texttt{true}$ if $p \leq q$, and $\mathcal{X}_P(p, q) := \texttt{false}$ otherwise. How do we turn this back into a preorder? Following the proof of Theorem 2.32, we construct a preorder with underlying set $\mathrm{Ob}(\mathcal{X}_P) = P$, and with $p \leq q$ if and only if $\mathcal{X}_P(p, q) = \texttt{true}$. This is precisely the preorder $(P, \leq)$!
2. Let $\mathcal{X}$ be a **Bool**-category. By the proof of Theorem 2.32, we construct a preorder $(\mathrm{Ob}(\mathcal{X}), \leq)$, where $x \leq y$ if and only if $\mathcal{X}(x, y) = \texttt{true}$. Then, following our generalization of Example 2.31 in part 1, we construct a **Bool**-category $\mathcal{X}'$ whose set of objects is $\mathrm{Ob}(\mathcal{X})$, and such that $\mathcal{X}'(x, y) = \texttt{true}$ if and only if $x \leq y$ in $(\mathrm{Ob}(\mathcal{X}), \leq)$. But, by construction, this means $\mathcal{X}'(x, y) = \mathcal{X}(x, y)$. So we get back the **Bool**-category we started with.

Solution to Exercise 2.35

The distance $d(\mathrm{US}, \mathrm{Spain})$ is bigger: the distance from, for example, San Diego to anywhere in Spain is bigger than the distance from anywhere in Spain to New York City.

Solution to Exercise 2.38

The difference between a Lawvere metric space – that is, a category enriched over $([0, \infty], \geq, 0, +)$ – and a category enriched over $(\mathbb{R}_{\geq 0}, \geq, 0, +)$ is that, in the latter, infinite distances are not allowed between points. You might thus call the latter a finite-distance Lawvere metric space.

Solution to Exercise 2.39
The table of distances for X is

$d(\nearrow)$	A	B	C	D
A	0	6	3	11
B	2	0	5	5
C	5	3	0	8
D	11	9	6	0

Solution to Exercise 2.40
The matrix of edge weights of X is

$$M_X = \begin{array}{c|cccc} \nearrow & A & B & C & D \\ \hline A & 0 & \infty & 3 & \infty \\ B & 2 & 0 & \infty & 5 \\ C & \infty & 3 & 0 & \infty \\ D & \infty & \infty & 6 & 0 \end{array}$$

Solution to Exercise 2.41
A **NMY**-category $\mathcal{X}$ is a set X together with, for all pairs of elements x, y in X, a value $\mathcal{X}(x, y)$ equal to no, maybe, or yes. Moreover, we must have $\mathcal{X}(x, x) = $ yes and $\min(\mathcal{X}(x, y), \mathcal{X}(y, z)) \leq \mathcal{X}(x, z)$ for all x, y, z. So a **NMY**-category can be thought of as set of points together with an statement – no, maybe, or yes – of whether it is possible to get from one point to another. In particular, it's always possible to get to a point if you're already there, and it's at least as possible to get from x to z as it is to get from x to y and then y to z.

Solution to Exercise 2.42
Here is one way to do this task.

1.

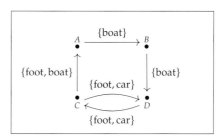

2. The corresponding $\mathcal{M}$-category, call it $\mathcal{X}$, has hom-objects:

$\mathcal{X}(\nearrow)$	A	B	C	D
A	M	{boat}	$\varnothing$	{boat}
B	$\varnothing$	M	$\varnothing$	{boat}
C	{foot, boat}	{boat}	M	M
D	{foot}	$\varnothing$	{foot, car}	M

For example, to compute the hom-object $X(C, D)$, we notice that there are two paths: $C \to A \to B \to D$ and $C \to D$. For the first path, the intersection is the set {boat}. For the second path, the intersection in the set {foot, car}. Their union, and thus the hom-object $X(C, D)$, is the entire set M.

This computation contains the key for why X is a M-category: by taking the union over all paths, we ensure that $X(x, y) \cap X(y, z) \subseteq X(x, z)$ for all x, y, z.

3. The person's interpretation looks right to us.

Solution to Exercise 2.43

1.

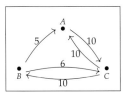

2. The matrix M with (x, y)th entry equal to the maximum, taken over all paths p from x to y, of the minimum edge label in p is

$M(\nearrow)$	A	B	C
A	∞	10	10
B	6	∞	6
C	10	10	∞

3. This is a matrix of hom-objects for a W-category since the diagonal values are all equal to the monoidal unit ∞, and because $\min(M(x, y), M(y, z)) \leq M(x, y)$ for all $x, y, z \in \{A, B, C\}$.

4. One interpretation is as a weight limit (not to be confused with "weighted limit," a more advanced categorical notion), for example for trucking cargo between cities. The hom-object indexed by a pair of points (x, y) describes the maximum cargo weight allowed on that route. There is no weight limit on cargo that remains at some point x, so the hom-object from x to x is always infinite. The maximum weight that can be trucked from x to z is always at least the minimum of that that can be trucked from x to y and then y to z. (It is "at least" this much because there may be some other, better route that does not pass through y.)

Solution to Exercise 2.46

This preorder describes the "is a part of" relation. That is, $x \leq y$ when $d(x, y) = 0$, which happens when x is a part of y. So Boston is a part of the US, and Spain is a part of Spain, but the US is not a part of Boston.

Solution to Exercise 2.47

1. Recall the monoidal monotones d and u from Exercise 2.28. The function f is equal to d; let g be equal to u.

2. Let X be the Lawvere metric space with two objects A and B, such that $d(A, B) = d(B, A) =$

5. Then we have $X_f = \boxed{\overset{A}{\bullet} \quad \overset{B}{\bullet}}$ while $X_g = \boxed{\overset{A}{\bullet} \longrightarrow \overset{B}{\bullet}}$.

Solution to Exercise 2.52

1. An extended metric space is a Lawvere metric space that obeys in addition the properties (b) if $d(x, y) = 0$ then $x = y$, and (c) $d(x, y) = d(y, x)$ of Definition 2.34. Let's consider the dagger condition first. It says that the identity function to the opposite **Cost**-category is a functor, and so for all x, y we must have $d(x, y) \leq d(y, x)$. But this means also that $d(y, x) \leq d(x, y)$, and so $d(x, y) = d(y, x)$. This is exactly property (c).

   Now let's consider the skeletality condition. This says that if $d(x, y) = 0$ and $d(y, x) = 0$, then $x = y$. Thus when we have property (c), $d(x, y) = d(y, x)$, this is equivalent to property (b). Thus skeletal dagger **Cost**-categories are the same as extended metric spaces!

2. Recall from Exercise 1.68 that skeletal dagger preorders are sets. The analogy "preorders are to sets as Lawvere metric spaces are to extended metric spaces" is thus the observation that just as extended metric spaces are skeletal dagger **Cost**-categories, sets are skeletal dagger **Bool**-categories.

Solution to Exercise 2.54

1. Let $(x, y) \in \mathcal{X} \times \mathcal{Y}$. Since $\mathcal{X}$ and $\mathcal{Y}$ are $\mathcal{V}$-categories, we have $I \leq \mathcal{X}(x, x)$ and $I \leq \mathcal{Y}(y, y)$. Thus $I = I \otimes I \leq \mathcal{X}(x, x) \otimes \mathcal{Y}(y, y) = (\mathcal{X} \times \mathcal{Y})\big((x, y), (x, y)\big)$.

2. Using the definition of product hom-objects, and the symmetry and monotonicity of $\otimes$ we have

$$(\mathcal{X} \times \mathcal{Y})\big((x_1, y_1), (x_2, y_2)\big) \otimes (\mathcal{X} \times \mathcal{Y})\big((x_2, y_2), (x_3, y_3)\big)$$

$$= \mathcal{X}(x_1, x_2) \otimes \mathcal{Y}(y_1, y_2) \otimes \mathcal{X}(x_2, x_3) \otimes \mathcal{Y}(y_2, y_3)$$

$$= \mathcal{X}(x_1, x_2) \otimes \mathcal{X}(x_2, x_3) \otimes \mathcal{Y}(y_1, y_2) \otimes \mathcal{Y}(y_2, y_3)$$

$$\leq \mathcal{X}(x_1, x_3) \otimes \mathcal{Y}(y_1, y_3)$$

$$= (\mathcal{X} \times \mathcal{Y})\big((x_1, y_1), (x_3, y_3)\big).$$

3. In particular, we use the symmetry to conclude that $\mathcal{Y}(y_1, y_2) \otimes \mathcal{X}(x_2, x_3) = \mathcal{X}(x_2, x_3) \otimes \mathcal{Y}(y_1, y_2)$.

Solution to Exercise 2.56

We just apply Definition 2.53(ii): $(\mathbb{R} \times \mathbb{R})\big((5, 6), (-1, 4)\big) = \mathbb{R}(5, -1) + \mathbb{R}(6, 4) = 6 + 2 = 8$.

Solution to Exercise 2.59

1. The function $- \otimes v : V \to V$ is monotone, because if $u \leq u'$ then $u \otimes v \leq u' \otimes v$ by the monotonicity condition (a) in Definition 2.1.

2. Let $a := (v \multimap w)$ in Eq. (2.23). The right-hand side is thus $(v \multimap w) \leq (v \multimap w)$, which is true by reflexivity. Thus the left-hand side is true too. This gives $((v \multimap w) \otimes v) \leq w$.

3. Let $u \leq u'$. Then, using part 2, $(v \multimap u) \otimes v \leq u \leq u'$. Applying Eq. (2.23), we thus have $(v \multimap u) \leq (v \multimap u')$. This shows that the map $(v \multimap -) : V \to V$ is monotone.

4. Eq. (2.23) is exactly the adjointness condition from Definition 1.90, except for the fact that we do not know $(- \otimes v)$ and $(v \multimap -)$ are monotone maps. We proved this, however, in parts 1 and 3 above.

Solution to Exercise 2.61

We need to find the hom-element. This is given by implication. That is, define the function $x \Rightarrow y$ by the table

$\Rightarrow$	false	true
false	true	true
true	false	true

Then $(a \wedge v) \leq w$ if and only if $a \leq (v \Rightarrow w)$. Indeed, if $v = \mathtt{false}$ then $a \wedge \mathtt{false} = \mathtt{false}$, and so the left-hand side is always true. But $(\mathtt{false} \Rightarrow w) = \mathtt{true}$, so the right-hand side is always true too. If $v = \mathtt{true}$, then $a \wedge \mathtt{true} = a$ so the left-hand side says $a \leq w$. But $(\mathtt{true} \Rightarrow w) = w$, so the right-hand side is the same. Thus $\Rightarrow$ defines a hom-element as per Eq. (2.23).

Solution to Exercise 2.68

1a. In **Bool**, $(\bigvee \varnothing) = \mathtt{false}$, the least element.
1b. In **Cost**, $(\bigvee \varnothing) = \infty$. This is because we use the opposite order $\geq$ on $[0, \infty]$, so $\bigvee \varnothing$ is the greatest element of $[0, \infty]$. Note that in this case our convention from Definition 2.66, where we denote $(\bigvee \varnothing) = 0$, is a bit confusing! Beware!
2a. In **Bool**, $x \vee y$ is the usual join, OR.
2b. In **Cost**, $x \vee y$ is the minimum $\min(x, y)$. Again because we use the opposite order on $[0, \infty]$, the join is the greatest number less than or equal to x and y.

Solution to Exercise 2.69

We showed in Exercise 2.61 that **Bool** is symmetric monoidal closed, and in Exercise 1.4 and Example 1.83 that the join is given by the OR operation $\vee$. Thus **Bool** is a quantale.

Solution to Exercise 2.70

Yes, the power-set monoidal preorder $(\mathsf{P}(S), \subseteq, S, \cap)$ is a quantale. The hom-object $B \multimap C$ is given by $\overline{B} \cup C$, where $\overline{B}$ is the *complement* of B: it contains all elements of S not contained in B. To see that this satisfies Eq. (2.23), note that if $(A \cap B) \subseteq C$, then

$$A = (A \cap \overline{B}) \cup (A \cap B) \subseteq \overline{B} \cup C.$$

On the other hand, if $A \subseteq (\overline{B} \cup C)$, then

$$A \cap B \subseteq (\overline{B} \cup C) \cap B = (\overline{B} \cap B) \cup (C \cap B) = C \cap B \subseteq C.$$

So $(\mathsf{P}(S), \subseteq, S, \cap)$ is monoidal closed. Furthermore, joins are given by union of subobjects, so it is a quantale.

Solution to Exercise 2.77

The (2×2)-identity matrices for $(\mathbb{N}, \leq, 1, *)$, **Bool**, and **Cost** are respectively

$$\begin{pmatrix} 1 & 0 \\ 0 & 1 \end{pmatrix}, \quad \begin{pmatrix} \mathtt{true} & \mathtt{false} \\ \mathtt{false} & \mathtt{true} \end{pmatrix}, \quad \text{and} \quad \begin{pmatrix} 0 & \infty \\ \infty & 0 \end{pmatrix}.$$

Solution to Exercise 2.78

1. We first use Proposition 2.64 (2) and symmetry to show that, for all $v \in V$, $0 \otimes v = 0$,

$$0 \otimes v \cong v \otimes 0 \cong \left(v \otimes \bigvee_{a \in \varnothing} a \right) = \bigvee_{a \in \varnothing} (v \otimes a) = 0.$$

Then we may just follow the definition in Eq. (2.26):

$$I_X * M(x, y) = \bigvee_{x' \in X} I_X(x, x') \otimes M(x', y)$$

$$= (I_X(x, x) \otimes M(x, y)) \vee \left(\bigvee_{x' \in X, x' \neq x} I_X(x, x') \otimes M(x', y) \right)$$

$$= (I \otimes M(x, y)) \vee \left(\bigvee_{x' \in X, x' \neq x} 0 \otimes M(x', y) \right)$$

$$= M(x, y) \vee 0 = M(x, y).$$

2. This again follows from Proposition 2.64 (2) and symmetry, making use also of the associativity of $\otimes$:

$$((M * N) * P)(w, z) = \bigvee_{y \in Y} \left(\bigvee_{x \in X} M(w, x) \otimes N(x, y) \right) \otimes P(y, z)$$

$$\cong \bigvee_{y \in Y, x \in X} M(w, x) \otimes N(x, y) \otimes P(y, z)$$

$$\cong \bigvee_{x \in X} M(w, x) \otimes \left(\bigvee_{y \in Y} N(x, y) \otimes P(y, z) \right)$$

$$= (M * (N * P))(w, z).$$

Solution to Exercise 2.79
We have the matrices

$$M_X = \begin{pmatrix} 0 & \infty & 3 & \infty \\ 2 & 0 & \infty & 5 \\ \infty & 3 & 0 & \infty \\ \infty & \infty & 6 & 0 \end{pmatrix}, \quad M_X^2 = \begin{pmatrix} 0 & 6 & 3 & \infty \\ 2 & 0 & 5 & 5 \\ 5 & 3 & 0 & 8 \\ \infty & 9 & 6 & 0 \end{pmatrix},$$

$$M_X^3 = M_X^4 = \begin{pmatrix} 0 & 6 & 3 & 11 \\ 2 & 0 & 5 & 5 \\ 5 & 3 & 0 & 8 \\ 11 & 9 & 6 & 0 \end{pmatrix}.$$

A.3 Solutions for Chapter 3

Solution to Exercise 3.1
There are five non-ID columns in Eq. (3.1) and five arrows in Eq. (3.2). This is not a coincidence: there is always one arrow for every non-ID column.

Solution to Exercise 3.4
To do this precisely, we should define concatenation technically. If $G = (V, A, s, t)$ is a graph, define a path in G to be a tuple of the form $(v, a_1, \dots, a_n)$ where $v \in V$ is a vertex, $s(a_1) = v$,

and $t(a_i) = s(a_{i+1})$ for all $i \in \{1, \ldots, n-1\}$; the length of this path is n, and this definition makes sense for any $n \in \mathbb{N}$. We say that the source of p is $s(p) := v$ and the target of p is defined to be

$$t(p) := \begin{cases} v & \text{if } n = 0, \\ t(a_n) & \text{if } n \geq 1. \end{cases}$$

Two paths $p = (v, a_1, \ldots, a_m)$ and $q = (w, b_1, \ldots, b_n)$ can be concatenated if $t(p) = s(q)$, in which case the concatenated path $p \, \mathring{9} \, q$ is defined to be

$$(p \, \mathring{9} \, q) := (v, a_1, \ldots, a_m, b_1, \ldots, b_n).$$

We are now ready to check unitality and associativity. A path p is an identity in **Free**(G) iff $p = (v)$ for some $v \in V$. It is easy to see from the above that (v) and $(w, b_1, \ldots, b_n)$ can be concatenated iff $v = w$, in which case the result is $(w, b_1, \ldots, b_n)$. Similarly $(v, a_1, \ldots, a_m)$ and (w) can be concatenated iff $w = t(a_m)$, in which case the result is $(v, a_1, \ldots, a_m)$. Finally, for associativity with p and q as above and $r = (x, c_1, \ldots, c_o)$, the formula readily reads that whichever way they are concatenated, $(p \, \mathring{9} \, q) \, \mathring{9} \, r$ or $p \, \mathring{9} \, (q \, \mathring{9} \, r)$, the result is

$$(v, a_1, \ldots, a_m, b_1, \ldots, b_n, c_1, \ldots, c_o).$$

Solution to Exercise 3.5

We often like to name identity morphisms by the objects they're on, and we do that here: v_2 means id_{v_2}. We write $\boxtimes$ when the composite does not make sense (i.e. when the target of the first morphism does not agree with the source of the second).

$\nearrow$	v_1	f_1	$f_1 \, \mathring{9} \, f_2$	v_2	f_2	v_3
v_1	v_1	f_1	$f_1 \, \mathring{9} \, f_2$	$\boxtimes$	$\boxtimes$	$\boxtimes$
f_1	$\boxtimes$	$\boxtimes$	$\boxtimes$	f_1	$f_1 \, \mathring{9} \, f_2$	$\boxtimes$
$f_1 \, \mathring{9} \, f_2$	$\boxtimes$	$\boxtimes$	$\boxtimes$	$\boxtimes$	$\boxtimes$	$f_1 \, \mathring{9} \, f_2$
v_2	$\boxtimes$	$\boxtimes$	$\boxtimes$	v_2	f_2	$\boxtimes$
f_2	$\boxtimes$	$\boxtimes$	$\boxtimes$	$\boxtimes$	$\boxtimes$	f_2
v_3	$\boxtimes$	$\boxtimes$	$\boxtimes$	$\boxtimes$	$\boxtimes$	v_3

Solution to Exercise 3.6

1. The category **1** has one object v_1 and one morphism, the identity id_{v_1}.
2. The category **0** is empty; it has no objects and no morphisms.
3. The pattern for number of morphisms in **0**, **1**, **2**, **3** is 0, 1, 3, 6; does this pattern look familiar? These are the first few "triangle numbers," so we could guess that the number of morphisms in **n**, the free category on the following graph

$$\overset{v_1}{\bullet} \xrightarrow{f_1} \overset{v_2}{\bullet} \xrightarrow{f_2} \cdots \xrightarrow{f_{n-1}} \overset{v_n}{\bullet}$$

is $1 + 2 + \cdots + n$. This makes sense because (and the proof strategy would be to verify that) the above graph has n paths of length 0, it has $n - 1$ paths of length 1, and so on: it has $n - i$ paths of length i for every $0 \leq i \leq n$.

Solution to Exercise 3.8

The correspondence was given by sending a path to its length. Concatenating a path of length m with a path of length n results in a path of length $m + n$.

Solution to Exercise 3.9

$$
\text{Free\_square} :=
\begin{array}{|c|}
\hline
\begin{array}{ccc}
A & \xrightarrow{\;f\;} & B \\
\Big\downarrow{\scriptstyle g} & & \Big\downarrow{\scriptstyle h} \\
C & \xrightarrow[\;i\;]{} & D
\end{array} \\
\hline
\textit{no equations} \\
\hline
\end{array}
$$

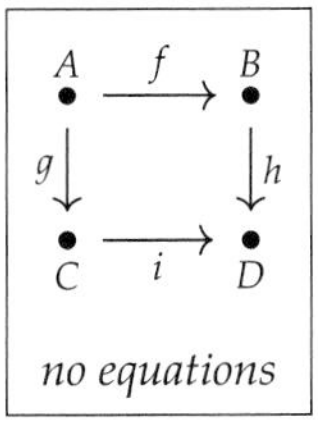

1. The ten paths are as follows

$$ A, \quad A ⨾ f, \quad A ⨾ g, \quad A ⨾ f ⨾ h, \quad A ⨾ g ⨾ i, \quad B, \quad B ⨾ h, \quad C, \quad C ⨾ i, \quad D $$

2. $A ⨾ f ⨾ h$ is parallel to $A ⨾ g ⨾ i$, in that they both have the same domain and both have the same codomain.

3. A is not parallel to any of the other nine paths.

Solution to Exercise 3.10

The morphisms in the given diagram are as follows:

$$ A, \quad A ⨾ f, \quad A ⨾ g, \quad A ⨾ j, \quad B, \quad B ⨾ h, \quad C, \quad C ⨾ i, \quad D $$

Note that $A ⨾ f ⨾ h = j = A ⨾ g ⨾ i$.

Solution to Exercise 3.12

There are four morphisms in $\mathcal{D}$, shown below, namely z, s, $s ⨾ s$, and $s ⨾ s ⨾ s$:

$$
\mathcal{D} :=
\begin{array}{|c|}
\hline
\begin{array}{c}
s \\
\circlearrowleft \\
\bullet \\
z
\end{array} \\
\hline
s ⨾ s ⨾ s ⨾ s = s ⨾ s \\
\hline
\end{array}
$$

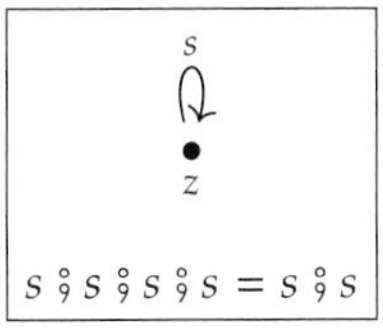

Solution to Exercise 3.14

The equations that make the graphs into preorders are shown below

$$
G_1 =
\begin{array}{|c|}
\hline
\bullet \underset{g}{\overset{f}{\rightrightarrows}} \bullet \\
\hline
f = g \\
\hline
\end{array}
\qquad
G_2 =
\begin{array}{|c|}
\hline
\begin{array}{c} f \\ \circlearrowleft \\ \bullet \\ a \end{array} \\
\hline
f = a \\
\hline
\end{array}
\qquad
G_3 =
\begin{array}{|c|}
\hline
\begin{array}{ccc}
\bullet & \xrightarrow{f} & \bullet \\
{\scriptstyle g}\big\downarrow & & \big\downarrow{\scriptstyle h} \\
\bullet & \xrightarrow[i]{} & \bullet
\end{array} \\
\hline
f ⨾ h = g ⨾ i \\
\hline
\end{array}
\qquad
G_4 =
\begin{array}{|c|}
\hline
\begin{array}{ccc}
\bullet & \xrightarrow{f} & \bullet \\
{\scriptstyle g}\big\downarrow & & \big\downarrow{\scriptstyle h} \\
\bullet & & \bullet
\end{array} \\
\hline
\textit{no equations} \\
\hline
\end{array}
$$

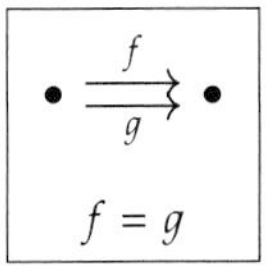

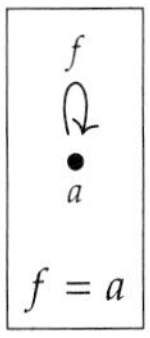

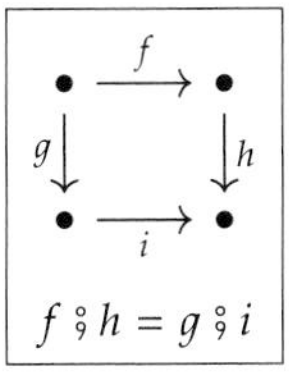

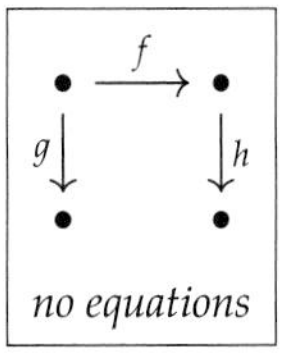

Solution to Exercise 3.15

The preorder reflection of a category $\mathcal{C}$ has the same objects and either one morphism or none between two objects, depending on whether or not a morphism between them exists in $\mathcal{C}$. So the preorder reflection of $\mathbb{N}$ has one object and one morphism from it to itself, which must be the identity. In other words, the preorder reflection of $\mathbb{N}$ is **1**.

Solution to Exercise 3.18

A function $f : \underline{2} \to \underline{3}$ can be described as an ordered pair $(f(1), f(2))$. The nine such functions are given by the following ordered pairs, which we arrange into a two-dimensional grid with three entries in each dimension, just for "funzies":[1]

$$
\begin{array}{ccc}
(1,1) & (1,2) & (1,3) \\
(2,1) & (2,2) & (2,3) \\
(3,1) & (3,2) & (3,3)
\end{array}
$$

Solution to Exercise 3.23

1. The inverse to $f(a) = 2$, $f(b) = 1$, $f(c) = 3$ is given by

$$
f^{-1}(1) = b, \quad f^{-1}(2) = a, \quad f^{-1}(3) = c.
$$

2. There are six distinct isomorphisms. In general, if A and B are sets, each with n elements, then the number of isomorphisms between them is n-factorial, often denoted $n!$. So for example there are $5 * 4 * 3 * 2 * 1 = 120$ isomorphisms between $\{1, 2, 3, 4, 5\}$ and $\{a, b, c, d, e\}$.

Solution to Exercise 3.24

We have to show that, for any object $c \in \mathcal{C}$, the identity id_c has an inverse, i.e. a morphism $f : c \to c$ such that $f \,\mathring{\,}\, \mathrm{id}_c = \mathrm{id}_c$ and $\mathrm{id}_c \,\mathring{\,}\, f = \mathrm{id}_c$. Take $f = \mathrm{id}_c$; this works.

Solution to Exercise 3.25

1. The monoid in Example 3.7 is not a group, because the morphism s has no inverse. Indeed each morphism is of the form s^n for some $n \in \mathbb{N}$ and composing it with s gives s^{n+1}, which is never s^0.

2. $\mathcal{C}$ from Example 3.11 is a group: the identity is always an isomorphism, and the other morphism s has inverse s.

Solution to Exercise 3.26

You may have found a person whose mathematical claims you can trust! Whenever you compose two morphisms in **Free**(G), their lengths add, and the identities are exactly those morphisms whose length is 0. In order for p to be an isomorphism, there must be some q such that $p \,\mathring{\,}\, q = \mathrm{id}$ and $q \,\mathring{\,}\, p = \mathrm{id}$, in which case the length of p (or q) must be 0.

Solution to Exercise 3.30

The other three functors $\mathbf{2} \to \mathbf{3}$ are shown here:

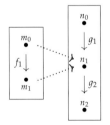 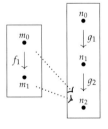 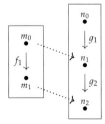

---

[1] Of course, this is not *mere funzies*; this is category theory!

Solution to Exercise 3.32

There are ten morphisms in $\mathcal{F}$; as usual we denote identities by the object they're on. These morphisms are sent to the following morphisms in $\mathcal{C}$:

$$A' \mapsto A, \quad f' \mapsto f, \quad g' \mapsto g, \quad f' \mathbin{\mathring{\,}} h' \mapsto f \mathbin{\mathring{\,}} h, \quad g' \mathbin{\mathring{\,}} i' \mapsto f \mathbin{\mathring{\,}} h,$$

$$B' \mapsto B, \quad h' \mapsto h, \quad C' \mapsto C, \quad i' \mapsto i, \quad D' \mapsto D.$$

If one of these seems different from the rest, it's probably $g' \mathbin{\mathring{\,}} i' \mapsto f \mathbin{\mathring{\,}} h$. But note that in fact also $g' \mathbin{\mathring{\,}} i' \mapsto g \mathbin{\mathring{\,}} i$ because $g \mathbin{\mathring{\,}} i = f \mathbin{\mathring{\,}} h$, so it's not an outlier after all.

Solution to Exercise 3.33

We need to give two functors F, G from $\overset{a}{\bullet} \overset{f}{\to} \overset{b}{\bullet}$ to $\overset{a'}{\bullet} \overset{f_1}{\underset{f_2}{\rightrightarrows}} \overset{b'}{\bullet}$ whose on-objects parts are the same and whose on-morphisms parts are different. There are only two ways to do this, and we choose one of them:

$$F(a) := a', \quad G(a) := a', \quad F(b) := b', \quad G(b) := b', \quad F(f) := f_1, \text{ and } G(f) := f_2.$$

The other way reverses f_1 and f_2.

Solution to Exercise 3.36

1. Let $\mathcal{C}$ be a category. Then defining $\mathrm{id}_{\mathcal{C}} : \mathcal{C} \to \mathcal{C}$ by $\mathrm{id}_{\mathcal{C}}(x) = x$ for every object and morphism in $\mathcal{C}$ is a functor because it preserves identities $\mathrm{id}_{\mathcal{C}}(\mathrm{id}_c) = \mathrm{id}_c = \mathrm{id}_{\mathrm{id}_{\mathcal{C}}(c)}$ for each object $c \in \mathrm{Ob}(\mathcal{C})$, and it preserves composition $\mathrm{id}_{\mathcal{C}}(f \mathbin{\mathring{\,}} g) = f \mathbin{\mathring{\,}} g = \mathrm{id}_{\mathcal{C}}(f) \mathbin{\mathring{\,}} \mathrm{id}_{\mathcal{C}}(g)$ for each pair of composable morphisms f, g in $\mathcal{C}$.

2. Given functors $F : \mathcal{C} \to \mathcal{D}$ and $G : \mathcal{D} \to \mathcal{E}$, we need to show that $F \mathbin{\mathring{\,}} G$ is a functor, i.e. that it preserves identities and compositions. If $c \in \mathcal{C}$ is an object then $(F \mathbin{\mathring{\,}} G)(\mathrm{id}_c) = G(F(\mathrm{id}_c)) = G(\mathrm{id}_{F(c)}) = \mathrm{id}_{G(F(c))}$ because F and G preserve identities. If f, g are composable morphisms in $\mathcal{C}$ then

$$(F \mathbin{\mathring{\,}} G)(f \mathbin{\mathring{\,}} g) = G(F(f) \mathbin{\mathring{\,}} F(g)) = G(F(f)) \mathbin{\mathring{\,}} G(F(g))$$

because F and G preserve composition.

3. We have proposed objects, morphisms, identities, and a composition formula for a category **Cat**: they are categories, functors, and the identities and compositions given above. We need to check that the two properties, unitality and associativity, hold. So suppose $F : \mathcal{C} \to \mathcal{D}$ is a functor and we pre-compose it as above with $\mathrm{id}_{\mathcal{C}}$; it is easy to see that the result will again be F, and similarly if we post-compose F with $\mathrm{id}_{\mathcal{D}}$. This gives unitality, and associativity is just as easy, though more wordy. Given F as above and $G : \mathcal{D} \to \mathcal{E}$ and $H : \mathcal{E} \to \mathcal{F}$, we need to show that $(F \mathbin{\mathring{\,}} G) \mathbin{\mathring{\,}} H = F \mathbin{\mathring{\,}} (G \mathbin{\mathring{\,}} H)$. It's a simple application of the definition: for any $x \in \mathcal{C}$, be it an object or morphism, we have

$$((F \mathbin{\mathring{\,}} G) \mathbin{\mathring{\,}} H)(c) = H((F \mathbin{\mathring{\,}} G)(c)) = H(G(F(c))) = (G \mathbin{\mathring{\,}} H)(F(c)) = (F \mathbin{\mathring{\,}} (G \mathbin{\mathring{\,}} H))(c).$$

Solution to Exercise 3.38

Let $S \in \mathbf{Set}$ be a set. Define $F_S : \mathbf{1} \to \mathbf{Set}$ by $F_S(1) = S$ and $F_S(\mathrm{id}_1) = \mathrm{id}_S$. With this definition, F_S preserves identities and compositions (the only compositions in $\mathbf{1}$ is the composite of the identity with itself), so F_S is a functor with $F_S(1) = S$ as desired.

Solution to Exercise 3.40

We are asked what sort of data "makes sense" for the schemas below.

1.
$$
\begin{array}{c}
s \\
\circlearrowleft \\
\bullet \\
z
\end{array}
$$
$$s \mathbin{\fatsemi} s = z$$

2.
$$
a \xrightarrow{f} b \underset{h}{\overset{g}{\rightrightarrows}} c
$$
$$f \mathbin{\fatsemi} g = f \mathbin{\fatsemi} h$$

This is a subjective question, so we propose an answer for your consideration.

1. Data on this schema, i.e. a set-valued functor, assigns a set $D(z)$ and a function $D(s)\colon D(z) \to D(z)$, such that applying that function twice is the identity. This sort of function is called an *involution* of the set D_z:

It's a do-si-do, a "partner move," where everyone picks a partner (possibly themselves) and exchanges with them. For example one could take D to be the set of pixels in a photograph, and take s to be the function sending each pixel to its mirror image across the vertical center line of the photograph.

2. We could make $D(c)$ the set of people at a "secret Santa" Christmas party, where everyone gives a gift to someone, possibly themselves. Take $D(b)$ to be the set of gifts, g the giver function (each gift is given by a person), and h the receiver function (each gift is received by a person), $D(a)$ is the set of people who give a gift to themselves, and $d(f)\colon D(a) \to D(b)$ is the inclusion.

Solution to Exercise 3.46

1. The expert packs so much information in so little space! Suppose we are given three objects $F, G, H \in \mathcal{D}^{\mathcal{C}}$; these are functors $F, G, H\colon \mathcal{C} \to \mathcal{D}$. Morphisms $\alpha\colon F \to G$ and $\beta\colon G \to H$ are natural transformations. Most beginners seem to think about a natural transformation in terms of its naturality squares, but the main thing to keep in mind is its components; the naturality squares constitute a check that comes later.

So for each $c \in \mathcal{C}$, α has a component $\alpha_c\colon F(c) \to G(c)$ and β has a component $\beta_c\colon G(c) \to H(c)$ in $\mathcal{D}$. The expert has told us to define $(\alpha \mathbin{\fatsemi} \beta)_c := (\alpha_c \mathbin{\fatsemi} \beta_c)$, and indeed that is a morphism $F(c) \to H(c)$.

Now we do the check. For any $f\colon c \to c'$ in $\mathcal{C}$, the inner squares of the following diagram commute because α and β are natural; hence the outer rectangle does too:

$$
\begin{array}{ccccc}
F(c) & \xrightarrow{\;\alpha_c\;} & G(c) & \xrightarrow{\;\beta_c\;} & H(c) \\
{\scriptstyle F(f)}\downarrow & & {\scriptstyle G(f)}\downarrow & & {\scriptstyle H(f)}\downarrow \\
F(c') & \xrightarrow[\;\alpha_c\;]{} & G(c') & \xrightarrow[\;\beta_c\;]{} & H(c')
\end{array}
$$

2. We propose that the identity natural transformation id_F on a functor $F \colon \mathcal{C} \to \mathcal{D}$ has as its c-component the morphism $(\text{id}_F)_c := \text{id}_{F(c)}$ in $\mathcal{D}$, for any c. The naturality square

$$
\begin{array}{ccc}
F(c) & \xrightarrow{\;\text{id}_{F(c)}\;} & F(c) \\
{\scriptstyle F(f)}\downarrow & & \downarrow{\scriptstyle F(f)} \\
F(c') & \xrightarrow[\;\text{id}_{F(c')}\;]{} & F(c')
\end{array}
$$

obviously commutes for any $f \colon c \to c'$. And it is unital: post-composing id_F with any $\beta \colon F \to G$ (and similarly for precomposing with any $\alpha \colon E \to F$) results in a natural transformation $\text{id}_F \,\mathbin{\fatsemi}\, \beta$ with components $(\text{id}_F)_c \,\mathbin{\fatsemi}\, \beta_c = (\text{id}_{F(c)} \,\mathbin{\fatsemi}\, \beta_c) = \beta_c$, and this is just β as desired.

Solution to Exercise 3.49
We have a category $\mathcal{C}$ and a preorder $\mathcal{P}$, considered as a category.

1. Suppose that $F, G \colon \mathcal{C} \to \mathcal{P}$ are functors and $\alpha, \beta \colon F \to G$ are natural transformations; we need to show that $\alpha = \beta$. It suffices to check that $\alpha_c = \beta_c$ for each object $c \in \mathrm{Ob}(\mathcal{C})$. But α_c and β_c are morphisms $F(c) \to G(c)$ in $\mathcal{P}$, which is a preorder, and the definition of a preorder – considered as a category – is that it has at most one morphism between any two objects. Thus $\alpha_c = \beta_c$, as desired.

2. This is false. Let $\mathcal{P} := \mathbf{1}$, let $\mathcal{C} := \boxed{\;\overset{a}{\bullet} \mathrel{\substack{\xrightarrow{f_1}\\[-4pt]\xrightarrow[f_2]{}}} \overset{b}{\bullet}\;}$, let $F(1) := a$, let $G(1) := b$, let $\alpha_1 := f_1$, and let $\beta_1 := g_2$.

Solution to Exercise 3.52
We need to write down the following

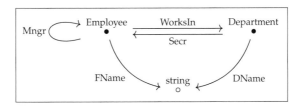

as a **Gr**-instance, as in Eq. (3.10). The answer is as follows:

Arrow	source	target
Mngr	Employee	Employee
WorksIn	Employee	Department
Secr	Department	Employee
FName	Employee	string
DName	Department	string

Vertex
Department
Employee
string

Solution to Exercise 3.54

Let G, H be the following graphs:

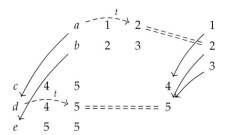

$$G := \boxed{\overset{1}{\bullet} \overset{a}{\longrightarrow} \overset{2}{\bullet} \overset{b}{\longrightarrow} \overset{3}{\bullet}}$$

$$H := \boxed{\overset{4}{\bullet} \overset{d}{\underset{c}{\rightrightarrows}} \overset{5}{\bullet} \circlearrowright e}$$

and let's believe the authors that there is a unique graph homomorphism $\alpha : G \to H$ for which $\alpha_{\text{Arrow}}(a) = d$.

1. We have $\alpha_{\text{Arrow}}(b) = e$ and $\alpha_{\text{Vertex}}(1) = 4$, $\alpha_{\text{Vertex}}(2) = 5$, and $\alpha_{\text{Vertex}}(3) = 5$.
2. We roughly copy the tables and then draw the lines (shown in black; ignore the dashed lines for now):

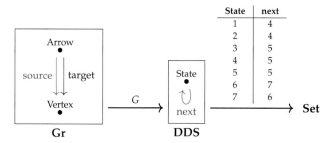

3. It works! One example of the naturality is shown with the help of dashed blue lines above. See how both paths starting at a end at 5?

Solution to Exercise 3.55

We just need to write out the composite of the following functors

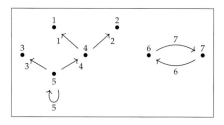

State	next
1	4
2	4
3	5
4	5
5	5
6	7
7	6

in the form of a database, and then draw the graph. The results are given below.

Arrow	source	target		Vertex
1	4	1		1
2	4	2		2
3	5	3		3
4	5	4		4
5	5	5		5
6	7	6		6
7	6	7		7

Solution to Exercise 3.60

We are interested in how the functors $- \times B$ and $(-)^B$ should act on morphisms for a given set B. We didn't specify this in the text – we only specified $- \times B$ and $(-)^B$ on objects – so in some sense this exercise is open: you can make up anything you want, under the condition that it is functorial. However, the authors cannot think of any such answers except the one we give below.

1. Given an arbitrary function $f : X \to Y$, we need a function $X \times B \to Y \times B$. We suggest the function which might be denoted $f \times B$; it sends (x, b) to $(f(x), b)$. This assignment is functorial: applied to id_X it returns $\mathrm{id}_{X \times B}$ and it preserves composition.
2. Given a function $f : X \to Y$, we need a function $X^B \to Y^B$. The canonical function would be denoted f^B; it sends a function $g : B \to X$ to the composite $(g \, \mathring{\,}\, f) : B \to X \to Y$. This is functorial: applied to id_X it sends g to g, i.e. $f^B(\mathrm{id}_X) = \mathrm{id}_{X^B}$, and applied to the composite $(f_1 \, \mathring{\,}\, f_2) : X \to Y \to Z$, we have

$$(f_1 \, \mathring{\,}\, f_2)^B(g) = g \, \mathring{\,}\, (f_1 \, \mathring{\,}\, f_2) = (g \, \mathring{\,}\, f_1) \, \mathring{\,}\, f_2 = (f_1^B \, \mathring{\,}\, f_2^B)(g)$$

   for any $g \in X^B$.
3. If $p : \mathbb{N} \to \mathbb{N}^{\mathbb{N}}$ is the result of currying $+ : \mathbb{N} \times \mathbb{N} \to \mathbb{N}$, then $p(3)$ is an element of $\mathbb{N}^{\mathbb{N}}$, i.e. we have $p(3) : \mathbb{N} \to \mathbb{N}$; what function is it? It is the function that adds 3. That is $p(3)(n) := n + 3$.

Solution to Exercise 3.62

The functor $! : \mathcal{C} \to \mathbf{1}$ from Eq. (3.14) sends each object $c \in \mathcal{C}$ to the unique object $1 \in \mathbf{1}$ and sends each morphism $f : c \to d$ in $\mathcal{C}$ to the unique morphism $\mathrm{id}_1 : 1 \to 1$ in $\mathbf{1}$.

Solution to Exercise 3.63

We want to draw the graph corresponding to the instance $I : \mathcal{G} \to \mathbf{Set}$ shown below:

Email	sent_by	received_by		Address
Em_1	Bob	Grace		Bob
Em_2	Grace	Pat		Doug
Em_3	Bob	Emmy		Emmy
Em_4	Sue	Doug		Grace
Em_5	Doug	Sue		Pat
Em_6	Bob	Bob		Sue

Here it is, with names and emails shortened (e.g. B=Bob, 3=Em_3):

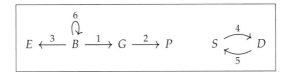

Solution to Exercise 3.66

An object z is terminal in some category $\mathcal{C}$ if, for every $c \in \mathcal{C}$ there exists a unique morphism $c \to z$. When $\mathcal{C}$ is the category underlying a preorder, there is at most one morphism between any two objects, so the condition simplifies: an object z is terminal iff, for every $c \in \mathcal{C}$ there exists a morphism $c \to z$. The morphisms in a preorder are written with $\leq$ signs, so z is terminal iff, for every $c \in P$, we have $c \leq z$, and this is the definition of top element.

Solution to Exercise 3.67

The terminal object in **Cat** is **1** because by Exercise 3.62 there is a unique morphism (functor) $\mathcal{C} \to \mathbf{1}$ for any object (category) $\mathcal{C} \in \mathbf{Cat}$.

Solution to Exercise 3.68

Consider the graph $2V := \boxed{\bullet \ \bullet}$ with two vertices and no arrows, and let $\mathcal{C} = \mathbf{Free}(2V)$; it has two objects and two morphisms (the identities). This category does not have a terminal object because it does not have any morphisms from one object to the other.

Solution to Exercise 3.73

A product of x and y in $\mathcal{P}$ is an object $z \in \mathcal{P}$ equipped with maps $z \to x$ and $z \to y$ such that for any other object z' and maps $z' \to x$ and $z' \to y$, there is a unique morphism $z' \to z$ making the evident triangles commute. But in a preorder, the maps are denoted $\leq$, they are unique if they exist, and all diagrams commute. Thus the above becomes: a product of x and y in $\mathcal{P}$ is an object z with $z \leq x$ and $z \leq y$ such that for any other z', if $z' \leq x$ and $z' \leq y$ then $z' \leq z$. This is exactly the definition of meet, $z = x \wedge y$.

Solution to Exercise 3.75

1. The identity morphism on the object (c, d) in the product category $\mathcal{C} \times \mathcal{D}$ is $(\mathrm{id}_c, \mathrm{id}_d)$.
2. Suppose we are given three composable morphisms in $\mathcal{C} \times \mathcal{D}$

$$(c_1, d_1) \xrightarrow{(f_1, g_1)} (c_2, d_2) \xrightarrow{(f_2, g_2)} (c_3, d_3) \xrightarrow{(f_3, g_3)} (c_4, d_4).$$

We want to check that $((f_1, g_1) \,\mathring{,}\, (f_2, g_2)) \,\mathring{,}\, (f_3, g_3) = (f_1, g_1) \,\mathring{,}\, ((f_2, g_2) \,\mathring{,}\, (f_3, g_3))$. But composition in a product category is given component-wise. That means the left-hand side is $((f_1 \,\mathring{,}\, f_2) \,\mathring{,}\, f_3, (g_1 \,\mathring{,}\, g_2) \,\mathring{,}\, g_3)$, whereas the right-hand side is $(f_1 \,\mathring{,}\, (f_2 \,\mathring{,}\, f_3), g_1 \,\mathring{,}\, (g_2 \,\mathring{,}\, g_3))$, and these are equal because both $\mathcal{C}$ and $\mathcal{D}$ individually have associative composition.
3. The product category $\mathbf{1} \times \mathbf{2}$ has two objects $(1, 1)$ and $(1, 2)$ and one non-identity morphism $(1, 1) \to (1, 2)$. It is not hard to see that it looks the same as $\mathbf{2}$. In fact, for any $\mathcal{C}$ there is an isomorphism of categories $\mathbf{1} \times \mathcal{C} \cong \mathcal{C}$.
4. Let P and Q be preorders, let $X = P \times Q$ be their product preorder as defined in Example 1.51, and let $\mathcal{P}$, $\mathcal{Q}$, and $\mathcal{X}$ be the corresponding categories. Then $\mathcal{X} = \mathcal{P} \times \mathcal{Q}$.

Solution to Exercise 3.76

A product of X and Y is an object Z equipped with morphisms $X \xleftarrow{p_X} Z \xrightarrow{p_Y} Y$ such that, for any other object Z' equipped with morphisms $X \xleftarrow{p'_X} Z' \xrightarrow{p'_Y} Y$, there is a unique morphism $f \colon Z' \to Z$ making the triangles commute, $f \,\mathring{,}\, p_X = p'_X$ and $f \,\mathring{,}\, p_Y = p'_Y$. But "an object equipped with morphisms to X and Y" is exactly the definition of an object in $\mathbf{Cone}(X, Y)$, and a morphism f making the triangles commute is exactly the definition of a morphism in $\mathbf{Cone}(X, Y)$. So the definition above becomes: a product of X and Y is an object $Z \in \mathbf{Cone}(X, Y)$ such that for any other object Z' there is a unique morphism $Z' \to Z$ in $\mathbf{Cone}(X, Y)$. This is exactly the definition of Z being terminal in $\mathbf{Cone}(X, Y)$.

Solution to Exercise 3.82

Suppose $\mathcal{J}$ is the graph $\boxed{\overset{v_1 \ v_2}{\bullet \ \bullet}}$ and $D \colon \mathcal{J} \to \mathbf{Set}$ is given by two sets, $D(v_1) = X$ and $D(v_2) = Y$ for sets X, Y. The product of these two sets is $X \times Y$. Let's check that the limit formula in Theorem 3.80 gives the same answer. It says

$$\lim_{g} D := \big\{(d_1, \ldots, d_n) \mid d_i \in D(v_i) \text{ for all } 1 \le i \le n \text{ and}$$

$$\text{for all } a\colon v_i \to v_j \in A, \text{ we have } D(a)(d_i) = d_j\big\}.$$

But in our case $n = 2$, there are no arrows in the graph, and $D(v_1) = X$ and $D(v_2) = Y$. So the formula reduces to

$$\lim_{g} D := \big\{(d_1, d_2) \mid d_1 \in X \text{ and } d_2 \in Y\big\},$$

which is exactly the definition of $X \times Y$.

Solution to Exercise 3.86

Given a functor $F\colon \mathcal{C} \to \mathcal{D}$, we define its opposite $F^{\mathrm{op}}\colon \mathcal{C}^{\mathrm{op}} \to \mathcal{D}^{\mathrm{op}}$ as follows. For each object $c \in \mathrm{Ob}(\mathcal{C}^{\mathrm{op}}) = \mathrm{Ob}(\mathcal{C})$, put $F^{\mathrm{op}}(c) := F(c)$. For each morphism $f\colon c_1 \to c_2$ in $\mathcal{C}^{\mathrm{op}}$, we have a corresponding morphism $f'\colon c_2 \to c_1$ in $\mathcal{C}$ and thus a morphism $F(f')\colon F(c_2) \to F(c_1)$ in $\mathcal{D}$, and thus a morphism $F(f')'\colon F^{\mathrm{op}}(c_1) \to F^{\mathrm{op}}(c_2)$. Hence we can define $F^{\mathrm{op}}(f) := F(f')'$. Note that the primes $(-')$ are pretty meaningless, we only put them there to differentiate between things that are very closely related.

It is easy to check that our definition of F^{op} is functorial: it sends identities to identities and composites to composites.

A.4   Solutions for Chapter 4

Solution to Exercise 4.2

1. The Hasse diagram for $\mathcal{X}^{\mathrm{op}} \times \mathcal{Y}$ is shown here (ignore the colors):

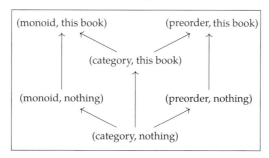

2. There is a profunctor $\Lambda\colon \mathcal{X} \nrightarrow \mathcal{Y}$, i.e. a functor $\mathcal{X}^{\mathrm{op}} \times \mathcal{Y} \to \mathbb{B}$, such that, in the picture above, blue is sent to true and black is sent to false, i.e.

$$\Lambda(\text{monoid, nothing}) = \Lambda(\text{monoid, this book})$$

$$= \Lambda(\text{preorder, this book}) \qquad = \Lambda(\text{category, this book}) = \texttt{true}$$

$$\Lambda(\text{preorder, nothing}) = \Lambda(\text{category, nothing}) = \texttt{false}.$$

The preorder $\mathcal{X}^{\mathrm{op}} \times \mathcal{Y}$ describes tasks in decreasing difficulty. For example, (we hope) it is easier for my aunt to explain a monoid given this book than for her to explain a monoid without this book. The profunctor Λ describes possible states of knowledge for my aunt: she can describe monoids without help, categories with help from the book, etc. It is an upper set because we assume that if she can do a task, she can also do any easier task.

Solution to Exercise 4.3
We've done this one before! We hope you remembered how to do it. If not, see Exercise 2.61.

Solution to Exercise 4.5
Recall from Definition 2.25 that a V-functor $\Phi\colon \mathcal{X}^{\mathrm{op}} \times \mathcal{Y} \to V$ is a function $\Phi\colon \mathrm{Ob}(\mathcal{X}^{\mathrm{op}} \times \mathcal{Y}) \to \mathrm{Ob}(V)$ such that for all (x, y) and (x', y') in $\mathcal{X}^{\mathrm{op}} \times \mathcal{Y}$ we have

$$(\mathcal{X}^{\mathrm{op}} \times \mathcal{Y})\big((x, y), (x', y')\big) \leq V\big(\Phi(x, y), \Phi(x', y')\big).$$

Using the definitions of product V-category (Definition 2.53) and opposite V-category (Exercise 2.52) on the left-hand side, and using Remark 2.65, which describes how we are viewing the quantale V as enriched over itself, on the right-hand side, this unpacks to

$$\mathcal{X}(x', x) \otimes \mathcal{Y}(y, y') \leq \Phi(x, y) \multimap \Phi(x', y').$$

Using symmetry of $\otimes$ and the definition of hom-element, Eq. (2.23), we see that Φ is a profunctor if and only if

$$\mathcal{X}(x', x) \otimes \Phi(x, y) \otimes \mathcal{Y}(y, y') \leq \Phi(x', y').$$

Solution to Exercise 4.6
Yes, since a **Bool**-functor is exactly the same as a monotone map, the definition of **Bool**-profunctor and that of feasibility relation line up perfectly!

Solution to Exercise 4.8
The feasibility matrix for Φ is

Φ	a	b	c	d	e
N	true	false	true	false	true
E	true	true	true	true	true
W	true	false	true	false	true
S	true	true	true	true	true

Solution to Exercise 4.10
The **Cost**-matrix for Φ is

Φ	x	y	z
A	17	20	20
B	11	14	14
C	14	17	17
D	12	9	15

Solution to Exercise 4.12

$$\Phi = M_X^3 * M_\Phi * M_Y^2 = \begin{pmatrix} 0 & 6 & 3 & 11 \\ 2 & 0 & 5 & 5 \\ 5 & 3 & 0 & 8 \\ 11 & 9 & 6 & 0 \end{pmatrix} \begin{pmatrix} \infty & \infty & \infty \\ 11 & \infty & \infty \\ \infty & \infty & \infty \\ \infty & 9 & \infty \end{pmatrix} \begin{pmatrix} 0 & 4 & 3 \\ 3 & 0 & 6 \\ 7 & 4 & 0 \end{pmatrix}$$

$$= \begin{pmatrix} 17 & 20 & \infty \\ 11 & 14 & \infty \\ 14 & 17 & \infty \\ 20 & 9 & \infty \end{pmatrix} \begin{pmatrix} 0 & 4 & 3 \\ 3 & 0 & 6 \\ 7 & 4 & 0 \end{pmatrix}$$

$$
= \begin{pmatrix} 17 & 20 & 20 \\ 11 & 14 & 14 \\ 14 & 17 & 17 \\ 12 & 9 & 15 \end{pmatrix}.
$$

Solution to Exercise 4.13

Yes, this is valid: it just means that the profunctor $\Phi: (T \times E) \nrightarrow \$$ does not relate (good-natured, funny) to any element of $\$$. More formally, it means that $\Phi((\text{good-natured, funny}), p) = \texttt{false}$ for all $p \in \$ = \{\$100\text{K}, \$500\text{K}, \$1\text{M}\}$. We can interpret this to mean that it is not feasible to produce a good-natured, funny movie for any of the cost options presented – so at least not for less than a million dollars.

Solution to Exercise 4.15

There are a number of methods that can be used to get the correct answer. One way that works well for this example is to search for the shortest paths on the diagram: it so happens that all the shortest paths go through the bridges from D to y and y to r, so in this case $(\Phi \mathbin{\overset{\circ}{,}} \Psi)(-, -) = \mathfrak{X}(-, D) + 9 + \mathfrak{Z}(r, -)$. This gives:

$\Phi \mathbin{\overset{\circ}{,}} \Psi$	p	q	r	s
A	22	24	20	21
B	16	18	14	15
C	19	21	17	18
D	11	13	9	10

A more methodical way is to use matrix multiplication. Here's one way you might do the multiplication, using a few tricks.

$$
\begin{aligned}
\Phi \mathbin{\overset{\circ}{,}} \Psi &= (M_X^3 * M_\Phi * M_Y^2) * (M_Y^2 * M_\Psi * M_Z^3) \\
&= M_X^3 * M_\Phi * M_Y^4 * M_\Psi * M_Z^3 \\
&= (M_X^3 * M_\Phi * M_Y^2) * M_\Psi * M_Z^3 \\
&= \Phi * M_\Psi * M_Z^3 \\
&= \begin{pmatrix} 17 & 20 & 20 \\ 11 & 14 & 14 \\ 14 & 17 & 17 \\ 12 & 9 & 15 \end{pmatrix} \begin{pmatrix} \infty & \infty & \infty & \infty \\ \infty & \infty & 0 & \infty \\ 4 & \infty & \infty & 4 \end{pmatrix} \begin{pmatrix} 0 & 2 & 4 & 5 \\ 4 & 0 & 2 & 3 \\ 2 & 4 & 0 & 1 \\ 1 & 3 & 5 & 0 \end{pmatrix} \\
&= \begin{pmatrix} 17 & 20 & 20 \\ 11 & 14 & 14 \\ 14 & 17 & 17 \\ 12 & 9 & 15 \end{pmatrix} \begin{pmatrix} \infty & \infty & \infty & \infty \\ 2 & 4 & 0 & 1 \\ 4 & 6 & 8 & 4 \end{pmatrix} \\
&= \begin{pmatrix} 22 & 24 & 20 & 21 \\ 16 & 18 & 14 & 15 \\ 19 & 21 & 17 & 18 \\ 11 & 13 & 9 & 10 \end{pmatrix}.
\end{aligned}
$$

Solution to Exercise 4.18

We choose the **Cost**-category $\mathcal{X}$ from Eq. (2.18). The unit profunctor $U_{\mathcal{X}}$ on $\mathcal{X}$ is described by the bridge diagram

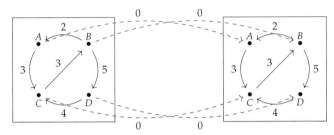

Solution to Exercise 4.20

1. The first equality is the unitality of $\mathcal{V}$ (Definition 2.1(b)). The second step uses the monotonicity of $\otimes$ (Definition 2.1(a)) applied to the inequalities $I \leq \mathcal{P}(p, p)$ (the identity law for $\mathcal{P}$ at p, Definition 2.30(a)) and $\Phi(p, q) \leq \Phi(p, q)$ (reflexivity of preorder $\mathcal{V}$, Definition 1.25(a)). The third step uses the definition of join: for any set I element $i_0 \in I$, a subset $s_{i \in I}$ at a poset, we have $s_{i_0} \leq \bigvee_{i \in I} s$. In our case, $I i_0 = p$, and $s_i = \mathcal{P}(p, i) \otimes \Phi$. The final equality is just the definition of profunctor composition, Definition 4.14.

2. Note that, in **Bool**, $I = $ true. Since the identity law at p says true $\leq \mathcal{P}(p, p)$, and true is the largest element of the preorder **Bool**, we thus have $\mathcal{P}(p, p) = $ true for all p. This shows that the first inequality in Eq. (4.9) is an equality.

   The second inequality is more involved. Note that, by the above, we can assume the left-hand side of the inequality is equal to $\Phi(p, q)$. We split into two cases. Suppose $\Phi(p, q) = $ true. Then, again since true is the largest element of $\mathbb{B}$, we must have equality.

   Next, suppose $\Phi(p, q) = $ false. Note that since Φ is a monotone map $\mathcal{P}^{\mathrm{op}} \times \mathcal{Q} \to $ **Bool**, if $p \leq p_1$ in $\mathcal{P}$, then $\Phi(p_1, q) \leq \Phi(p, q)$ in **Bool**. Thus if $\mathcal{P}(p, p_1) = $ true then $\Phi(p_1, q) = \Phi(p, q) = $ false. This implies that for all $p_1 \in \mathcal{P}$, we have either $\mathcal{P}(p, p_1) = $ false or $\Phi(p_1, q) = $ false, and hence $\bigvee_{p_1 \in \mathcal{P}} \mathcal{P}(p, p_1) \wedge \Phi(p_1, q) = \bigvee_{p_1 \in \mathcal{P}}$ false $= $ false. Thus, in either case, we see that $\Phi(p, q) = \bigvee_{p_1 \in \mathcal{P}} \mathcal{P}(p, p_1) \wedge \Phi(p_1, q)$, as required.

3. The first equation is unitality in monoidal categories, $v \otimes I = v$ for any $v \in V$. The second is that $I \leq \mathcal{Q}(q, q)$ by unitality of enriched categories, see Definition 2.30, together with monotonicity of monoidal product: $v_1 \leq v_2$ implies $v \otimes v_1 \leq v \otimes v_2$. The third was shown in Exercise 4.5.

Solution to Exercise 4.22

This is very similar to Exercise 2.78: we exploit the associativity of $\otimes$. Note, however, we also require $\mathcal{V}$ to be symmetric monoidal closed, since this implies the distributivity of $\otimes$ over $\vee$ (Proposition 2.64 2), and $\mathcal{V}$ to be skeletal, so we can turn equivalences into equalities.

$$((\Phi \,\mathring{,}\, \Psi) \,\mathring{,}\, \Upsilon)(p, s) = \bigvee_{r \in \mathcal{R}} \left(\bigvee_{q \in \mathcal{Q}} \Phi(p, q) \otimes \Psi(q, r) \right) \otimes \Upsilon(r, s)$$

$$= \bigvee_{r \in \mathcal{R}, q \in \mathcal{Q}} \Phi(p, q) \otimes \Psi(q, r) \otimes \Upsilon(r, s)$$

$$= \bigvee_{q \in \mathcal{Q}} \Phi(p, q) \otimes \bigvee_{r \in \mathcal{R}} \left(\Psi(q, r) \otimes \Upsilon(r, s) \right)$$

$$= (\Phi \,\mathring{,}\, (\Psi \,\mathring{,}\, \Upsilon))(p, s).$$

Solution to Exercise 4.26

This is very straightforward. We wish to check $\widehat{\mathrm{id}} \colon \mathcal{P} \nrightarrow \mathcal{P}$ has the formula $\widehat{\mathrm{id}}(p, q) = \mathcal{P}(p, q)$. By Definition 4.24, $\widehat{\mathrm{id}}(p, q) := \mathcal{P}(\mathrm{id}(p), q) = \mathcal{P}(p, q)$. So they're the same.

Solution to Exercise 4.28

The conjoint $\check{+} \colon \mathbb{R} \nrightarrow \mathbb{R} \times \mathbb{R} \times \mathbb{R}$ sends (a, b, c, d) to $\mathbb{R}(a, b + c + d)$, which is `true` if $a \le b + c + d$, and false otherwise.

Solution to Exercise 4.30

1. By Definition 4.24, $\widehat{F}(p, q) = \mathcal{Q}(F(p), q)$ and $\check{G}(p, q) = \mathcal{Q}(p, G(q))$. Since $\mathcal{V}$ is skeletal, F and G are $\mathcal{V}$ adjoints if and only if $\mathcal{Q}(F(p), q) = \mathcal{Q}(p, G(q))$. Thus F and G are adjoints if and only if $\widehat{F} = \check{G}$.
2. Note that id $\colon \mathcal{P} \to \mathcal{P}$ is $\mathcal{V}$-adjoint to itself, since both sides of Eq. (4.11) then equal $\mathcal{P}(p, q)$. Thus $\widehat{\mathrm{id}} = \check{\mathrm{id}}$.

Solution to Exercise 4.33

The Hasse diagram for the collage of the given profunctor is quite simply this:

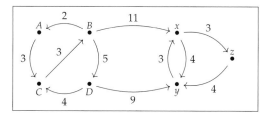

Solution to Exercise 4.37

Since we only have a rough definition, we can only roughly check this: we won't bother with the notion of well behaved. Nonetheless, we can still compare Definition 2.1 with Definition 4.34.

First, recall from Section 3.2.3 that a preorder is a category $\mathcal{P}$ such that for every $p, q \in \mathcal{P}$, the set $\mathcal{P}(p, q)$ has at most one element.

On the surface, all looks promising: both definitions have two constituents and four properties. In constituent (i), both Definition 2.1 and Definition 4.34 call for the same: an element, or object, of the preorder $\mathcal{P}$. So far so good. Constituent (ii), however, is where it gets interesting: Definition 2.1 calls for merely a function $\otimes \colon \mathcal{P} \times \mathcal{P} \to \mathcal{P}$, while Definition 4.34 calls for a *functor*.

Recall from Example 3.35 that functors between preorders are exactly monotone maps. So we need for the function $\otimes$ in Definition 2.1 to be a monotone map. This is exactly property (a) of Definition 2.1: it says that if $(x_1, x_2) \le (y_1, y_2)$ in $\mathcal{P} \otimes \mathcal{P}$, then we must have $x_1 \otimes x_2 \le y_1 \otimes y_2$ in $\mathcal{P}$. So it is also the case that in Definition 2.1 $\otimes$ is a functor.

The remaining properties compare easily, taking the natural isomorphisms to be equality or equivalence in $\mathcal{P}$. Indeed, property (b) of Definition 2.1 corresponds to *both* properties (a) and (b) of Definition 4.34, and then the respective properties (c) and (d) similarly correspond.

Solution to Exercise 4.39

1. $g_E(5, 3) = $ `false`, $g_F(5, 3) = 2$.
2. $g_E(3, 5) = $ `true`, $g_F(3, 5) = -2$.

3. $h(5, \texttt{true}) = 5$.
4. $h(-5, \texttt{true}) = -5$.
5. $h(-5, \texttt{false}) = 6$.
6. $q_G(-2, 3) = 2, q_F(-2, 3) = -13$.
7. $q_G(2, 3) = -1, q_F(2, 3) = 7$.

Solution to Exercise 4.41

Yes, the rough definition roughly agrees: plain old categories are **Set**-categories! In detail, we need to compare Definition 4.40 when $V = (\textbf{Set}, \{1\}, \times)$ with Definition 3.2. In both cases, we see that (i) asks for a collection of objects and (ii) asks for, for all pairs of objects x, y, a *set* $\mathcal{C}(x, y)$ of morphisms. Moreover, recall that the monoidal unit I is the one-element set $\{1\}$. This means a morphism $\mathrm{id}_x : I \to \mathcal{C}(x, x)$ is a function $\mathrm{id}_x : \{1\} \to \mathcal{C}(x, x)$. This is the same data as simply an element $\mathrm{id}_x = \mathrm{id}_x(1) \in \mathcal{C}(x, x)$; we call this data the identity morphism on x. Finally, a morphism $\mathring{,} : \mathcal{C}(x, y) \otimes \mathcal{C}(y, z) \to \mathcal{C}(x, z)$ is a function $\mathring{,} : \mathcal{C}(x, y) \times \mathcal{C}(y, z) \to \mathcal{C}(x, z)$; this is exactly the composite required in Definition 3.2 (iv).

So in both cases the data agrees. In Definition 3.2 we also require this data to satisfy two conditions, unitality and associativity. This is what is meant by the last sentence of Definition 4.40.

Solution to Exercise 4.43

An identity element in a **Cost**-category $\mathcal{X}$ is a morphism $I \to \mathcal{X}(x, x)$ in $\textbf{Cost} = ([0, \infty], \geq, 0, +)$, and hence the condition that $0 \geq \mathcal{X}(x, x)$. This implies that $\mathcal{X}(x, x) = 0$. In terms of distances, we interpret this to mean that the distance from any point to itself is equal to 0. We think this is a pretty sensible condition for a notion of distance to obey.

Solution to Exercise 4.47

1. Here is a picture of the unit corelation $\varnothing \to \underline{3} \sqcup \underline{3}$, where we have drawn the empty set with an empty dotted rectangle on the left:

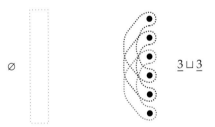

$\varnothing$ $\underline{3} \sqcup \underline{3}$

2. Here is a picture of the counit corelation $\underline{3} \sqcup \underline{3} \to \varnothing$:

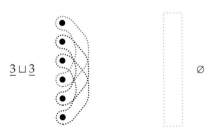

$\underline{3} \sqcup \underline{3}$ $\varnothing$

3. Here is a picture of the snake equation on the left of Eq. (4.15).

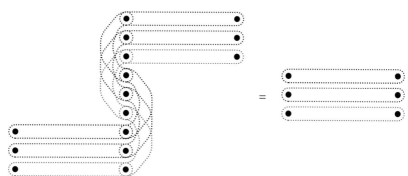

Solution to Exercise 4.49

Given two resource preorders $\mathcal{X}$ and $\mathcal{Y}$, the preorder $\mathcal{X} \times \mathcal{Y}$ represents the set of all pairs of resources, $x \in \mathcal{X}$ and $y \in \mathcal{Y}$, with $(x, y) \leq (x', y')$ iff $x \leq x'$ and $y \leq y'$. That is, if x is available given x' and y is available given y', then (x, y) is available given (x', y').

Given two profunctors $\Phi \colon \mathcal{X}_1 \nrightarrow \mathcal{X}_2$ and $\Psi \colon \mathcal{Y}_1 \nrightarrow \mathcal{Y}_2$, the profunctor $\Phi \times \Psi$ represents their conjunction, i.e. AND. In other words, if y_1 can be obtained given x_1 AND y_2 can be obtained given x_2, then (y_1, y_2) can be obtained given (x_1, x_2).

Solution to Exercise 4.50

The profunctor $\mathcal{X} \times \mathbf{1} \nrightarrow \mathcal{X}$ defined by the functor $\alpha \colon (\mathcal{X} \times \mathbf{1})^{\mathrm{op}} \times \mathcal{X} \to \mathcal{V}$ that maps $\alpha((x, 1), y) \coloneqq \mathcal{X}(x, y)$ is an isomorphism. It has inverse $\alpha^{-1} \colon \mathcal{X} \nrightarrow \mathcal{X} \times \mathbf{1}$ defined by $\alpha^{-1}(x, (y, 1)) \coloneqq \mathcal{X}(x, y)$. To see that $\alpha^{-1} \, \mathring{,} \, \alpha = \mathrm{U}_{\mathcal{X}}$, note first that the unit law for $\mathcal{X}$ at z and the definition of join imply

$$\mathcal{X}(x, z) = \mathcal{X}(x, z) \otimes I \leq \mathcal{X}(x, z) \otimes \mathcal{X}(z, z) \leq \bigvee_{y \in \mathcal{X}} \mathcal{X}(x, y) \otimes \mathcal{X}(y, z),$$

while composition says $\mathcal{X}(x, y) \otimes \mathcal{X}(y, z) \leq \mathcal{X}(x, z)$ and hence

$$\bigvee_{y \in \mathcal{X}} \mathcal{X}(x, y) \otimes \mathcal{X}(y, z) \leq \bigvee_{y \in \mathcal{X}} \mathcal{X}(x, z) = \mathcal{X}(x, z).$$

Thus, unpacking the definition of composition of profunctors, we have

$$(\alpha^{-1} \, \mathring{,} \, \alpha)(x, z) = \bigvee_{(y,1) \in \mathcal{X} \times \mathbf{1}} \alpha(x, (y, 1)) \otimes \alpha^{-1}((y, 1), z) = \bigvee_{y \in \mathcal{X}} \mathcal{X}(x, y) \otimes \mathcal{X}(y, z) = \mathcal{X}(x, z).$$

Similarly we can show $\alpha \, \mathring{,} \, \alpha^{-1} = \mathrm{U}_{\mathcal{X} \times \mathbf{1}}$, and hence that α is an isomorphism $\mathcal{X} \times \mathbf{1} \nrightarrow \mathcal{X}$.

Moreover, we can similarly show that $\beta((1, x), y) \coloneqq \mathcal{X}(x, y)$ defines an isomorphism $\beta \colon \mathbf{1} \times \mathcal{X} \nrightarrow \mathcal{X}$.

Solution to Exercise 4.51

We check the first snake equation, the one on the left-hand side of Eq. (4.15). The proof of the one on the right-hand side is analogous.

We must show that the composite Φ of profunctors

$$\mathcal{X} \xrightarrow{\alpha^{-1}} \mathcal{X} \times \mathbf{1} \xrightarrow{\mathrm{U}_{\mathcal{X}} \times \eta_{\mathcal{X}}} \mathcal{X} \times \mathcal{X}^{\mathrm{op}} \times \mathcal{X} \xrightarrow{\epsilon_{\mathcal{X}} \times \mathrm{U}_{\mathcal{X}}} \mathbf{1} \times \mathcal{X} \xrightarrow{\alpha} \mathcal{X}$$

is itself the identity (i.e. the unit profunctor on $\mathcal{X}$), where α and α^{-1} are the isomorphisms defined in the solution to Exercise 4.50 above.

Freely using the distributivity of $\otimes$ over $\vee$, the value $\Phi(x, y)$ of this composite at $(x, y) \in \mathcal{X}^{\text{op}} \times \mathcal{X}$ is given by

$$
\bigvee_{a,b,c,d,e \in \mathcal{X}} \alpha^{-1}(x, (a, 1)) \otimes (U_{\mathcal{X}} \times \eta_{\mathcal{X}})((a, 1), (b, c, d))
$$
$$
\otimes (\epsilon_{\mathcal{X}} \times U_{\mathcal{X}})((b, c, d), (1, e)) \otimes \alpha((1, e), y)
$$
$$
= \bigvee_{a,b,c,d,e \in \mathcal{X}} \alpha^{-1}(x, (a, 1)) \otimes U_{\mathcal{X}}(a, b) \otimes \eta_{\mathcal{X}}(1, c, d)
$$
$$
\otimes \epsilon_{\mathcal{X}}(b, c, 1) \otimes U_{\mathcal{X}}(d, e) \otimes \alpha((1, e), y)
$$
$$
= \bigvee_{a,b,c,d,e \in \mathcal{X}} \mathcal{X}(x, a) \otimes \mathcal{X}(a, b) \otimes \mathcal{X}(c, d) \otimes \mathcal{X}(b, c) \otimes \mathcal{X}(d, e) \otimes \mathcal{X}(e, y)
$$
$$
= \mathcal{X}(x, y),
$$

where in the final step we repeatedly use the argument the Lemma 4.19 that shows that composing with the unit profunctor $U_{\mathcal{X}}(a, b) = \mathcal{X}(a, b)$ is the identity.

This shows that $\Phi(x, y)$ is the identity profunctor, and hence shows the first snake equation holds. Again, checking the other snake equation is analogous.

A.5   Solutions for Chapter 5

Solution to Exercise 5.3

1. Below we draw a morphism $f : 3 \to 2$ and a morphism $g : 2 \to 4$ in **FinSet**:

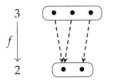 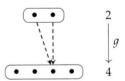

2. Here is a picture of $f + g$:

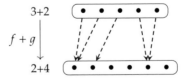

3. The composite of morphisms $f : m \to n$ and $g : n \to p$ in **FinSet** is the function $(f \mathbin{\mathring{,}} g) : m \to p$ given by $(f \mathbin{\mathring{,}} g)(i) = g(f(i))$ for all $1 \le i \le m$.

4. The identity $\text{id}_m : m \to m$ is given by $\text{id}_m(i) = i$ for all $1 \le i \le m$. Here is a picture of id_2 and id_8:

 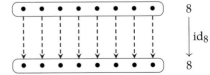

5. Here is a picture of the symmetry $\sigma_{3,5} \colon 8 \to 8$:

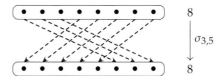

Solution to Exercise 5.7

We need to give examples of posetal props, i.e. each will be a poset whose set of objects is $\mathbb{N}$, whose order is denoted $m \preceq n$, and with the property that whenever $m_1 \preceq n_1$ and $m_2 \preceq n_2$ hold then $m_1 + m_2 \preceq n_1 + n_2$ does too.

   The question only asks for three, but we will additionally give a quasi-example and a non-example.

1. Take $\preceq$ to be the discrete order: $m \preceq n$ iff $m = n$.
2. Take $\preceq$ to be the usual order, $m \preceq n$ iff there exists $d \in \mathbb{N}$ with $d + m = n$.
3. Take $\preceq$ to be the reverse of the usual order, $m \preceq n$ iff there exists $d \in \mathbb{N}$ with $m = n + d$.
4. Take $\preceq$ to be the codiscrete order $m \preceq n$ for all m, n. Some may object that this is a preorder, not a poset, so we call it a quasi-example.
5. (Non-example.) Take $\preceq$ to be the division order, $m \preceq n$ iff there exists $q \in \mathbb{N}$ with $m * q = d$. This is a perfectly good poset, but it does not satisfy the monotonicity property: we have $2 \preceq 4$ and $3 \preceq 3$ but not $5 \preceq^? 7$.

Solution to Exercise 5.8

Example 5.4: The prop **Bij** has

1.   **Bij**$(m, n) := \{f \colon \underline{m} \to \underline{n} \mid f$ is a bijection$\}$. Note that **Bij**$(m, n) = \varnothing$ if $m \neq n$ and it has $n!$ elements if $m = n$.
2.   The identity map $n \to n$ is the bijection $\underline{n} \to \underline{n}$ sending $i \mapsto i$.
3.   The symmetry map $m + n \to n + m$ is the bijection $\sigma_{m,n} \colon \underline{m + n} \to \underline{n + m}$ given by

$$\sigma_{m,n}(i) := \begin{cases} i + n & \text{if } i \leq m, \\ i - m & \text{if } m + 1 \leq i. \end{cases}$$

4.   Composition of bijections $m \to n$ and $n \to p$ is just their composition as functions, which is again a bijection.
5.   Given bijections $f \colon m \to m'$ and $g \colon n \to n'$, their monoidal product $(f + g) \colon (m + n) \to (m' + n')$ is given by

$$(f + g)(i) := \begin{cases} f(i) & \text{if } i \leq m, \\ g(i - m) + m' & \text{if } m + 1 \leq i. \end{cases}$$

Example 5.5: The prop **Corel** has

1.   **Corel**(m, n) is the set of equivalence relations on $\underline{m + n}$.
2.   The identity map $n \to n$ is the smallest equivalence relation, which is the smallest reflexive relation, i.e. where $i \sim j$ iff $i = j$.

3. The symmetry map $\sigma_{m,n}$, as an equivalence relation on $\underline{m+n+n+m}$ is "the obvious thing," namely "equating corresponding m's together and also equating corresponding ns together." To be pedantic, $i \sim j$ iff either
    - $|i - j| = m + n + n$, or
    - $m + 1 \le i \le m + n + n$ and $m + 1 \le j \le m + n + n$ and $|i - j| = n$.
4. The composition rule for correlations is explicitly described in footnote 2 on page 143.
5. Given equivalence relations $\sim$ on $\underline{m+n}$ and $\sim'$ on $\underline{m'+n'}$, we need an equivalence relation $(\sim + \sim')$ on $\underline{m+n+m'+n'}$. We take it to be "the obvious thing," namely "using $\sim$ on the unprimed stuff and using $\sim'$ on the primed stuff, with no other interaction." To be pedantic, $i \sim j$ iff either
    - $i \le m + n$ and $j \le m + n$ and $i \sim j$, or
    - $m + n + 1 \le i$ and $m + n + 1 \le j$ and $i \sim' j$.

Example 5.6: The prop **Rel** has

1. **Rel**(m, n) is the set of relations on the set $\underline{m} \times \underline{n}$, i.e. the set of subsets of $\underline{m} \times \underline{n}$, i.e. its power set.
2. The identity map $n \to n$ is the subset $\{(i, j) \in \underline{n} \times \underline{n} \mid i = j\}$.
3. The symmetry map $m + n \to n + m$ is the subset of pairs $(i, j) \in (\underline{m+n}) \times (\underline{n+m})$ such that either
    - $i \le m$ and $m + 1 \le j$ and $i + m = j$, or
    - $m + 1 \le i$ and $j \le m$ and $j + m = i$.
4. Composition of relations is as in Example 5.6.
5. Given a relation $R \subseteq \underline{m} \times \underline{n}$ and a relation $R' \subseteq \underline{m'} \times \underline{n'}$, we need a relation $(R + R') \subseteq \underline{m+m'} \times \underline{n+n'}$. As stated in the example (footnote), this can be given by a universal property: The monoidal product $R_1 + R_2$ of relations $R_1 \subseteq \underline{m_1} \times \underline{n_1}$ and $R_2 \subseteq \underline{m_2} \times \underline{n_2}$ is given by $R_1 \sqcup R_2 \subseteq (\underline{m_1} \times \underline{n_1}) \sqcup (\underline{m_2} \times \underline{n_2}) \subseteq (\underline{m_1} \sqcup \underline{m_2}) \times (\underline{n_1} \sqcup \underline{n_2})$.

Solution to Exercise 5.13

Composition of an (m, n)-port graph G and an (n, p)-port graph H looks visually like sticking them end to end, connecting the wires in order, removing the two outer boxes, and adding a new outer box.

For example, suppose we want to compose the following in the order shown:

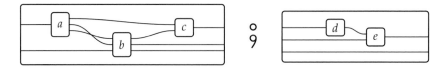

The result is:

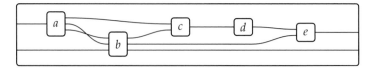

Solution to Exercise 5.14

The monoidal product of two morphisms is drawn by stacking the corresponding port graphs. For this problem, we just stack the left-hand picture on top of itself to obtain the right-hand picture:

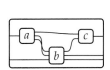 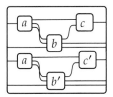

Solution to Exercise 5.16

We have a relation $R \subseteq P \times P$ which generates a preorder $\leq_P$ on P, we have an arbitrary preorder $(Q, \leq_Q)$ and a function $f : P \to Q$, not necessarily monotonic.

1. Assume that, for every $x, y \in P$, if $R(x, y)$ then $f(x) \leq f(y)$; we want to show that f is monotone, i.e. that for every $x \leq_P y$ we have $f(x) \leq_Q f(y)$. By definition of P being the reflexive, transitive closure of R, we have $x \leq_P y$ iff there exists $n \in \mathbb{N}$ and $x_0, \dots, x_n$ in P with $x_0 = x$ and $x_n = y$ and $R(x_i, x_{i+1})$ for each $0 \leq i \leq n - 1$. (The case $n = 0$ handles reflexivity.) But then, by assumption, $R(x_i, x_{i+1})$ implies $f(x_i) \leq_Q f(x_{i+1})$ for each i. By induction on i we show that $f(x_0) \leq_Q f(x_i)$ for all $0 \leq i \leq n$, at which point we are done.
2. Suppose now that f is monotone, and take $x, y \in P$ for which $R(x, y)$ holds. Then $x \leq_P y$ because $\leq_P$ is the smallest preorder relation containing R. (Another way to see this based on the above description is with $n = 1$, $x_0 = x$, and $x_n = y$, which we said implies $x \leq_P y$.) Since f is monotone, we indeed have $f(x) \leq_Q f(y)$.

Solution to Exercise 5.17

Suppose that P, Q, and R are as in Exercise 5.16 and we have a function $g : Q \to P$.

1. If $R(g(a), g(b))$ holds for all $a \leq_Q b$ then g is monotone, because $R(x, y)$ implies $x \leq_P y$.
2. It is possible for $g : (Q, \leq_Q) \to (P, \leq_P)$ to be monotone and yet have some $a, b \in Q$ with $a \leq_Q b$ and $(g(a), g(b)) \notin R$. Indeed, take $Q := \{1\}$ to be the free preorder on one element, and take $P := \{1\}$ with $R = \varnothing$. Then the unique function $g : Q \to P$ is monotone (because $\leq_P$ is reflexive even though R is empty), and yet $(g(1), g(1)) \notin R$.

Solution to Exercise 5.19

Let $G = (V, A, s, t)$ be a graph, let $\mathcal{G}$ be the free category on G, and let $\mathcal{C}$ be another category, whose set of morphisms is denoted $\mathrm{Mor}(\mathcal{C})$.

1. To give a function $\mathrm{Mor}(\mathcal{C}) \to \mathrm{Ob}(\mathcal{C})$ means that for every element $\mathrm{Mor}(\mathcal{C})$ we need to give exactly one element of $\mathrm{Ob}(\mathcal{C})$. So for dom we take any $q \in \mathrm{Mor}(\mathcal{C})$, view it as a morphism $q : y \to z$, and send it to its domain y. Similarly for cod: we put $\mathrm{cod}(q) := z$.
2. Suppose first that we are given a functor $F : \mathcal{G} \to \mathcal{C}$. On objects we have a function $\mathrm{Ob}(\mathcal{G}) \to \mathrm{Ob}(\mathcal{C})$, and this defines f since $\mathrm{Ob}(\mathcal{G}) = V$. On morphisms, first note that the arrows of graph G are exactly the length=1 paths in G, whereas $\mathrm{Mor}(\mathcal{G})$ is the set of all paths in G, so we have an inclusion $A \subseteq \mathrm{Mor}(\mathcal{G})$. The functor F provides a function

$\mathrm{Mor}(\mathcal{G}) \to \mathrm{Mor}(\mathcal{C})$, which we can restrict to A to obtain $g\colon A \to \mathrm{Mor}(\mathcal{C})$. All functors satisfy $\mathrm{dom}(F(r)) = F(\mathrm{dom}(r))$ and $\mathrm{cod}(F(r)) = F(\mathrm{cod}(r))$ for any $r\colon w \to x$. In particular when $r \in A$ is an arrow we have $\mathrm{dom}(r) = s(r)$ and $\mathrm{cod}(r) = t(r)$. Thus we have found (f, g) with the required properties.

Suppose second that we are given a pair of functions (f, g) where $f\colon V \to \mathrm{Ob}(\mathcal{C})$ and $g\colon A \to \mathrm{Mor}(\mathcal{C})$ such that $\mathrm{dom}(g(a)) = f(s(a))$ and $\mathrm{cod}(g(a)) = f(t(a))$ for all $a \in A$. Define $F\colon \mathcal{G} \to \mathcal{C}$ on objects by f. An arbitrary morphism in $\mathcal{G}$ is a path $p := (v_0, a_1, a_2, \ldots, a_n)$ in G, where $v_0 \in V$, $a_i \in A$, $v_0 = s(a_1)$, and $t(a_i) = s(a_{i+1})$ for all $1 \leq i \leq n - 1$. Then $g(a_i)$ is a morphism in $\mathcal{C}$ whose domain is $f(v_0)$ and the morphisms $g(a_i)$ and $g(a_{i+1})$ are composable for every $1 \leq i \leq n - 1$. We then take $F(p) := \mathrm{id}_{f(v_0)} \mathbin{\raise1pt\hbox{$\scriptstyle\circ$}\kern-1pt\raise-1pt\hbox{$\scriptstyle\circ$}} g(a_1) \mathbin{\raise1pt\hbox{$\scriptstyle\circ$}\kern-1pt\raise-1pt\hbox{$\scriptstyle\circ$}} \cdots \mathbin{\raise1pt\hbox{$\scriptstyle\circ$}\kern-1pt\raise-1pt\hbox{$\scriptstyle\circ$}} g(a_n)$ to be the composite. It is easy to check that this is indeed a functor (preserves identities and compositions).

Third, we want to see that the two operations we just gave are mutually inverse. On objects this is straightforward, and on morphisms it is straightforward to see that, given (f, g), if we turn them into a functor $F\colon \mathcal{G} \to \mathcal{C}$ and then extract the new pair of functions (f', g'), then $f = f'$ and $g = g'$. Finally, given a functor $F\colon \mathcal{G} \to \mathcal{C}$, we extract the pair of functions (f, g) as above and then turn them into a new functor $F'\colon \mathcal{G} \to \mathcal{C}$. It is clear that F and F' act the same on objects, so what about on morphisms? The formula says that F' acts the same on morphisms of length 1 in $\mathcal{G}$ (i.e. on the elements of A). But an arbitrary morphism in $\mathcal{G}$ is just a path, i.e. a sequence of composable arrows, and so by functoriality, both F and F' must act the same on arbitrary paths.

3. $(\mathrm{Mor}(\mathcal{C}), \mathrm{Ob}(\mathcal{C}), \mathrm{dom}, \mathrm{cod})$ is a graph; let's denote it $U(\mathcal{C}) \in \mathbf{Grph}$. We have functors **Free**: $\mathbf{Grph} \leftrightarrows \mathbf{Cat} : U$, and **Free** is left adjoint to U.

Solution to Exercise 5.20

1. The elements of the free monoid on the set $\{a\}$ are:

$$a^0, a^1, a^2, a^3, \ldots, a^{2019}, \ldots$$

with monoid multiplication $*$ given by the usual natural number addition on the exponents, $a^i * a^j = a^{i+j}$.

2. This is isomorphic to $\mathbb{N}$, by sending $a^i \mapsto i$.

3. The elements of the free monoid on the set $\{a, b\}$ are 'words in a and b,' each of which we will represent as a list whose entries are either a or b. Here are some:

$$[\,], \quad [a], \quad [b], \quad [a, a], \quad [a, b], \quad \ldots, \quad [b, a, b, b, a, b, a, a, a, a], \quad \ldots$$

Solution to Exercise 5.23

We have two props: the prop of port graphs and the free prop **Free**(G, s, t) where

$$G := \{\rho_{m,n}\colon m \to n \mid m, n \in \mathbb{N}\}, \qquad s(\rho_{m,n}) := m, \quad t(\rho_{m,n}) := n;$$

we want to show they are the same prop. As categories they have the same set of objects (in both cases, $\mathbb{N}$), so we need to show that, for every $m, n \in \mathbb{N}$, they have the same set of morphisms (and that their composition formulas and monoidal product formulas agree).

By Definition 5.21, a morphism $m \to n$ in **Free**(G) is a G-labeled port graph, i.e. a pair (Γ, ℓ), where $\Gamma = (V, in, out, \iota)$ is an (m, n)-port graph and $\ell\colon V \to G$ is a function, such that the "arities agree." What does this mean? Recall that every vertex $v \in V$ is drawn as a box with some left-hand ports and some right-hand ports – an arity – and $\ell(v) \in G$ is supposed to have the correct arity; precisely, $s(\ell(v)) = in(v)$ and $t(\ell(v)) = out(v)$. But G was chosen so that it has exactly one element with any given arity, so the function ℓ has only one choice, and thus contributes nothing: it neither increases nor decreases the freedom. In other words, a morphism in our particular **Free**(G) can be identified with an (m, n)-port graph Γ, as desired.

Again by Definition 5.21, the "composition and the monoidal structure are just those for port graphs **PG** (see Eq. (5.4)); the labelings (the ℓs) are just carried along." So we are done.

Solution to Exercise 5.27

Here is a picture of $(f + \mathrm{id}_1 + \mathrm{id}_1) \mathbin{\overset{\circ}{,}} (\sigma + \mathrm{id}_1) \mathbin{\overset{\circ}{,}} (\mathrm{id}_1 + h) \mathbin{\overset{\circ}{,}} \sigma \mathbin{\overset{\circ}{,}} g$, in the free prop on generators $G = \{f : 1 \to 1, g : 2 \to 2, h : 2 \to 1\}$:

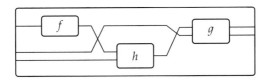

Solution to Exercise 5.30

The free prop on generators (G, s, t), defined in Definition 5.21, is – for all intents and purposes – the same thing as the prop presented by $(G, s, t, \varnothing)$, having no relations. The only possible "subtle difference" we might have to admit is if someone said that a set S is "subtly different" from its quotient by the trivial equivalence relation. In the latter, the elements are the singleton subsets of S. So for example the quotient of $S = \{1, 2, 3\}$ by the trivial equivalence relation is the set $\{\{1\}, \{2\}, \{3\}\}$. It is subtly different from S, but the two are naturally isomorphic, and category-theoretically, the difference will never make a difference.

Solution to Exercise 5.36

1. If $(R, 0, +, 1, *)$ is a rig, then the multiplicative identity $1 \in \mathrm{Mat}_n(R)$ is the usual n-by-n identity matrix: 1's on the diagonal and 0's everywhere else (where by "1" and "0," we mean those elements of R). So for $n = 4$ it is:

$$
\begin{pmatrix}
1 & 0 & 0 & 0 \\
0 & 1 & 0 & 0 \\
0 & 0 & 1 & 0 \\
0 & 0 & 0 & 1
\end{pmatrix}.
$$

2. We choose $n = 2$ and hence need to find two elements $A, B \in \mathrm{Mat}_2(\mathbb{N})$ such that $A * B \neq B * A$.

$$
A * B = \begin{pmatrix} 0 & 1 \\ 0 & 0 \end{pmatrix} * \begin{pmatrix} 0 & 1 \\ 1 & 0 \end{pmatrix} \neq \begin{pmatrix} 0 & 1 \\ 1 & 0 \end{pmatrix} * \begin{pmatrix} 0 & 1 \\ 0 & 0 \end{pmatrix} = B * A.
$$

One can calculate from the multiplication formula (recalled in Example 5.35) that $(A * B)(1, 1) = 0 * 0 + 1 * 1 = 1$ and $(B * A)(1, 1) = 0 * 0 + 0 * 0 = 0$, which are not equal.

Solution to Exercise 5.38

Semantically, if we apply the flow graph below to the input signal (x, y)

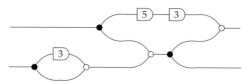

the resulting output signal is $(16x + 4y, x + 4y)$.

Solution to Exercise 5.44

The monoidal product of $A = \begin{pmatrix} 3 & 3 & 1 \\ 2 & 0 & 4 \end{pmatrix}$ and $B = \begin{pmatrix} 2 & 5 & 6 & 1 \end{pmatrix}$ is

$$A + B = \begin{pmatrix} 3 & 3 & 1 & 0 & 0 & 0 & 0 \\ 2 & 0 & 4 & 0 & 0 & 0 & 0 \\ 0 & 0 & 0 & 2 & 5 & 6 & 1 \end{pmatrix}.$$

Solution to Exercise 5.47

1. The signal flow graph on the left represents the matrix on the right:

 $\begin{pmatrix} 1 & 1 & 1 \end{pmatrix}$

2. The signal flow graph on the left represents the matrix on the right:

 $\begin{pmatrix} 1 & 1 & 1 \end{pmatrix}$

3. They are equal.

Solution to Exercise 5.49

1.

$$\begin{pmatrix} 0 \\ 1 \\ 2 \end{pmatrix} \sim$$

2.

$$\begin{pmatrix} 0 & 0 \\ 0 & 0 \end{pmatrix} \sim$$

3.

$$\begin{pmatrix} 1 & 2 & 3 \\ 4 & 5 & 6 \end{pmatrix} \sim$$

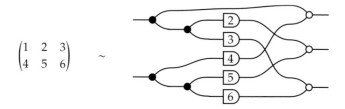

Solution to Exercise 5.50

- For the first layer g_1, take the monoidal product of m copies of c_n,

$$g_1 := c_n + \cdots + c_n : m \to (m \times n),$$

where c_n is the signal flow diagram that makes n copies of a single input:

$$c_n := \prec \, \S \, (1 + \prec) \, \S \, (1 + 1 + \prec) \, \S \ldots \S \, (1 + \cdots + 1 + \prec) : 1 \to n.$$

- Next, define

$$
\begin{aligned}
g_2 := \; & s_{M(1,1)} + \cdots + s_{M(1,n)} \\
& + s_{M(2,1)} + \cdots + s_{M(2,n)} \\
& + \cdots \\
& + s_{M(m,1)} + \cdots + s_{M(m,n)} : (m \times n) \to (m \times n),
\end{aligned}
$$

where $s_a : 1 \to 1$ is the signal flow graph generator "scalar multiplication by a." This layer amplifies each copy of the input signal by the relevant rig element.

- The third layer rearranges wires. We will not write this down explicitly, but simply say it is the signal flow graph $g_3 : m \times n \to m \times n$, i.e. the composite and monoidal product of swap and identity maps, such that the $(i - 1)m + j$th input is sent to the $(j - 1)n + i$th output, for all $1 \le i \le n$ and $1 \le j \le m$.

- Finally, the fourth layer is similar to the first, but instead adds the amplified input signals. We define

$$g_4 := a_m + \cdots + a_m : (m \times n) \to n,$$

where a_m is the signal flow graph that adds m inputs to produce a single output:

$$a_m := (1 + \cdots + 1 + \succ) \, \S \cdots \S \, (1 + 1 + \succ) \, \S \, (1 + \succ) \, \S \succ : m \to 1.$$

Using Proposition 5.46, it is a straightforward but tedious calculation to show that $g = g_1 \, \S \, g_2 \, \S \, g_3 \, \S \, g_4 : m \to n$ has the property that $S(g) = M$.

Solution to Exercise 5.53

1. The matrices in Example 5.49 may also be drawn as the following signal flow graphs:

(a)

$$\begin{pmatrix} 0 \\ 1 \\ 2 \end{pmatrix} \quad \sim$$

(b)

$$\begin{pmatrix} 0 & 0 \\ 0 & 0 \end{pmatrix} \quad \sim$$

(c)

$$\begin{pmatrix} 1 & 2 & 3 \\ 4 & 5 & 6 \end{pmatrix} \quad \sim$$

2. Here are graphical proofs that the representations we chose in our solution to Example 5.49 agree with those chosen in part 1 above.

(a)

(b)

(c)

Solution to Exercise 5.54

1. The signal flow graphs

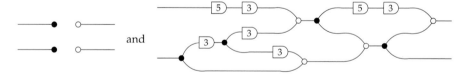

cannot represent the same morphism because one has a path from a vertex on the left to one on the right, and the other does not. To prove this, observe that the only graphical equation in Theorem 5.51 that breaks a path from left to right is the equation

So a 0 scalar must occur within a path from left to right before we could rewrite the diagram to break that path. No such 0 scalar can appear, however, because the diagram does not contain any, and the sum and product of any two nonzero natural numbers is always nonzero.

2. Replacing each of the 3's with 0 allows us to rewrite the diagram to

Solution to Exercise 5.57

The three conditions of Definition 5.55 are

(a) $(\mu \otimes \mathrm{id}) \,\overset{\circ}{,}\, \mu = (\mathrm{id} \otimes \mu) \,\overset{\circ}{,}\, \mu$,
(b) $(\eta \otimes \mathrm{id}) \,\overset{\circ}{,}\, \mu = \mathrm{id} = (\mathrm{id} \otimes \eta) \,\overset{\circ}{,}\, \mu$,
(c) $\sigma_{M,M} \,\overset{\circ}{,}\, \mu = \mu$,

where $\sigma_{M,M}$ is the swap map on M in $\mathcal{C}$.

1. Suppose $\mu \colon \mathbb{R} \times \mathbb{R} \to \mathbb{R}$ is defined by $\mu(a, b) = a * b$ and $\eta \in \mathbb{R}$ is defined to be $\eta = 1$. The conditions, written diagrammatically, say that starting in the upper left of each diagram below, the result in the lower right is the same regardless of which path you take:

$$
\begin{array}{ccc}
(a,b,c) & \xrightarrow{(\mu \otimes \mathrm{id})} & (a * b, c) \\
{\scriptstyle(\mathrm{id} \otimes \mu)} \downarrow & & \downarrow {\scriptstyle \mu} \\
(a, b * c) & \xrightarrow{\ \ \mu \ \ } & a * b * c
\end{array}
\qquad
\begin{array}{ccc}
a & \xrightarrow{(\eta \otimes \mathrm{id})} & (1, a) \\
{\scriptstyle(\mathrm{id} \otimes \eta)} \downarrow & {\scriptstyle \mathrm{id}_a} & \downarrow {\scriptstyle \mu} \\
(a, 1) & \xrightarrow{\ \ \mu \ \ } & a
\end{array}
\qquad
\begin{array}{ccc}
(a,b) & \xrightarrow{\sigma} & (b, a) \\
& {\scriptstyle \mu} \searrow & \downarrow {\scriptstyle \mu} \\
& & a * b
\end{array}
$$

and this is true for $(\mathbb{R}, *, 1)$.

2. The same reasoning works for $(\mathbb{R}, +, 0)$, shown below:

$$
\begin{array}{ccc}
(a,b,c) & \xrightarrow{(\mu \otimes \mathrm{id})} & (a+b, c) \\
{\scriptstyle(\mathrm{id} \otimes \mu)} \downarrow & & \downarrow {\scriptstyle \mu} \\
(a, b+c) & \xrightarrow{\ \ \mu \ \ } & a+b+c
\end{array}
\qquad
\begin{array}{ccc}
a & \xrightarrow{(\eta \otimes \mathrm{id})} & (0, a) \\
{\scriptstyle(\mathrm{id} \otimes \eta)} \downarrow & {\scriptstyle \mathrm{id}_a} & \downarrow {\scriptstyle \mu} \\
(a, 0) & \xrightarrow{\ \ \mu \ \ } & a
\end{array}
\qquad
\begin{array}{ccc}
(a,b) & \xrightarrow{\sigma} & (b, a) \\
& {\scriptstyle \mu} \searrow & \downarrow {\scriptstyle \mu} \\
& & a+b
\end{array}
$$

Solution to Exercise 5.59

The functor $U \colon \mathbf{Mat}(R) \to \mathbf{Set}$ is given on objects by sending n to the set R^n, and on morphisms by matrix-vector multiplication. Here R^n means the set of n-tuples or n-dimensional vectors in R. In particular, $R^0 = \{()\}$ consists of a single vector of dimension 0.

1. U preserves the monoidal unit because 0 is the monoidal unit of any prop ($\mathbf{Mat}(R)$ is a prop), $\{1\}$ is the monoidal unit of $\mathbf{Set}$, and R^0 is canonically isomorphic to $\{1\}$. U also preserves the monoidal product because there is a canonical isomorphism $R^m \times R^n \cong R^{m+n}$.

2. A monoid object in $\mathbf{Mat}(R)$ is a tuple (m, μ, η) where $m \in \mathbb{N}$, $\mu \colon m+m \to m$, and $\eta \colon 0 \to m$ satisfy the properties $\mu(\eta, x) = x = \mu(x, \eta)$ and $\mu(x, \mu(y, z)) = \mu(\mu(x, y), z)$. Note that there is only one morphism $0 \to m$ in $\mathbf{Mat}(R)$ for any m. It is not hard to show that for any $m \in \mathbb{N}$ there is only one monoid structure. For example, when $m = 2$, μ must be the following matrix

$$
\mu := \begin{pmatrix} 1 & 0 \\ 0 & 1 \\ 1 & 0 \\ 0 & 1 \end{pmatrix}
$$

Anyway, for any monoid (m, μ, η), the morphism $U(\eta) \colon R^0 \to R^m$ is given by $U(\eta)(1) := (0, \dots, 0)$, and the morphism $U(\mu) \colon R^m \times R^m \to R^m$ is given by

$$
U(\mu)((a_1, \dots, a_m), (b_1, \dots, b_m)) := (a_1 + b_1, \dots, a_m + b_m).
$$

These give R^m the structure of a monoid.

3. The triple $(1, -, -\!\!\!<)$ corresponds to the additive monoid structure on $\mathbb{R}$, e.g. with $(5, 3) \mapsto 8$.

Solution to Exercise 5.65

1. The behavior $B(\prec)$ of the reversed addition icon $\prec: 1 \to 2$ is the relation $\{(x, y, z) \in R^3 \mid x = y + z\}$.
2. The behavior $B(\succ)$ of the reversed copy icon, $\succ: 2 \to 1$ is the relation $\{(x, y, z) \in R^3 \mid x = y = z\}$.

Solution to Exercise 5.67

If $B \subseteq R^m \times R^n$ and $C \subseteq R^p \times R^q$ are morphisms in $\mathbf{Rel}_R$, then take $B + C \subseteq R^{m+p} \times R^{n+q}$ to be the set

$$B + C := \{(w, y, x, z) \in R^{m+p} \times R^{n+q} \mid (w, x) \in B \text{ and } (y, z) \in C\}.$$

Solution to Exercise 5.68

The behavior of $g: m \to n$ and $h^{\mathrm{op}}: n \to \ell$ are

$$B(g) = \{(x, z) \in R^m \times R^n \mid S(g)(x) = z\},$$
$$B(h^{\mathrm{op}}) = \{(z, y) \in R^n \times R^\ell \mid z = S(h)(y)\},$$

respectively, and, by Eq. (5.13), the composite $B(g \mathbin{\fatsemi} (h^{\mathrm{op}})) = B(g) \mathbin{\fatsemi} B(h^{\mathrm{op}})$ is:

$$\{(x, y) \mid \text{ there exists } z \in R^n \text{ such that } S(g)(x) = z \text{ and } z = S(h)(y)\}.$$

Since $S(g)$ and $S(h)$ are functions, the above immediately reduces to the desired formula:

$$B(g \mathbin{\fatsemi} (h^{\mathrm{op}})) = \{(x, y) \mid S(g)(x) = S(h)(y)\}.$$

Solution to Exercise 5.69

The behavior of $g^{\mathrm{op}}: n \to m$ and $h: m \to p$ are

$$B(g^{\mathrm{op}}) = \{(y, x) \in R^n \times R^m \mid y = S(g)(x)\},$$
$$B(h) = \{(x, z) \in R^m \times R^p \mid S(h)(x) = z\},$$

respectively, and, by Eq. (5.13), the composite $B((g^{\mathrm{op}}) \mathbin{\fatsemi} h) = B(g^{\mathrm{op}}) \mathbin{\fatsemi} B(h)$ is:

$$\{(y, z) \mid \text{ there exists } x \in R^m \text{ such that } y = S(g)(x) \text{ and } S(h)(x) = z\}.$$

This immediately reduces to the desired formula:

$$B((g^{\mathrm{op}}) \mathbin{\fatsemi} h) = \{(S(g)(x), S(h)(x)) \mid x \in R^m\}.$$

Solution to Exercise 5.70

1. The behavior of the 0-reverse $\multimap$ is the subset $\{y \in R \mid y = 0\}$, and its n-fold tensor is similarly $\{y \in R^n \mid y = 0\}$. Composing this relation with $S(g) \subseteq R^m \times R^n$ gives $\{x \in R^m \mid S(g) = 0\}$, which is the kernel of $S(g)$.
2. The behavior of the discard-inverse $\multimapdotbothA$ is the subset $\{x \in R\}$, i.e. the largest subset of R, and similarly its m-fold tensor is $R^n \subseteq R^n$. Composing this relation with $S(g) \subseteq R^m \times R^n$ gives $\{y \in R^n \mid \text{ there exists } x \in R^m \text{ such that } S(g)(x) = y\}$, which is exactly the image of $S(g)$.

3. For any $g: m \to n$, we first claim that the behavior $B(g) = \{(x, y) \mid S(g)(x) = y\}$ is linear, i.e. it is closed under addition and scalar multiplication. Indeed, $S(g)$ is multiplication by a matrix, so if $S(g)(x) = y$ then $S(g)(rx) = ry$ and $S(g)(x_1 + x_2) = S(g)(x_1) + S(g)(x_2)$. Thus we conclude that $(x, y) \in B(g)$ implies $(rx, ry) \in B(g)$, so it's closed under scalar multiplication, and $(x_1, y_1), (x_2, y_2) \in B(g)$ implies $(x_1 + x_2, y_1 + y_2) \in B(g)$, so it's closed under addition. Similarly, the behavior $B(g^{\mathrm{op}})$ is also linear; the proof is similar.

Finally, we need to show that the composite of any two linear relations is linear. Suppose that $B \subseteq R^m \times R^n$ and $C \subseteq R^n \times R^p$ are linear. Take $(x_1, z_1), (x_2, z_2) \in B \,\mathring{\circ}\, C$ and take $r \in R$. By definition, there exist $y_1, y_2 \in R^n$ such that $(x_1, y_1), (x_2, y_2) \in B$ and $(y_1, z_1), (y_2, z_2) \in C$. Since B and C are linear, $(rx_1, ry_1) \in B$ and $(ry_1, rz_1) \in C$, and also $(x_1 + x_2, y_1 + y_2) \in B$ and $(y_1 + y_2, z_1 + z_2) \in C$. Hence $(rx_1, rz_1) \in (B \,\mathring{\circ}\, C)$ and $(x_1 + x_2, z_1 + z_2) \in (B \,\mathring{\circ}\, C)$, as desired.

Solution to Exercise 5.71

Suppose that $B \subseteq R^m \times R^n$ and $C \subseteq R^n \times R^p$ are linear. Their composite is the relation $(B \,\mathring{\circ}\, C) \subseteq R^m \times R^p$ consisting of all (x, z) for which there exists $y \in R^n$ with $(x, y) \in B$ and $(y, z) \in C$. We want to show that the set $(B \,\mathring{\circ}\, C)$ is linear, i.e. closed under scalar multiplication and addition.

For scalar multiplication, take an $(x, z) \in (B \,\mathring{\circ}\, C)$ and any $r \in R$. Since B is linear, we have $(r * x, r * y) \in B$ and since C is linear we have $(r * y, r * z) \in C$, so then $(r * x, r * z) \in (B \,\mathring{\circ}\, C)$. For addition, if we also have $(x', z') \in (B \,\mathring{\circ}\, C)$ then there is some $y' \in R^n$ with $(x', y') \in B$ and $(y', z') \in C$, so since B and C are linear we have $(x + x', y + y') \in B$ and $(y + y', z + z') \in C$, hence $(x + x', z + z') \in (B \,\mathring{\circ}\, C)$.

A.6   Solutions for Chapter 6

Solution to Exercise 6.3

Let $A = \{a, b\}$, and consider the preorders shown here: $\boxed{\begin{smallmatrix} a & b \\ \bullet & \bullet \end{smallmatrix}}$, $\boxed{\begin{smallmatrix} a & & b \\ \bullet & \to & \bullet \end{smallmatrix}}$, $\boxed{\begin{smallmatrix} a & & b \\ \bullet & \leftrightarrows & \bullet \end{smallmatrix}}$.

1. The left-most (the discrete preorder on A) has no initial object, because $a \not\leq b$ and $b \not\leq a$.
2. The middle one has one initial object, namely a.
3. The right-most (the codiscrete preorder on A) has two initial objects.

Solution to Exercise 6.6

Recall that the objects of a free category on a graph are the vertices of the graph, and the morphisms are paths. Thus the free category on a graph G has an initial object if there exists a vertex v that has a unique path to every object. In graphs 1 and 2, the vertex a has this property, so the free categories on graphs 1 and 2 have initial objects. In graph 3, neither a nor b has a path to each other, and so there is no initial object. In graph 4, the vertex a has many paths to itself, and hence its free category does not have an initial object either.

Solution to Exercise 6.7

1. The remaining conditions are that $f(1_R) = 1_S$, and that $f(r_1 *_R r_2) = f(r_1) *_S f(r_2)$.
2. The initial object in the category **Rig** is the natural numbers rig $(\mathbb{N}, 0, +, 1, *)$. The fact that it is initial means that, for any other rig $R = (R, 0_R, +_R, 1_R, *_R)$, there is a unique rig homomorphism $f: \mathbb{N} \to R$.

What is this homomorphism? Well, to be a rig homomorphism, f must send 0 to 0_R, 1 to 1_R. Furthermore, we must also have $f(m + n) = f(m) +_R f(n)$, and hence

$$f(m) = f(\underbrace{1 + 1 + \cdots + 1}_{m \text{ summands}}) = \underbrace{f(1) + f(1) + \cdots + f(1)}_{m \text{ summands}} = \underbrace{1_R + 1_R + \cdots + 1_R}_{m \text{ summands}}.$$

So if there is a rig homomorphism $f : \mathbb{N} \to R$, it must be given by the above formula. But does this formula work correctly for multiplication?

It remains to check $f(m * n) = f(m) *_R f(n)$, and this will follow from distributivity. Noting that $f(m * n)$ is equal to the sum of mn copies of 1_R, we have

$$f(m) *_R f(n) = (\underbrace{1_R + \cdots + 1_R}_{m \text{ summands}}) *_R (\underbrace{1_R + \cdots + 1_R}_{n \text{ summands}})$$

$$= \underbrace{\overbrace{1_R * (\underbrace{1_R + \cdots + 1_R}_{n \text{ summands}}) + \cdots + 1_R * (\underbrace{1_R + \cdots + 1_R}_{n \text{ summands}})}^{m \text{ summands}}}$$

$$= \underbrace{1_R + \cdots + 1_R}_{mn \text{ summands}} = f(m * n).$$

Thus $(\mathbb{N}, 0, +, 1, *)$ is the initial object in **Rig**.

Solution to Exercise 6.8

In Definition 6.1, it is the initial object $\varnothing \in \mathcal{C}$ that is universal. In this case, all objects $c \in \mathcal{C}$ are "comparable objects." So the universal property of the initial object is that to any *object* $c \in \mathcal{C}$, there is a unique map $\varnothing \to c$ coming from the initial object.

Solution to Exercise 6.10

If c_1 is initial then by the universal property, for any c there is a unique morphism $c_1 \to c$; in particular, there is a unique morphism $c_1 \to c_2$, call it f. Similarly, if c_2 is initial then there is a unique morphism $c_2 \to c_1$, call it g. But how do we know that f and g are mutually inverse? Well since c_1 is initial there is a unique morphism $c_1 \to c_1$. But we can think of two: id_{c_1} and $f \, \mathring{,} \, g$. Thus they must be equal. Similarly for c_2, so we have $f \, \mathring{,} \, g = \text{id}_{c_1}$ and $g \, \mathring{,} \, f = \text{id}_{c_2}$, which is the definition of f and g being mutually inverse.

Solution to Exercise 6.12

Let $(P, \leq)$ be a preorder, and $p, q \in P$. Recall that a preorder is a category with at most one morphism, denoted $\leq$, between any two objects. Also recall that all diagrams in a preorder commute, since this means any two morphisms with the same domain and codomain are equal.

Translating Definition 6.11 to this case, a coproduct $p + q$ in P is an element of P such that $p \leq p + q$ and $q \leq p + q$, and such that for all elements $x \in P$ with maps $p \leq x$ and $q \leq x$, we have $p + q \leq x$. But this says exactly that $p + q$ is a join: it is a least element above both p and q. Thus coproducts in preorders are exactly the same as joins.

Solution to Exercise 6.14

The function $[f, g]$ is defined by

$$[f, g] : A \sqcup B \longrightarrow T$$

$$\text{apple1} \longmapsto a$$

$$\text{banana1} \longmapsto b$$
$$\text{pear1} \longmapsto p$$
$$\text{cherry1} \longmapsto c$$
$$\text{orange1} \longmapsto o$$
$$\text{apple2} \longmapsto e$$
$$\text{tomato2} \longmapsto o$$
$$\text{mango2} \longmapsto o.$$

Solution to Exercise 6.15

1. The equation $\iota_A \,\mathbin{\raise0.2ex\hbox{$\scriptstyle\circ$}\kern-0.1em\raise-0.3ex\hbox{$\scriptstyle\circ$}}\, [f, g] = f$ is the commutativity of the left-hand triangle in the commutative diagram (6.1) defining $[f, g]$.

2. The equation $\iota_B \,\mathbin{\raise0.2ex\hbox{$\scriptstyle\circ$}\kern-0.1em\raise-0.3ex\hbox{$\scriptstyle\circ$}}\, [f, g] = g$ is the commutativity of the right-hand triangle in the commutative diagram (6.1) defining $[f, g]$.

3. The equation $[f, g] \,\mathbin{\raise0.2ex\hbox{$\scriptstyle\circ$}\kern-0.1em\raise-0.3ex\hbox{$\scriptstyle\circ$}}\, h = [f \,\mathbin{\raise0.2ex\hbox{$\scriptstyle\circ$}\kern-0.1em\raise-0.3ex\hbox{$\scriptstyle\circ$}}\, h, g \,\mathbin{\raise0.2ex\hbox{$\scriptstyle\circ$}\kern-0.1em\raise-0.3ex\hbox{$\scriptstyle\circ$}}\, h]$ follows from the universal property of the coproduct. Indeed, the diagram

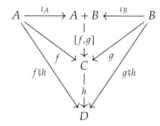

commutes, and the universal property says there is a unique map $[f \,\mathbin{\raise0.2ex\hbox{$\scriptstyle\circ$}\kern-0.1em\raise-0.3ex\hbox{$\scriptstyle\circ$}}\, h, g \,\mathbin{\raise0.2ex\hbox{$\scriptstyle\circ$}\kern-0.1em\raise-0.3ex\hbox{$\scriptstyle\circ$}}\, h]: A + B \to D$ for which this occurs. Hence we must have $[f, g] \,\mathbin{\raise0.2ex\hbox{$\scriptstyle\circ$}\kern-0.1em\raise-0.3ex\hbox{$\scriptstyle\circ$}}\, h = [f \,\mathbin{\raise0.2ex\hbox{$\scriptstyle\circ$}\kern-0.1em\raise-0.3ex\hbox{$\scriptstyle\circ$}}\, h, g \,\mathbin{\raise0.2ex\hbox{$\scriptstyle\circ$}\kern-0.1em\raise-0.3ex\hbox{$\scriptstyle\circ$}}\, h]$.

4. Similarly, to show $[\iota_A, \iota_B] = \mathrm{id}_{A+B}$, observe that the diagram

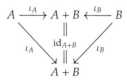

trivially commutes. Hence by the uniqueness in (6.1), $[\iota_A, \iota_B] = \mathrm{id}_{A+B}$.

Solution to Exercise 6.16

This exercise is about showing that coproducts and an initial object give a symmetric monoidal category. Since all we have are coproducts and an initial object, and since these are defined by their universal properties, the solution is to use these universal properties over and over, to prove that all the data of Definition 4.34 can be constructed.

1. To define a functor $+ \colon \mathcal{C} \times \mathcal{C} \to \mathcal{C}$ we must define its action on objects and morphisms. In both cases, we just take the coproduct. If (A, B) is an object of $\mathcal{C} \times \mathcal{C}$, its image $A + B$ is, as usual, the coproduct of the two objects of $\mathcal{C}$. If $(f, g) \colon (A, B) \to (C, D)$ is a morphism, then we can form a morphism $f + g = [f \,\mathbin{\raise0.2ex\hbox{$\scriptstyle\circ$}\kern-0.1em\raise-0.3ex\hbox{$\scriptstyle\circ$}}\, \iota_C, g \,\mathbin{\raise0.2ex\hbox{$\scriptstyle\circ$}\kern-0.1em\raise-0.3ex\hbox{$\scriptstyle\circ$}}\, \iota_D] \colon A + B \to C + D$, where $\iota_C \colon C \to C + D$ and $\iota_D \colon D \to C + D$ are the canonical morphisms given by the definition of the coproduct $A + B$.

Note that this construction sends identity morphisms to identity morphisms, since by Exercise 6.15 part 4 we have

$$\mathrm{id}_A + \mathrm{id}_B = [\mathrm{id}_A \,\mathring{\,}\, \iota_A, \mathrm{id}_B \,\mathring{\,}\, \iota_B] = [\iota_A, \iota_B] = \mathrm{id}_{A+B}.$$

To show that $+$ is a functor, we need to also show it preserves composition. Suppose we also have a morphism $(h, k) \colon (C, D) \to (E, F)$ in $\mathcal{C} \times \mathcal{C}$. We need to show that $(f + g) \,\mathring{\,}\, (h + k) = (f \,\mathring{\,}\, h) + (g \,\mathring{\,}\, k)$. This is a slightly more complicated version of the argument in Exercise 6.15 3. It follows from the fact that the diagram below commutes:

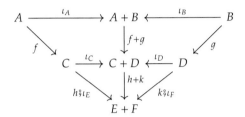

Indeed, we again use the uniqueness of the copairing in (6.1), this time to show that $(f \,\mathring{\,}\, h) + (g \,\mathring{\,}\, k) = [f \,\mathring{\,}\, h \,\mathring{\,}\, \iota_E, g \,\mathring{\,}\, k \,\mathring{\,}\, \iota_F] = (f + g) \,\mathring{\,}\, (h + k)$, as required.

2. Recall that the universal property of the initial object gives a unique map $!_A \colon \varnothing \to A$. Then the copairing $[\mathrm{id}_A, !_A]$ is a map $A + \varnothing \to A$. Moreover, it is an isomorphism with inverse $\iota_A \colon A \to A + \varnothing$. Indeed, using the properties in Exercise 6.15 and the universal property of the initial object, we have $\iota_A \,\mathring{\,}\, [\mathrm{id}_A, !_A] = \mathrm{id}_A$, and

$$[\mathrm{id}_A, !_A] \,\mathring{\,}\, \iota_A = [\mathrm{id}_A \,\mathring{\,}\, \iota_A, !_A \,\mathring{\,}\, \iota_A] = [\iota_A, !_{A+\varnothing}] = [\iota_A, \iota_\varnothing] = \mathrm{id}_{A+\varnothing}.$$

An analogous argument shows $[!_A, \mathrm{id}_A] \colon \varnothing + A \to A$ is an isomorphism.

3. We'll just write down the maps and their inverses; we leave it to you, if you like, to check that they indeed are inverses.

   (a) The map $[\mathrm{id}_A + \iota_B, \iota_C] = [[\iota_A, \iota_B \,\mathring{\,}\, \iota_{B+C}], \iota_C \,\mathring{\,}\, \iota_{B+C}] \colon (A + B) + C \to A + (B + C)$ is an isomorphism, with inverse $[\iota_A, \iota_B + \mathrm{id}_C] \colon A + (B + C) \to (A + B) + C$.

   (b) The map $[\iota_A, \iota_B] \colon A + B \to B + A$ is an isomorphism. Note our notation here is slightly confusing: there are two maps named ι_A, (i) $\iota_A \colon A \to A + B$, and (ii) $\iota_A \colon A \to B + A$, and similarly for ι_B. In the above we mean the map (ii). It has inverse $[\iota_B, \iota_A] \colon B + A \to A + B$, where in this case we mean the map (i).

Solution to Exercise 6.20

1. Suppose we are given an arbitrary diagram of the form $B \leftarrow A \to C$ in $\mathbf{Disc}_S$; we need to show that it has a pushout. The only morphisms in $\mathbf{Disc}_S$ are identities, so in particular $A = B = C$, and the square consisting of all identities is its pushout.

2. Suppose $\mathbf{Disc}_S$ has an initial object s. Then S cannot be empty! But it also cannot have more than one object, because if s' is another object then there is a morphism $s \to s'$, but the only morphisms in S are identities so $s = s'$. Hence the set S must consist of exactly one element.

Solution to Exercise 6.22

The pushout is the set $\underline{4}$, as depicted in the top right in the diagram below, equipped also with the depicted functions:

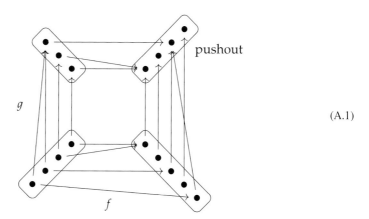

(A.1)

We want to see that this checks out with the description from Example 6.21, i.e. that it is the set of equivalence classes in $\underline{5} \sqcup \underline{3}$ generated by the relation $\{f(a) \sim g(a) \mid a \in \underline{4}\}$. If we denote elements of $\underline{5}$ as $\{1, \dots, 5\}$ and those of $\underline{3}$ as $\{1', 2', 3'\}$, we can redraw the functions f, g:

$$
\begin{array}{ccc}
1 & \longrightarrow \bullet \longrightarrow & 1' \\
2 & \bullet & 2' \\
3 & \bullet & 3' \\
4 & \bullet & \\
5 & &
\end{array}
$$

which says we take the equivalence relation on $\underline{5} \sqcup \underline{3}$ generated by: $1 \sim 1'$, , $3 \sim 1'$, $5 \sim 2'$, and $5 \sim 3'$. The equivalence classes are $\{1, 1', 3\}$, $\{2\}$, $\{4\}$, and $\{5, 2', 3'\}$. These four are exactly the four elements in the set labeled "pushout" in Eq. (A.1).

Solution to Exercise 6.24

1. The diagram to the left commutes because $\varnothing$ is initial, and so has a unique map $\varnothing \to X + Y$. This implies we must have $f \mathring{\,}_9 \iota_X = g \mathring{\,}_9 \iota_Y$.
2. There is a unique map $X + Y \to T$ making the diagram in (6.4) commute simply by the universal property of the coproduct (6.1) applied to the maps $x \colon X \to T$ and $y \colon Y \to T$.
3. Suppose $X +_\varnothing Y$ exists. By the universal property of $\varnothing$, given any pair of arrows $x \colon X \to T$ and $y \colon Y \to T$, the diagram

$$
\begin{array}{ccc}
\varnothing & \xrightarrow{\ f\ } & X \\
{\scriptstyle g}\downarrow & & \downarrow{\scriptstyle x} \\
Y & \xrightarrow[\ y\ ]{} & T
\end{array}
$$

commutes. This means, by the universal property of the pushout $X +_\varnothing Y$, there exists a unique map $t \colon X +_\varnothing Y \to T$ such that $\iota_X \mathring{\,}_9 t = x$ and $\iota_Y \mathring{\,}_9 t = y$. Thus $X +_\varnothing Y$ is the coproduct $X + Y$.

Solution to Exercise 6.30

We have to check that the colimit of the diagram shown left really is given by taking three pushouts as shown right:

That is, we need to show that S, together with the maps from A, B, X, Y, and Z, has the required universal property. So suppose given an object T with two commuting diagrams as shown:

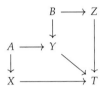

We need to show there is a unique map $S \to T$ making everything commute. Since Q is a pushout of $X \leftarrow A \to Y$, there is a unique map $Q \to T$ making a commutative triangle with Y, and since R is the pushout of $Y \leftarrow B \to Z$, there is a unique map $R \to T$ making a commutative triangle with Y. This implies that there is a commuting (Y, Q, R, T) square, and hence a unique map $S \to T$ from its pushout making everything commute. This is what we wanted to show.

Solution to Exercise 6.36

The formula in Theorem 6.32 says that the pushout $X +_N Y$ is given by the set of equivalence classes of $X \sqcup N \sqcup Y$ under the equivalence relation generated by $x \sim n$ if $x = f(n)$, and $y \sim n$ if $y = g(n)$, where $x \in X$, $y \in Y$, $n \in N$. Since for every $n \in N$ there exists an $x \in X$ such that $x = f(n)$, this set is isomorphic to the set of equivalence classes of $X \sqcup Y$ under the equivalence relation generated by $x \sim y$ if there exists n such that $x = f(n)$ and $y = g(n)$. This is exactly the description of Example 6.21.

Solution to Exercise 6.41

The monoidal product is

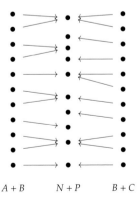

$$A + B \qquad N + P \qquad B + C$$

Solution to Exercise 6.42

Let x and y be composable cospans in **Cospan$_{\text{FinSet}}$**. In terms of wires and connected components, the composition rule in **Cospan$_{\text{FinSet}}$** says that (i) the composite cospan has a unique element in the apex for every connected component of the concatenation of the wire diagrams x and y, and (ii) in the wire diagram for $x \, \mathring{9} \, y$, each element of the feet is connected by a wire to the element representing the connected component to which it belongs.

Solution to Exercise 6.47

Morphisms 1, 4, and 6 are equal, and morphisms 3 and 5 are equal. Morphism 2 is not equal to any other depicted morphism. This is an immediate consequence of Theorem 6.45.

Solution to Exercise 6.49

1. The input to h should be labeled B.
2. The output of g should be labeled D, since we know from the labels in the top right that h is a morphism $B \to D \otimes D$.
3. The fourth output wire of the composite should be labeled D too!

Solution to Exercise 6.52

We draw the function depictions above, and the wiring depictions below. Note that we depict the empty set with blank space.

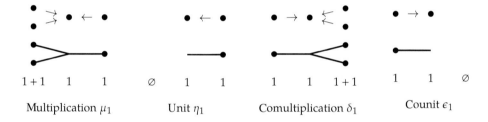

$1+1$	1	1	$\varnothing$	1	1	1	1	$1+1$	1	1	$\varnothing$

Multiplication μ_1 Unit η_1 Comultiplication δ_1 Counit ϵ_1

Solution to Exercise 6.53

The special law says that the composite of cospans

$$\multimap\!\!\bigcirc\!\!\multimap = X \xrightarrow{\text{id}} X \xleftarrow{[\text{id},\text{id}]} X + X \xrightarrow{[\text{id},\text{id}]} X \xleftarrow{\text{id}} X$$

is the identity. This comes down to checking that the square

$$\begin{array}{ccc} X + X & \xrightarrow{[\text{id},\text{id}]} & X \\ {\scriptstyle [\text{id},\text{id}]}\downarrow & & \downarrow{\scriptstyle \text{id}} \\ X & \xrightarrow{\text{id}} & X \end{array} \qquad (\text{A.2})$$

is a pushout square. It is trivial to see that the square commutes. Suppose now that we have maps $f : X \to Y$ and $g : X \to Y$ such that

$$\begin{array}{ccc} X + X & \xrightarrow{[\text{id},\text{id}]} & X \\ {\scriptstyle [\text{id},\text{id}]}\downarrow & & \downarrow{\scriptstyle f} \\ X & \xrightarrow{g} & T \end{array}$$

Write $\iota_1 : X \to X + X$ for the map into the first copy of X in $X + X$, given by the definition of coproduct. Then, using the fact that $\iota_1 \,\fatsemi\, [\mathrm{id}, \mathrm{id}] = \mathrm{id}$ from Exercise 6.15 1, and the commutativity of the above square, we have $f = \iota_1 \,\fatsemi\, [\mathrm{id}, \mathrm{id}] \,\fatsemi\, f = \iota_1 \,\fatsemi\, [\mathrm{id}, \mathrm{id}] \,\fatsemi\, g = g$. This means that $f : X \to T$ is the unique map such that

$$
\begin{array}{ccc}
X & \xrightarrow{[\mathrm{id},\mathrm{id}]} & X \\
{\scriptstyle[\mathrm{id},\mathrm{id}]}\downarrow & & \downarrow{\scriptstyle\mathrm{id}} \quad\searrow{\scriptstyle f} \\
X & \xrightarrow[\mathrm{id}]{} & X \\
& \searrow{\scriptstyle f} & \\
& g=f & T
\end{array}
$$

commutes, and so (A.2) is a pushout square.

Solution to Exercise 6.57
The missing diagram is

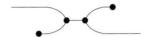

Solution to Exercise 6.60
Let $A \subseteq S$ and $B \subseteq T$. Then

$$
\begin{aligned}
\varphi_{S',T'}\big((\mathrm{im}_f \times \mathrm{im}_g)(A \times B)\big) &= \varphi_{S',T'}(\{f(a) \mid a \in A\} \times \{g(b) \mid b \in B\}) \\
&= \{(f(a), g(b)) \mid a \in A,\ b \in B\} \\
&= \mathrm{im}_{f \times g}(A \times B) \\
&= \mathrm{im}_{f \times g}(\varphi_{S,T}(A, B)).
\end{aligned}
$$

Thus the required square commutes.

Solution to Exercise 6.63
They mean that every category $\mathbf{Cospan}_{\mathcal{C}}$ is equal to a category $\mathbf{Cospan}_F$, for some well-chosen F. They also tell you how to choose this F: take the functor $F \colon \mathcal{C} \to \mathbf{Set}$ that sends every object of $\mathcal{C}$ to the set $\{*\}$, and every morphism of $\mathcal{C}$ to the identity function on $\{*\}$. Of course, you will have to check this functor is a lax symmetric monoidal functor, but in fact this is not hard to do.

To check that $\mathbf{Cospan}_{\mathcal{C}}$ is equal to $\mathbf{Cospan}_F$, first observe that they have the same objects: the objects of $\mathcal{C}$. Next, observe that a morphism in $\mathbf{Cospan}_F$ is a cospan $X \leftarrow N \to Y$ in $\mathcal{C}$ together with an element of $FN = \{*\}$. But FN also has a unique element, $*$! So there's no choice here, and we can consider morphisms of $\mathbf{Cospan}_F$ just to be cospans in $\mathcal{C}$. Moreover, composition of morphisms in $\mathbf{Cospan}_F$ is simply the usual composition of cospans via pushout, so $\mathbf{Cospan}_F = \mathbf{Cospan}_{\mathcal{C}}$.

(More technically, we might say that $\mathbf{Cospan}_{\mathcal{C}}$ and $\mathbf{Cospan}_F$ are isomorphic, where the isomorphism is the identity-on-objects functor $\mathbf{Cospan}_{\mathcal{C}} \to \mathbf{Cospan}_F$ that simply decorates each cospan with $*$, and its inverse is the one that forgets this $*$. But this is close enough to equal that many category theorists, us included, don't mind saying equal in this case.)

Solution to Exercise 6.64

We can represent the circuit in Eq. (6.11) by the tuple (V, A, s, t, ℓ) where $V = \{ul, ur, dl, dr\}$, $A = \{r1, r2, r3, c1, i1\}$, and $s, t,$ and ℓ are defined by the table

	r1	r2	r3	c1	i1
$s(-)$	dl	ul	ur	ul	dl
$t(-)$	ul	ur	dr	ur	dr
$\ell(-)$	1Ω	2Ω	1Ω	$3F$	$1H$

Solution to Exercise 6.65

The circuit $\mathrm{Circ}(f)(c)$ is

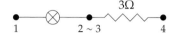

Solution to Exercise 6.66

The circuit $\psi_{2,2}(b, s)$ is the disjoint union of the two labeled graphs b and s:

Solution to Exercise 6.67

The cospan is the cospan $\underline{1} \xrightarrow{f} \underline{2} \xleftarrow{g} \underline{1}$, where $f(1) = 1$ and $g(1) = 2$. The decoration is the C-circuit $(\underline{2}, \{a\}, s, t, \ell)$, where $s(a) = 1, t(a) = 2$ and $\ell(a) = \texttt{battery}$.

Solution to Exercise 6.68

Recall the circuit $C := (V, A, s, t, \ell)$ from the solution to Exercise 6.64. Then the first decorated cospan is given by the cospan $\underline{1} \xrightarrow{f} V \xleftarrow{g} \underline{2}$, $f(1) = ul$, $g(1) = ur$, and $g(2) = ur$, decorated by circuit C. The second decorated cospan is given by the cospan $\underline{2} \xrightarrow{f'} V' \xleftarrow{g'} \underline{2}$ and the circuit $C' := (V', A', s', t', \ell')$, where $V' = \{l, r, d\}$, $A' = \{r1', r2'\}$, and the functions are given by the tables

	1	2
$f'(-)$	1	d
$g'(-)$	r	r

	r1'	r2'
$s(-)$	l	r
$t(-)$	r	d
$\ell(-)$	5Ω	8Ω

To compose these, we first take the pushout of $V \xleftarrow{g} \underline{2} \xrightarrow{f'} V'$. This gives the a new apex $V'' = \{ul, dl, dr, m, r\}$ with five elements, and composite cospan $\underline{1} \xrightarrow{h} V'' \xleftarrow{k} \underline{2}$ given by $h(1) = ul$, $k(1) = r$ and $k(2) = m$. The new circuit is given by $(V,'' A + A', s,'' t,'' \ell'')$ where the functions are given by

	r1	r2	r3	c1	i1	r1'	r2'
$s''(-)$	dl	ul	m	ul	dl	m	r
$t''(-)$	ul	m	dr	m	dr	r	m
$\ell''(-)$	1Ω	2Ω	1Ω	$3F$	$1H$	5Ω	8Ω

This is exactly what is depicted in Eq. (6.14).

Solution to Exercise 6.69

Composing η and x we have

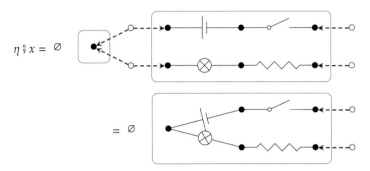

and composing the result with ϵ gives

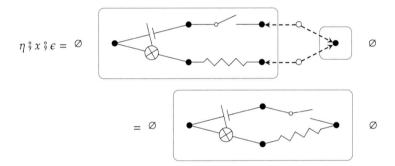

Solution to Exercise 6.74

1. The cospan shown left corresponds to the wiring diagram shown right:

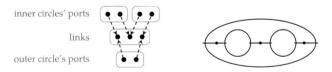

It has two inner circles, each with two ports. One port of the first is wired to a port of the second. One port of the first is wired to the outside circle, and one port of the second is wired to the outside circle. This is exactly what the cospan says to do.

2. The cospan shown left corresponds to the wiring diagram shown right:

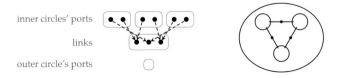

3. The composite $g \circ_1 f$ has arity $(2, 2, 2, 2; 0)$; there is a depiction on the left:

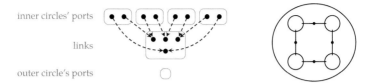

4. The associated wiring diagram is shown on the right above. One can see that one diagram has been substituted into a circle of the other.

A.7   Solutions for Chapter 7

Solution to Exercise 7.2

In the commutative diagram below, suppose the (B, C, B', C') square is a pullback:

$$
\begin{array}{ccccc}
A & \xrightarrow{f} & B & \xrightarrow{g} & C \\
\downarrow{h_1} & & \downarrow{h_2} & & \downarrow{h_3} \\
A' & \xrightarrow{f'} & B' & \xrightarrow{g'} & C'
\end{array}
$$

We need to show that the (A, B, A', B') square is a pullback iff the (A, C, A', C') rectangle is a pullback.

Suppose first that (A, B, A', B') is a pullback, and take any (X, p, q) as in the following diagram:

$$
\begin{array}{ccc}
X & \xrightarrow{\quad p \quad} & \\
& A \xrightarrow{f} B \xrightarrow{g} C & \\
q & \downarrow{h_1} \quad h_2\downarrow \quad h_3\downarrow & \\
& A' \xrightarrow{f'} B' \xrightarrow{g'} C' &
\end{array}
$$

where $q \,\fatsemi\, f' \,\fatsemi\, g' = p \,\fatsemi\, h_3$. Then by the universal property of the (B, C, B', C') pullback, we get a unique dotted arrow r making the left-hand diagram below commute:

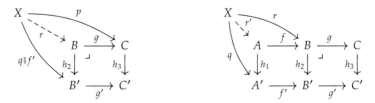

In other words $r \,\fatsemi\, h_2 = g \,\fatsemi\, f'$ and $r \,\fatsemi\, g = p$. Then by the universal property of the (A, B, A', B') pullback, we get a unique dotted arrow $r' : X \to A$ making the right-hand diagram commute, i.e. $r' \,\fatsemi\, f = r$ and $r' \,\fatsemi\, h_1 = q$. This gives the existence of an r with the required property, $r' \,\fatsemi\, f = r$

and $r' \fatsemi f \fatsemi g = r \fatsemi g = p$. To see uniqueness, suppose given another morphisms r_0 such that $r_0 \fatsemi f \fatsemi g = p$ and $r_0 \fatsemi h_1 = q$:

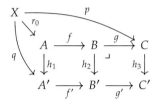

Then by the uniqueness of r, we must have $r_0 \fatsemi f = r$, and then by the uniqueness of r', we must have $r_0 = r'$. This proves the first result.

The second is similar. Suppose that (A, C, A', C') and (B, C, B', C') are pullbacks and suppose we are given a commutative diagram of the following form:

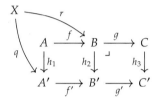

i.e. where $r \fatsemi h_2 = q \fatsemi f'$. Then letting $p := r \fatsemi g$, we have

$$p \fatsemi h_3 = r \fatsemi g \fatsemi h_3 = r \fatsemi h_2 \fatsemi g' = q \fatsemi f' \fatsemi g',$$

so by the universal property of the (A, C, A', C') pullback, there is a unique morphism $r' : X \rightarrow A$ such that $r' \fatsemi f \fatsemi g = p$ and $r_0 \fatsemi h_1 = q$, as shown:

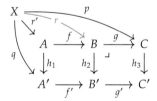

But now let $r_0 := r' \fatsemi f$. It satisfies $r_0 \fatsemi g = p$ and $r_0 \fatsemi h_2 = q \fatsemi f'$, and r satisfies the same equations: $r \fatsemi g = p$ and $r \fatsemi h_2 = q \fatsemi f'$. Hence by the universal property of the (B, C, B', C') pullback $r_0 = r'$. It follows that r' is a pullback of the (A, B, A', B') square, as desired.

Solution to Exercise 7.4

A function $f : A \rightarrow B$ is injective iff for all $a_1, a_2 \in A$, if $f(a_1) = f(a_2)$ then $a_1 = a_2$. It is a monomorphism iff for all sets X and functions $g_1, g_2 : X \rightarrow A$, if $g_1 \fatsemi f = g_2 \fatsemi f$ then $g_1 = g_2$. Indeed, this comes directly from the universal property of the pullback from Definition 7.3, because the dashed arrow is forced to equal both g_1 and g_2, thus forcing $g_1 = g_2$.

1. Suppose f is a monomorphism, let $a_1, a_2 \in A$ be elements, and suppose $f(a_1) = f(a_2)$. Let $X = \{*\}$ be a one-element set, and let $g_1, g_2 \colon X \to A$ be given by $g_1(*) := a_1$ and $g_2(*) := a_2$. Then $g_1 \,\fatsemi\, f = g_2 \,\fatsemi\, f$, so $g_1 = g_2$, so $a_1 = a_2$.
2. Suppose that f is an injection, let X be any set, and let $g_1, g_2 \colon X \to A$ be such that $g_1 \,\fatsemi\, f = g_2 \,\fatsemi\, f$. We will have $g_1 = g_2$ if we can show that $g_1(x) = g_2(x)$ for every $x \in X$. So take any $x \in X$; since $f(g_1(x)) = f(g_2(x))$ and f is injective, we have $g_1(x) = g_2(x)$ as desired.

Solution to Exercise 7.5
1. Suppose we have a pullback as shown, where i is an isomorphism:

$$
\begin{array}{ccc}
A' & \xrightarrow{\;f'\;} & B' \\
{\scriptstyle i'}\big\downarrow & \lrcorner & {\scriptstyle\cong}\big\downarrow{\scriptstyle i} \\
A & \xrightarrow[\;f\;]{} & B
\end{array}
$$

Let $j := i^{-1}$ be the inverse of i, and consider $g := (f \,\fatsemi\, j) \colon A \to B'$. Then $g \,\fatsemi\, i = f$, so by the existence part of the universal property, there is a map $j' \colon A \to A'$ such that $j' \,\fatsemi\, i' = \mathrm{id}_A$ and $j' \,\fatsemi\, f' = f \,\fatsemi\, j$. We will be done if we can show $i' \,\fatsemi\, j' = \mathrm{id}_{A'}$. One checks that $(i' \,\fatsemi\, j') \,\fatsemi\, i' = i'$ and that $(i' \,\fatsemi\, j') \,\fatsemi\, f' = i' \,\fatsemi\, f \,\fatsemi\, j = f' \,\fatsemi\, i \,\fatsemi\, j = f'$. But $\mathrm{id}_{A'}$ also satisfies those properties: $\mathrm{id}_{A'} \,\fatsemi\, i' = i'$ and $\mathrm{id}_{A'} \,\fatsemi\, f' = f'$, so by the uniqueness part of the universal property, $(i' \,\fatsemi\, j') = \mathrm{id}_{A'}$.
2. We need to show that the following diagram is a pullback:

$$
\begin{array}{ccc}
A & \xrightarrow{\;f\;} & B \\
\big\| & \lrcorner & \big\| \\
A & \xrightarrow[\;f\;]{} & B
\end{array}
$$

So take any object X and morphisms $g \colon X \to A$ and $h \colon X \to B$ such that $g \,\fatsemi\, f = h \,\fatsemi\, \mathrm{id}_B$. We need to show there is a unique morphism $r \colon X \to A$ such that $r \,\fatsemi\, \mathrm{id}_A = g$ and $r \,\fatsemi\, f = h$. That's easy: the first requirement forces $r = g$ and the second requirement is then fulfilled.

Solution to Exercise 7.6
Consider the diagram shown left, in which all three squares are pullbacks:

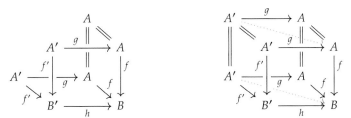

The front and bottom squares are the same – the assumed pullback – and the right-hand square is a pullback because f is assumed monic. We can complete it to the commutative diagram shown right, where the back square and top square are pullbacks by Exercise 7.5. Our goal is to show that the left-hand square is a pullback.

To do this, we use two applications of the pasting lemma, Exercise 7.2. Since the right-hand face is a pullback and the back face is a pullback, the diagonal rectangle (lightly drawn) is also a pullback. Since the front face is a pullback, the left-hand face is also a pullback.

Solution to Exercise 7.7

The following is an epi-mono factorization of f:

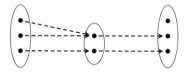

Solution to Exercise 7.8

1. If V is a quantale with the stated properties, then
    * I serves as a top element: $v \leq I$ for all $v \in V$,
    * $v \otimes w$ serves as a meet operation, i.e. it satisfies the same universal property as $\wedge$, namely $v \otimes w$ is a greatest lower bound for v and w.

    Now the $\multimap$ operation satisfies the same universal property as exponentiation (hom-object) does, namely $v \leq (w \multimap x)$ iff $v \otimes W \leq x$. So V is a cartesian closed category, and of course it is a preorder.

2. Not every cartesian closed preorder comes from a quantale with the stated properties, because quantales have all joins and cartesian closed preorders need not. Finding a counterexample – a cartesian closed preorder that is missing some joins – takes some ingenuity, but it can be done. Here's one we came up with:

$$
\begin{array}{llll}
(0,0) \leftarrow (0,1) \leftarrow (0,2) \leftarrow (0,3) & \cdots \\
\uparrow \qquad \uparrow \qquad \quad \uparrow \\
(1,0) \leftarrow (1,1) \leftarrow (1,2) & \cdots \\
\uparrow \qquad \uparrow \\
(2,0) \leftarrow (1,1) & \cdots \\
\uparrow \\
(3,0) & \cdots
\end{array}
$$

This is the product preorder $\mathbb{N}^{\text{op}} \times \mathbb{N}^{\text{op}}$: its objects are pairs $(a, b) \in \mathbb{N} \times \mathbb{N}$ with $(a, b) \leq (a', b')$ iff, in the usual ordering on $\mathbb{N}$, we have $a' \leq a$ and $b' \leq b$. But you can just look at the diagram.

It has a top element, $(0, 0)$, and it has binary meets, $(a, b) \wedge (a', b') = (\max(a, a'), \max(b, b'))$. But it has no bottom element, so it has no empty join. Thus we will be done if we can show that for each x, y, the hom-object $x \multimap y$ exists. The formula for it is $x \multimap y = \bigvee \{w \mid w \wedge x \leq y\}$, i.e. we need these particular joins to exist. Since $y \wedge x \leq y$, we have $y \leq x \multimap y$. So we can replace the formula with $x \multimap y = \bigvee \{w \mid y \leq w \text{ and } w \wedge x \leq y\}$. But the set of elements in $\mathbb{N}^{\text{op}} \times \mathbb{N}^{\text{op}}$ that are bigger than y is finite and nonempty.[2] So this is a finite nonempty join, and $\mathbb{N}^{\text{op}} \times \mathbb{N}^{\text{op}}$ has all finite nonempty joins: they are given by inf.

---

[2] If $y = (a, b)$ then there are exactly $(a + 1) * (b + 1)$ elements y' for which $y \leq y'$.

Solution to Exercise 7.10

Let $m: \mathbb{Z} \to \mathbb{B}$ be the characteristic function of the inclusion $\mathbb{N} \subseteq \mathbb{Z}$.

1. $\ulcorner m \urcorner(-5) = \texttt{false}$.
2. $\ulcorner m \urcorner(0) = \texttt{true}$.

Solution to Exercise 7.11

1. The characteristic function $\ulcorner \text{id}_\mathbb{N} \urcorner: \mathbb{N} \to \mathbb{B}$ sends each $n \in \mathbb{N}$ to $\texttt{true}$.
2. Let $!_\mathbb{N}: \varnothing \to \mathbb{N}$ be the inclusion of the empty set. The characteristic function $\ulcorner !_\mathbb{N} \urcorner: \mathbb{N} \to \mathbb{B}$ sends each $n \in \mathbb{N}$ to $\texttt{false}$.

Solution to Exercise 7.12

1. The sort of thing (*?*) we're looking for is a subobject of $\mathbb{B}$, say $A \subseteq \mathbb{B}$. This would have a characteristic function, and we're trying to find the A for which the characteristic function is $\neg: \mathbb{B} \to \mathbb{B}$.
2. The question now asks "what is A?" The answer is $\{\texttt{false}\} \subseteq \mathbb{B}$.

Solution to Exercise 7.13

1. Here is the truth table for $P = (P \wedge Q)$:

P	Q	$P \wedge Q$	$P = (P \wedge Q)$	
true	true	true	true	
true	false	false	false	(A.3)
false	true	false	true	
false	false	false	true	

2. Yes!
3. The characteristic function for $P \Rightarrow Q$ is the function $\ulcorner \Rightarrow \urcorner: \mathbb{B} \times \mathbb{B} \to \mathbb{B}$ given by the first, second, and fourth column of Eq. (A.3).
4. It classifies the subset $\{(\texttt{true}, \texttt{true}), (\texttt{false}, \texttt{true}), (\texttt{false}, \texttt{false})\} \subseteq \mathbb{B} \times \mathbb{B}$.

Solution to Exercise 7.14

Say that $\ulcorner E \urcorner, \ulcorner P \urcorner, \ulcorner T \urcorner: \mathbb{N} \to \mathbb{B}$ classify respectively the subsets $E := \{n \in \mathbb{N} \mid n \text{ is even}\}$, $P := \{n \in \mathbb{N} \mid n \text{ is prime}\}$, and $T := \{n \in \mathbb{N} \mid n \geq 10\}$ of $\mathbb{N}$.

1. $\ulcorner E \urcorner(17) = \texttt{false}$ because 17 is not even.
2. $\ulcorner P \urcorner(17) = \texttt{true}$ because 17 is prime.
3. $\ulcorner T \urcorner(17) = \texttt{true}$ because $17 \geq 10$.
4. The set classified by $(\ulcorner E \urcorner \wedge \ulcorner P \urcorner) \vee \ulcorner T \urcorner$ is that of all natural numbers that are either above 10 or an even prime. The smallest three elements of this set are 2, 10, 11.

Solution to Exercise 7.19

1. The one-dimensional analogue of an ϵ-ball around a point $x \in \mathbb{R}$ is $B(x, \epsilon) := \{x' \in \mathbb{R} \mid |x - x'| < \epsilon\}$, i.e. the set of all points within ϵ of x.
2. A subset $U \subseteq \mathbb{R}$ is open if, for every $x \in U$, there is some $\epsilon > 0$ such that $B(x, \epsilon) \subseteq U$.
3. Let $U_1 := \{x \in \mathbb{R} \mid 0 < x < 2\}$ and $U_2 := \{x \in \mathbb{R} \mid 1 < x < 3\}$. Then $U := U_1 \cup U_2 = \{x \in \mathbb{R} \mid 0 < x < 3\}$.

4. Let $I = \{1, 2, 3, 4, \ldots\}$ and for each $i \in I$ let $U_i := \{x \in \mathbb{R} \mid \frac{1}{i} < x < 1\}$, so we have $U_1 \subseteq U_2 \subseteq U_3 \subseteq \cdots$. Their union is $U := \bigcup_{i \in I} U_i = \{x \in \mathbb{R} \mid 0 < x < 1\}$.

Solution to Exercise 7.21

1. The coarse topology on X is the one whose only open sets are $X \subseteq X$ and $\varnothing \subseteq X$. This is a topology because it contains the top and bottom subsets, it is closed under finite intersection (the intersection $A \cap B$ is $\varnothing$ iff one or the other is $\varnothing$), and it is closed under arbitrary union (the union $\bigcup_{i \in I} A_i$ is X iff $A_i = X$ for some $i \in I$).
2. The fine topology on X is the one where every subset $A \subseteq X$ is considered open. All the conditions on a topology say "if such-and-such then such-and-such is open," but these are all satisfied because everything is open!
3. If $(X, \mathsf{P}(X))$ is discrete, $(Y, \mathbf{Op}_Y)$ is any topological space, and $f : X \to Y$ is any function then it is continuous. Indeed, this just means that for any open set $U \subseteq Y$ the preimage $f^{-1}(U) \subseteq X$ is open, and everything in X is open.

Solution to Exercise 7.23

1. The Hasse diagram for the Sierpinski topology is $\boxed{\varnothing \to \{1\} \to \{1, 2\}}$.
2. A set $(U_i)_{i \in I}$ covers U iff
   - either $I = \varnothing$ and $U = \varnothing$,
   - or $U_i = U$ for some $i \in I$.

   In other words, the only way that some collection of these sets could cover another set U is if that collection contains U or if U is empty and the collection is also empty.

Solution to Exercise 7.24

Let $(X, \mathbf{Op})$ be a topological space, suppose that $Y \subseteq X$ is a subset, and consider the subspace topology $\mathbf{Op}_{?\cap Y}$.

1. We want to show that $Y \in \mathbf{Op}_{?\cap Y}$. We need to find $B \in \mathbf{Op}$ such that $Y = B \cap Y$; this is easy, take $B = X$.
2. We still need to show that $\mathbf{Op}_{?\cap Y}$ contains $\varnothing$ and is closed under finite intersection and arbitrary union. Note $\varnothing = \varnothing \cap Y$, so according to the formula, $\varnothing \in \mathbf{Op}_{?\cap Y}$. Suppose that $A_1, A_2 \in \mathbf{Op}_{?\cap Y}$. Then there exist $B_1, B_2 \in \mathbf{Op}$ with $A_1 = B_1 \cap Y$ and $A_2 = B_2 \cap Y$. But then $A_1 \cap A_2 = (B_1 \cap Y) \cap (B_2 \cap Y) = (B_1 \cap B_2) \cap Y$, so it is in $\mathbf{Op}_{?\cap Y}$ since $B_1 \cap B_2 \in \mathbf{Op}$. The same idea works for arbitrary unions: given a set I and A_i for each $i \in I$, we have $A_i = B_i \cap Y$ for some $B_i \in \mathbf{Op}$, and

$$\bigcup_{i \in I} A_i = \bigcup_{i \in I} (B_i \cap Y) = \left(\bigcup_{i \in i} B_i \right) \cap Y \in \mathbf{Op}_{?\cap Y}.$$

Solution to Exercise 7.26

Let's imagine a $\mathcal{V}$-category $\mathcal{C}$, where $\mathcal{V}$ is the quantale corresponding to the open sets of a topological space $(X, \mathbf{Op})$. Its Hasse diagram would be a set of dots and some arrows between them, each labeled by an open set $U \subseteq \mathbf{Op}$. It might look something like this:

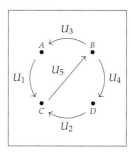

Recall from Section 2.3 that the "distance" between two points is computed by taking the join, over all paths between them, of the monoidal product of distances along that path. For example, $\mathcal{C}(B, C) = (U_3 \wedge U_1) \vee (U_4 \wedge U_2)$, because $\wedge$ is the monoidal product in $\mathcal{V}$.

In general, we can thus imagine the open set $\mathcal{C}(a, b)$ as a kind of "size restriction" for getting from a to b, like bridges that your truck needs to pass under. The size restriction for getting from a to itself is X: no restriction. In general, to go on any given route (path) from a to b, you have to fit under every bridge in the path, so we take their meet. But we can go along any path, so we take the join over all paths.

Solution to Exercise 7.29

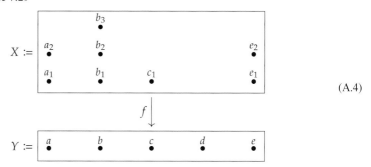

$$(A.4)$$

1. The fiber of f over a is $\{a_1, a_2\}$.
2. The fiber of f over c is $\{c_1\}$.
3. The fiber of f over d is $\varnothing$.
4. A function $f' \colon X \to Y$ for which every fiber has either one or two elements is shown below.

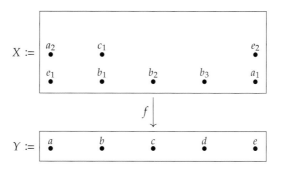

Solution to Exercise 7.30
Refer to Eq. (A.4).

1. Here is a drawing of all six sections over $V_1 = \{a, b, c\}$:

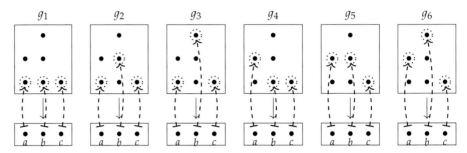

2. When $V_2 = \{a, b, c, d\}$, there are no sections: $\operatorname{Sec}_f(V_2) = \varnothing$.
3. When $V_3 = \{a, b, d, e\}$, the set $\operatorname{Sec}_f(V_3)$ has $2 * 3 * 1 * 2 = 12$ elements.

Solution to Exercise 7.31
$\operatorname{Sec}_f(\{a, b, c\})$ and $\operatorname{Sec}_f(\{a, c\})$ are drawn as the top row (six-element set) and bottom row (two-element set) below, and the restriction map is also shown:

$$(a_1, b_1, c_1) \quad (a_1, b_2, c_1) \quad (a_1, b_3, c_1) \quad (a_2, b_1, c_1) \quad (a_2, b_2, c_1) \quad (a_2, b_3, c_1)$$

$$(a_1, c_1) \qquad\qquad\qquad (a_2, c_1)$$

Solution to Exercise 7.32
1. Let $g_1 := (a_1, b_1)$ and $g_2 := (b_2, e_1)$; these do not agree on the overlap.
2. No, there's no section $g \in \operatorname{Sec}_f(U_1 \cup U_2)$ for which $g|_{U_1} = g_1$ and $g|_{U_2} = g_2$.

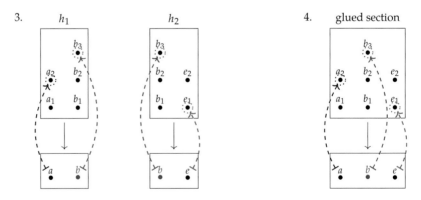

3. h_1 h_2 4. glued section

Solution to Exercise 7.35
No, there is not a one-to-one correspondence between sheaves on M and vector fields on M. The relationship between sheaves on M and vector fields on M is that the *set of all* vector fields on M corresponds to *one* sheaf, namely Sec_π, where $\pi : TM \to M$ is the tangent bundle as described in Example 7.34. There are so many sheaves on M that they don't even form a set (it's

just a "collection"); again, one member of this gigantic collection is the sheaf Sec_π of all possible vector fields on M.

Solution to Exercise 7.37

1. The Hasse diagram for the Sierpinski topology is $\boxed{\varnothing \to \{1\} \to \{1, 2\}}$.
2. A presheaf F on **Op** consists of any three sets and any two functions $F(\{1, 2\}) \to F(\{1\}) \to F(\varnothing)$ between them.
3. Recall from Exercise 7.23 that the only nontrivial covering (a covering of U is *nontrivial* if it does not contain U) occurs when $U = \varnothing$ in which case the empty family over U is a cover.
4. As explained in Example 7.28, F will be a sheaf iff $F(\varnothing) \cong \{1\}$. Thus the category of sheaves is equivalent to that of just two sets and one function $F(\{1, 2\}) \to F(\{1\})$.

Solution to Exercise 7.38

The one-point space $X = \{1\}$ has two open sets, $\varnothing$ and $\{1\}$, and every sheaf $S \in \mathbf{Shv}(X)$ assigns $S(\varnothing) = \{()\}$ by the sheaf condition (see Example 7.28). So the only data in a sheaf $S \in \mathbf{Shv}(X)$ is the set $S(\{1\})$. This is how we get the correspondence between sets and sheaves on the one-point space.

According to Eq. (7.13), the subobject classifier $\Omega \colon \mathbf{Op}(X)^{\mathrm{op}} \to \mathbf{Set}$ in $\mathbf{Shv}(X)$ should be the functor where $\Omega(\{1\})$ is the set of open sets of $\{1\}$. So we're hoping to see that there is a one-to-one correspondence between the set $\mathbf{Op}(\{1\})$ and the set $\mathbb{B} = \{\texttt{true}, \texttt{false}\}$ of booleans. Indeed there is: there are two open sets of $\{1\}$, as we said, $\varnothing$ and $\{1\}$, and these correspond to `false` and `true`, respectively.

Solution to Exercise 7.39

By Eqs. (7.13) and (7.14) the definition of $\Omega(U)$ is $\Omega(U) := \{U' \in \mathbf{Op} \mid U' \subseteq U\}$, and the definition of the restriction map for $V \subseteq U$ is $U' \mapsto U' \cap V$.

1. It is functorial: given $W \subseteq V \subseteq U$ and $U' \subseteq U$, we indeed have $(U' \cap V) \cap W = U' \cap W$, since $W \subseteq V$. For functoriality, we also need preservation of identities, and this amounts to $U' \cap U = U'$ for all $U' \subseteq U$.
2. Yes, a presheaf is just a functor; the above check is enough.

Solution to Exercise 7.41

We need a graph homomorphism of the following form:

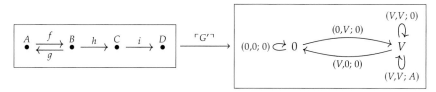

There is only one that classifies G', and here it is. Let's write $\gamma := \ulcorner G' \urcorner$.

- Since D is missing from G', we have $\gamma(D) = 0$ (vertex: missing).
- Since vertices A, B, C are present in G' we have $\gamma(A) = \gamma(B) = \gamma(C) = V$ (vertex: present).
- The above forces $\gamma(i) = (V, 0; 0)$ (arrow from present vertex to missing vertex: missing).
- Since the arrow f is in G', we have $\gamma(f) = (V, V; A)$ (arrow from present vertex to present vertex: present).

- Since the arrows g and h are missing in G', we have $\gamma(g) = \gamma(h) = (V, V; 0)$ (arrow from present vertex to present vertex: missing).

Solution to Exercise 7.43

With $U = \mathbb{R} - \{0\} \subseteq \mathbb{R}$, we have the following.

1. The complement of U is $\mathbb{R} - U = \{0\}$ and $\neg U$ is its interior, which is $\neg U = \varnothing$.
2. The complement of $\neg U$ is $\mathbb{R} - \varnothing = \mathbb{R}$, and this is open, so $\neg\neg U = \mathbb{R}$.
3. It is true that $U \subseteq \neg\neg U$.
4. It is false that $\neg\neg U \subseteq^? U$.

Solution to Exercise 7.44

1. If for any $V \in \mathbf{Op}$ we have $\top \wedge V = V$ then when $V = X$ we have $\top \wedge X := \top \cap X = X$, but anything intersected with X is itself, so $\top = \top \cap X = X$.
2. $(\top \vee V) := (X \cup V) = X$ holds, $(V \Rightarrow X) = \bigcup_{\{R \in \mathbf{Op}| R \cap V \subseteq X\}} R = X$ holds because $(X \cap V) \subseteq X$, and $(X \Rightarrow V) = V = \bigcup_{\{R \in \mathbf{Op}| R \cap X \leq V\}} R = Y$ holds because $R \cap X = R$.
3. If for any set $V \in \mathbf{Op}$ we have $(\bot \vee V) = V$, then when $V = \varnothing$ we have $(\bot \vee \varnothing) := (\bot \cup \varnothing) = \varnothing$, but anything in a union with $\varnothing$ is itself, so $\bot = \bot \cup \varnothing = \varnothing$.
4. $(\bot \wedge V) = (\varnothing \cap V) = \varnothing$ holds, and $(\bot \Rightarrow V) = \bigcup_{\{R \in \mathbf{Op}| R \cap \varnothing \subseteq V\}} R = X$ holds because $(X \cap \varnothing) \subseteq V$.

Solution to Exercise 7.46

S is the sheaf of people, the set of which changes over time: a section in S over any interval of time is a person who is alive throughout that interval. A section in the subobject $\{S \mid p\}$ over any interval of time is a person who is alive *and likes the weather* throughout that interval of time.

Solution to Exercise 7.47

We need an example of a space X, a sheaf $S \in \mathbf{Shv}(X)$, and two predicates $p, q \colon S \to \Omega$ for which $p(s) \vdash_{s:S} q(s)$ holds. Take X to be the one-point space, take S to be the sheaf corresponding to the set $S = \mathbb{N}$, let $p(s)$ be the predicate "$24 \leq s \leq 28$," and let $q(s)$ be the predicate "s is not prime." Then $p(s) \vdash_{s:S} q(s)$ holds.

As an informal example, take X to be the surface of the Earth, take S to be the sheaf of vector fields as in Example 7.34 thought of in terms of wind-blowing. Let p be the predicate "the wind is blowing due east at somewhere between 2 and 5 kilometers per hour" and let q be the predicate "the wind is blowing at somewhere between 1 and 5 kilometers per hour." Then $p(s) \vdash_{s:S} q(s)$ holds. This means that, for any open set U, if the wind is blowing due east at somewhere between 2 and 5 kilometers per hour throughout U, then the wind is blowing at somewhere between 1 and 5 kilometers per hour throughout U as well.

Solution to Exercise 7.49

We have the predicate $p \colon \mathbb{N} \times \mathbb{Z} \to \mathbb{B}$ given by $p(n, z)$ iff $n \leq |z|$.

1. The predicate $\forall(z : \mathbb{Z}).\, p(n, z)$ holds for $\{0\} \subseteq \mathbb{N}$.
2. The predicate $\exists(z : \mathbb{Z}).\, p(n, z)$ holds for $\mathbb{N} \subseteq \mathbb{N}$.
3. The predicate $\forall(n : \mathbb{N}).\, p(n, z)$ holds for $\varnothing \subseteq \mathbb{Z}$.
4. The predicate $\exists(n : \mathbb{N}).\, p(n, z)$ holds for $\mathbb{Z} \subseteq \mathbb{Z}$.

Solution to Exercise 7.50

Suppose s is a person alive throughout the interval U. Apply the above definition to the example $p(s, t) = $ "person s is worried about news t" from above.

1. The formula says that $\forall(t : T). p(s, t)$ "returns the largest open set $V \subseteq U$ for which $p(s|_V, t) = V$ for all $t \in T(V)$." Note that $T(V)$ is the set of items that are in the news throughout the interval V. Substituting, this becomes "the largest interval of time $V \subseteq U$ over which person s is worried about news t for every item t that is in the news throughout V." In other words, for V to be nonempty, the person s would have to be worried about *every single item of news* throughout V. My guess is that there's a festival happening or a happy kitten somewhere that person s is not worried about, but maybe I'm assuming that person s is sufficiently mentally "normal." There may be people who are sometimes worried about literally everything in the news; we ask you to please be kind to them.
2. Yes, it is exactly the same description.

Solution to Exercise 7.51

Suppose s is a person alive throughout the interval U. Apply the above definition to the example $p(s, t) = $ "person s is worried about news t" from above.

1. The formula says that $\exists(t : T). p(s, t)$ "returns the union $V = \bigcup_i V_i$ of all the open sets V_i for which there exists some $t_i \in T(V_i)$ satisfying $p(s|_{V_i}, t_i) = V_i$." Substituting, this becomes "the union of all time intervals V_i for which there is some item t_i in the news about which s is worried throughout V_i." In other words it is all the time that s is worried about at least one thing in the news. Perhaps when s is sleeping or concentrating on something, she is not worried about anything, in which case intervals of sleeping or concentrating would not be subsets of V. But if s said "there's been such a string of bad news this past year, it's like I'm always worried about something!," she is saying that it's like $V = $ "this past year."
2. This seems like a good thing for "there exists a piece of news that worries s" to mean: the news itself is allowed to change as long as the person's worry remains. Someone might disagree and think that the predicate should mean "there is one piece of news that worries s throughout the whole interval V." In that case, perhaps this person is working within a different topos, e.g. one where the site has fewer coverings. Indeed, it is the notion of covering that makes existential quantification work the way it does.

Solution to Exercise 7.53

It is clear that if $j(j(q)) = j(q)$ then $j(j(q)) \leq j(q)$ by reflexivity. On the other hand, assume the hypothesis, that $p \leq j(p)$ for all $U \subseteq X$ and $p \in \Omega(U)$. If $j(j(q)) \leq j(q)$, then letting $p := j(q)$ we have both $j(p) \leq p$ and $p \leq j(p)$. This means $p \cong j(p)$, but Ω is a poset (not just a preorder) so $p = j(p)$, i.e. $j(j(q)) = j(q)$ as desired.

Solution to Exercise 7.55

Let S be the sheaf of people and j be "assuming Bob is in San Diego ..."

1. Take $p(s)$ to be "s likes the weather."
2. Let U be the interval 2019/01/01 – 2019/02/01. For an arbitrary person $s \in S(U)$, $p(s)$ is a subset of U, and it means the subset of U throughout which s likes the weather.
3. Similarly $j(p(s))$ is a subset of U, and it means the subset of U throughout which, assuming Bob is in San Diego, s liked the weather. In other words, $j(p(s))$ is true whenever Bob is not in San Diego, and it is true whenever s likes the weather.
4. It is true that $p(s) \leq j(p(s))$, by the "in other words" above.

5. It is true that $j(j(p(s))) = j(p(s))$, because suppose we are given a time during which "if Bob is in San Diego then if Bob is in San Diego then s likes the weather." Then if Bob is in San Diego during this time then s likes the weather. But that is exactly what $j(p(s))$ means.

6. Take $q(s)$ to be "s is happy." Suppose "if Bob is in San Diego then both s likes the weather and s is happy." Then both "if Bob is in San Diego then s likes the weather" and "if Bob is in San Diego then s is happy" are true too. The converse is equally clear.

Solution to Exercise 7.58

We have $o_{[a,b]} := \{[d, u] \in \mathbb{IR} \mid a < d \leq u < b\}$.

1. Since $0 \leq 2 \leq 6 \leq 8$, we have $[2, 6] \in o_{[0,8]}$ by the above formula.

2. In order to have $[2, 6] \in^? o_{[0,5]} \cup o_{[4,8]}$, we would need to have either $[2, 6] \in^? o_{[0,5]}$ or $[2, 6] \in^? o_{[4,8]}$. But the formula does not hold in either case.

Solution to Exercise 7.59

A subset $U \subseteq \mathbb{R}$ is open in the subspace topology of $\mathbb{R} \subseteq \mathbb{IR}$ iff there is an open set $U' \subseteq \mathbb{IR}$ with $U = U' \cap \mathbb{R}$. We want to show that this is the case iff U is open in the usual topology.

Suppose that U is open in the subspace topology. Then $U = U' \cap \mathbb{R}$, where $U' \subseteq \mathbb{IR}$ is the union of some basic opens, $U' = \bigcup_{i \in I} o_{[a_i,b_i]}$, where $o_{[a_i,b_i]} = \{[d, u] \in \mathbb{IR} \mid a_i < d < u < b_i\}$. Since $\mathbb{R} = \{[x, x] \in \mathbb{IR}\}$, the intersection $U' \cap \mathbb{R}$ will then be

$$U = \bigcup_{i \in I} \{x \in \mathbb{R} \mid a_i < x < b_i\}$$

and this is just the union of open balls $B(m_i, r_i)$ where $m_i := \frac{a_i + b_i}{2}$ is the midpoint and $r_i := \frac{b_i - a_i}{2}$ is the radius of the interval (a_i, b_i). The open balls $B(m_i, r_i)$ are open in the usual topology on $\mathbb{R}$ and the union of opens is open, so U is open in the usual topology.

Suppose that U is open in the usual topology. Then $U = \bigcup_{j \in J} B(m_j, \epsilon_j)$ for some set J. Let $a_j := m_j - \epsilon_j$ and $b_j := m_j + \epsilon_j$. Then

$$U = \bigcup_{j \in J} \{x \in \mathbb{R} \mid a_j < x < b_j\} = \bigcup_{j \in J} (o_{[a_j,b_j]} \cap \mathbb{R}) = \left(\bigcup_{j \in J} o_{[a_j,b_j]} \right) \cap \mathbb{R}$$

which is open in the subspace topology.

Solution to Exercise 7.62

Fix any topological space $(X, \mathbf{Op}_X)$ and any subset $R \subseteq \mathbb{IR}$ of the interval domain. Define $H_X(U) := \{f : U \cap R \to X \mid f \text{ is continuous}\}$.

1. H_X is a presheaf: given $V \subseteq U$ the restriction map sends the continuous function $f : U \cap R \to X$ to its restriction along the subset $V \cap R \subseteq U \cap R$.

2. It is a sheaf: given any family U_i of open sets with $U = \bigcup_i U_i$ and a continuous function $f_i : U_i \cap R \to X$ for each i, agreeing on overlaps, they can be glued together to give a continuous function on all of $U \cap R$, since $U \cap R = (\bigcup_i U_i) \cap R = \bigcup_i (U_i \cap R)$.

References

[Ada17] Elie M. Adam. "Systems, Generativity and Interactional Effects." Available online: `www.mit.edu/~eadam/eadam_PhDThesis.pdf` PhD thesis. Massachusetts Institute of Technology, July 2017 (cit. on pp. 2, 26, 36).

[AGV71] Michael Artin, Alexander Grothendieck, and Jean-Louis Verdier. *Theorie de Topos et Cohomologie Etale des Schemas I, II, III.* Lecture Notes in Mathematics 269, 270, 305. Springer, 1971 (cit. on p. 256).

[AJ94] Samson Abramsky and Achim Jung. "Domain theory." In *Handbook of Logic in Computer Science.* Oxford University Press, 1994 (cit. on p. 257).

[AS05] Aaron D. Ames and Shankar Sastry. "Characterization of Zeno behavior in hybrid systems using homological methods." In *American Control Conference, 2005. Proceedings of the 2005.* IEEE. 2005, pp. 1160–1165 (cit. on p. 257).

[AV93] Samson Abramsky and Steven Vickers. "Quantales, observational logic and process semantics." *Mathematical Structures in Computer Science* 3.2 (1993), pp. 161–227 (cit. on p. 76).

[Awo10] Steve Awodey. *Category Theory*, 2nd edn. Oxford Logic Guides 52. Oxford University Press, 2010 (cit. on p. 114).

[BD98] John C. Baez and James Dolan. "Categorification." In Higher Category Theory, American Mathematical Society, 1998. eprint: `math/9802029` (cit. on p. 145).

[BE15] John C. Baez and Jason Erbele. "Categories in control." *Theory and Applications of Categories* 30 (2015), pp. 836–881 (cit. on pp. 170, 179).

[BF15] John C. Baez and Brendan Fong. "A compositional framework for passive linear networks." 2015. URL: `https://arxiv.org/abs/1504.05625` (cit. on p. 219).

[BFP16] John C. Baez, Brendan Fong, and Blake S. Pollard. "A compositional framework for Markov processes." *Journal of Mathematical Physics* 57.3 (2016) (cit. on p. 219).

[BH08] Philip A Bernstein and Laura M Haas. "Information integration in the enterprise." *Communications of the ACM* 51.9 (2008), pp. 72–79 (cit. on p. 77).

[Bor94] Francis Borceux. *Handbook of Categorical Algebra 1: Basic Category Theory.* Encyclopedia of Mathematics and its Applications 50. Cambridge University Press, 1994 (cit. on pp. 114, 192).

[BP17] John C. Baez and Blake S. Pollard. "A compositional framework for reaction networks." *Reviews in Mathematical Physics* 29.9 (2017) (cit. on p. 219).

[Bro61] Ronnie Brown. "Some problems of algebraic topology: a study of function spaces, function complexes, and FD-complexes." PhD thesis, University of Oxford, 1961 (cit. on p. 76).

[BSZ14] Filippo Bonchi, Paweł Sobociński, and Fabio Zanasi. "A categorical semantics of signal flow graphs." In *International Conference on Concurrency Theory.* Springer-Verlag, 2014, pp. 435–450 (cit. on p. 179).

[BSZ15] Filippo Bonchi, Paweł Sobociński, and Fabio Zanasi. "Full abstraction for signal flow graphs." *ACM SIGPLAN Notices.* 50.1. (2015), pp. 515–526 (cit. on p. 179).

[BSZ17] Filippo Bonchi, Paweł Sobociński, and Fabio Zanasi. "The calculus of signal flow diagrams I: Linear relations on streams." *Information and Computation* 252 (2017), pp. 2–29 (cit. on pp. 170, 179).

[BW90] Michael Barr and Charles Wells. *Category Theory for Computing Science.* Prentice Hall, 1990 (cit. on p. 114).

[Car91] Aurelio Carboni. "Matrices, relations, and group representations." *Journal of Algebra* 136.2 (1991), pp. 497–529. URL: www.sciencedirect.com/science/article/pii/002186939190057F (cit. on p. 219).

[CD95] Boris Cadish and Zinovy Diskin. "Algebraic graph-based approach to management of multibase systems, I: Schema integration via sketches and equations." In *Proceedings of Next Generation of Information Technologies and Systems, NGITS,* Vol. 95.1995 (cit. on p. 114).

[Cen15] Andrea Censi. "A mathematical theory of co-design." 2015. eprint: arXiv:1512.08055 (cit. on pp. 117, 145).

[Cen17] Andrea Censi. "Uncertainty in monotone co-design problems." *IEEE Robotics and Automation Letters* 2.3 (2017), pp. 1556–1563. URL: https://arxiv.org/abs/1609.03103 (cit. on p. 145).

[CFS16] Bob Coecke, Tobias Fritz, and Robert W. Spekkens. "A mathematical theory of resources." *Information and Computation* 250 (2016), pp. 59–86 (cit. on pp. 48, 75).

[CK17] Bob Coecke and Aleks Kissinger. *Picturing Quantum Processes.* Cambridge University Press, 2017 (cit. on p. iv).

[CP10] Bob Coecke and Eric O. Paquette. "Categories for the practising physicist." In *New Structures for Physics.* Springer, 2010, pp. 173–286 (cit. on p. iv).

[CW87] A. Carboni and R. F. C. Walters. "Cartesian bicategories I." *Journal of Pure and Applied Algebra* 49.1 (1987), pp. 11–32. URL: www.sciencedirect.com/science/article/pii/0022404987901216 (cit. on p. 219).

[CY96] Louis Crane and David N. Yetter. "Examples of categorification." 1996. eprint: q-alg/9607028 (cit. on p. 145).

[Dar70] Gabriele Darbo. "Aspetti algebrico-categoriali della teoria dei dispositivi." *Symposia Mathematica IV* (1970), Istituto Nazionale di Alta Mathematica, pp. 303–336 (cit. on p. 218).

[FGR03] Michael Fleming, Ryan Gunther, and Robert Rosebrugh. "A database of categories." *Journal of Symbolic Computation* 35 (2003), pp. 127–135 (cit. on p. 114).

[Fon15] Brendan Fong. "Decorated cospans." *Theory and Applications of Categories* 30.33 (2015), pp. 1096–1120 (cit. on p. 219).

[Fon16] Brendan Fong. "The algebra of open and interconnected systems." PhD thesis, University of Oxford, 2016 (cit. on p. 219).

[Fon18] Brendan Fong. "Decorated corelations." *Theory and Applications of Categories* 33.22 (2018), pp. 608–643 (cit. on p. 219).

[Fra67] John B. Fraleigh. *A First Course in Abstract Algebra.* Addison-Wesley, 1967 (cit. on p. 179).

[Fri17] Tobias Fritz. "Resource convertibility and ordered commutative monoids." *Mathematical Structures in Computer Science* 27.6 (2017), pp. 850–938 (cit. on pp. 48, 75).

[FS18a]   Brendan Fong and Maru Sarazola. "A recipe for black box functors." 2018. eprint: `arXiv:1812.03601` (cit. on p. 219).

[FS18b]   Brendan Fong and David I. Spivak. "Hypergraph categories." 2018. eprint: `arXiv: 1806.08304` (cit. on p. 219).

[FSR16]   Brendan Fong, Paweł Sobociński, and Paolo Rapisarda. "A categorical approach to open and interconnected dynamical systems." In *Proceedings of the 31st Annual ACM/IEEE Symposium on Logic in Computer Science.* Association for Computing Machinery, 2016, pp. 495–504 (cit. on pp. 168, 179).

[Gie+03] G. Gierz, K. H. Hofmann, K. Keimel, J. D. Lawson, M. Mislove, and D. S. Scott. *Continuous Lattices and Domains.* Encyclopedia of Mathematics and its Applications 93. Cambridge University Press, 2003 (cit. on p. 257).

[Gla13]   K Glazek. *A Guide to the Literature on Semirings and their Applications in Mathematics and Information Sciences: With Complete Bibliography.* Springer Science & Business Media, 2013 (cit. on p. 179).

[Gra18]   Marco Grandis. *Category Theory and Applications.* World Scientific, 2018 (cit. on p. 114).

[HMP98]   Claudio Hermida, Michael Makkai, and John Power. "Higher-dimensional multi-graphs." In *Proceedings of the 13th Annual IEEE Symposium on Logic in Computer Science (Indianapolis, IN, 1998).* IEEE Computer Society, 1998, pp. 199–206 (cit. on p. 214).

[HTP03]   Esfandiar Haghverdi, Paulo Tabuada, and George Pappas. "Bisimulation relations for dynamical and control systems." *Electronic Notes in Theoretical Computer Science* 69 (2003), pp. 120–136 (cit. on p. 257).

[IP94]   Amitavo Islam and Wesley Phoa. "Categorical models of relational databases I: Fibrational formulation, schema integration." In *International Symposium on Theoretical Aspects of Computer Software.* Springer, 1994, pp. 618–641 (cit. on p. 114).

[Jac99]   Bart Jacobs. *Categorical Logic and Type Theory.* Studies in Logic and the Foundations of Mathematics 141. North-Holland, 1999 (cit. on p. 257).

[JNW96]   André Joyal, Mogens Nielsen, and Glynn Winskel. "Bisimulation from open maps." *Information and Computation* 127.2 (1996), pp. 164–185 (cit. on p. 257).

[Joh02]   Peter T. Johnstone. *Sketches of an Elephant: A Topos Theory Compendium.* Oxford Logic Guides 43. Oxford University Press, 2002 (cit. on p. 257).

[Joh77]   Peter T. Johnstone. *Topos Theory.* London Mathematical Society Monographs 10. Academic Press, 1977 (cit. on p. 223).

[JR02]   Michael Johnson and Robert Rosebrugh. "Sketch data models, relational schema and data specifications." In: *Electronic Notes in Theoretical Computer Science CATS'02, Computing,* 61 (2002), pp. 51–63 (cit. on p. 114).

[JS93]   André Joyal and Ross Street. "Braided tensor categories." *Advances in Mathematics* 102.1 (1993), pp. 20–78 (cit. on pp. 40, 145).

[JSV96]   André Joyal, Ross Street, and Dominic Verity. "Traced monoidal categories." *Mathematical Proceedings of the Cambridge Philosophical Society* 119 (1996), 447–468 (cit. on p. 145).

[Kel05]   G. M. Kelly. "Basic concepts of enriched category theory." *Reprints in Theory and Applications of Categories* No. 10 (2005). URL: `www.tac.mta.ca/tac/ reprints/articles/10/tr0abs.html` (cit. on pp. 76, 139).

[Law04]   F. William Lawvere. "Functorial semantics of algebraic theories and some algebraic problems in the context of functorial semantics of algebraic theories." *Reprints in Theory and Applications of Categories* No. 5 (2004) (cit. on p. 179).

[Law73]   F. William Lawvere. "Metric spaces, generalized logic, and closed categories." *Rendiconti del seminario matematico efisico di Milano* 43.1 (1973), pp. 135–166 (cit. on pp. 76, 145).

[Law86]   F. William Lawvere. "State categories and response functors." 1986. URL: https://arxiv.org/pdf/1807.06000.pdf (cit. on p. 257).

[Lei04]   Tom Leinster. *Higher Operads, Higher Categories*. London Mathematical Society Lecture Note Series 298. Cambridge University Press, 2004 (cit. on pp. 213, 218, 219).

[Lei14]   Tom Leinster. *Basic Category Theory*. Cambridge Studies in Advanced Mathematics 14.3. Cambridge University Press, 2014 (cit. on p. 114).

[LS88]   J. Lambek and P. J. Scott. *Introduction to Higher Order Categorical Logic*. Cambridge Studies in Advanced Mathematics 7. Cambridge University Press, 1988 (cit. on p. 257).

[Mac98]   Saunders Mac Lane. *Categories for the Working Mathematician*, 2nd edn. Graduate Texts in Mathematics 5. Springer-Verlag, 1998 (cit. on pp. 114, 115, 145).

[May72]   J. Peter May. *The Geometry of Iterated Loop Spaces*. Lecture Notes in Mathematics 271. Springer, 1972 (cit. on p. 219).

[McL90]   Colin McLarty. "The uses and abuses of the history of topos theory." *British Journal for the Philosophy of Science* 41.3 (1990), pp. 351–375 (cit. on p. 256).

[McL92]   Colin McLarty. *Elementary Categories, Elementary Toposes*. Clarendon Press, 1992 (cit. on p. 256).

[MM92]   Saunders MacLane and Ieke Moerdijk. *Sheaves in Geometry and Logic: A First Introduction to Topos Theory*. Springer, 1992 (cit. on p. 257).

[nLa18]   nLab. "Symmetric monoidal category." 2018. URL: ncatlab.org/nlab/revision/symmetric+monoidal+category/30 (cit. on p. 139).

[NNH99]   Flemming Nielson, Hanne R. Nielson, and Chris Hankin. *Principles of Program Analysis*. Springer-Verlag, 1999 (cit. on p. 37).

[Pie91]   Benjamin C. Pierce. *Basic Category Theory for Computer Scientists*. MIT Press, 1991 (cit. on p. 114).

[PS95]   Frank Piessens and Eric Steegmans. "Categorical data specifications." *Theory and Applications of Categories* 1.8 (1995), pp. 156–173 (cit. on p. 114).

[Rie17]   Emily Riehl. *Category Theory in Context*. Courier Dover, 2017 (cit. on p. 114).

[Ros90]   Kimmo I. Rosenthal. *Quantales and their Applications*. Longman Scientific and Technical 234. Longman, 1990 (cit. on p. 76).

[RS13]   Dylan Rupel and David I. Spivak. "The operad of temporal wiring diagrams: formalizing a graphical language for discrete-time processes." 2013. eprint: arXiv:1307.6894 (cit. on p. 219).

[RW92]   Robert Rosebrugh and R. J. Wood. "Relational databases and indexed categories." In *Canadian Mathematical Society Conference Procedings*. International Summer Category Theory Meeting (June 23–30, 1991), ed. R. A. G. Seely. American Mathematical Society, 1992, pp. 391–407 (cit. on p. 114).

[Sch+17]   Patrick Schultz, David I. Spivak, Christina Vasilakopoulou, and Ryan Wisnesky. "Algebraic databases." *Theory and Applications of Categories* 32 (2017), pp. 547–619 (cit. on p. 114).

[Sel10]   Peter Selinger. "A survey of graphical languages for monoidal categories." In *New Structures for Physics*. Springer, 2010, pp. 289–355 (cit. on p. 146).

[Shu08]   Michael Shulman. "Framed bicategories and monoidal fibrations." *Theory and Applications of Categories* 20 (2008), pp. 650–738 (cit. on p. 145).

[Shu10]   Michael Shulman. "Constructing symmetric monoidal bicategories." 2010. eprint: `arXiv:1004.0993` (cit. on p. 145).

[SLG+15]   Eswaran Subrahmanian, Christopher Lee, Helen Granger, et al. "Managing and supporting product life cycle through engineering change management for a complex product." *Research in Engineering Design* 26.3 (2015), pp. 189–217 (cit. on p. 118).

[Sob]   Paweł Sobociński. *Graphical Linear Algebra*. URL: `https://graphicallin earalgebra.net/` (accessed on 03/11/2018) (cit. on p. 179).

[Spi+16]   David I. Spivak, Magdalen R. C. Dobson, Sapna Kumari, and Lawrence Wu. "Pixel arrays: A fast and elementary method for solving nonlinear systems." 2016. eprint: `arXiv:1609.00061` (cit. on p. 219).

[Spi12]   David I. Spivak. "Functorial data migration." *Information and Computation* 217 (2012), pp. 31–51 (cit. on p. 114).

[Spi13]   David I. Spivak. "The operad of wiring diagrams: formalizing a graphical language for databases, recursion, and plug-and-play circuits." 2013. eprint: `arXiv:1305 .0297` (cit. on p. 219).

[Spi14a]   David I. Spivak. *Category Theory for the Sciences*. MIT Press, 2014 (cit. on pp. 105, 114).

[Spi14b]   David I. Spivak. "Database queries and constraints via lifting problems." *Mathematical Structures in Computer Science* 24.6 (2014), e240602, 55. URL: `http://dx .doi.org/10.1017/S0960129513000479` (cit. on p. 94).

[SS18]   Patrick Schultz and David I. Spivak. *Temporal Type Theory: A Topos-Theoretic Approach to Systems and Behavior*. Springer, 2018 (cit. on p. 256).

[SSV18]   Alberto Speranzon, David I. Spivak, and Srivatsan Varadarajan. "Abstraction, composition and contracts: a sheaf theoretic approach." 2018. eprint: `arXiv:1802 .03080` (cit. on p. 255).

[SVS16]   David I. Spivak, Christina Vasilakopoulou, and Patrick Schultz. "Dynamical systems and sheaves." 2016. eprint: `arXiv:1609.08086`.

[SW15a]   Patrick Schultz and Ryan Wisnesky. "Algebraic data integration." 2015. eprint: `arXiv:1503.03571` (cit. on p. 114).

[SW15b]   David I. Spivak and Ryan Wisnesky. "Relational foundations for functorial data migration." In *Proceedings of the 15th Symposium on Database Programming Languages*. DBPL. Association for Computing Machinary, 2015, pp. 21–28 (cit. on p. 114).

[TG96]   Chris Tuijn and Marc Gyssens. "CGOOD, a categorical graph-oriented object data model." *Theoretical Computer Science* 160.1–2 (1996), pp. 217–239 (cit. on p. 114).

[Vig03]   Sebastiano Vigna. "A guided tour in the topos of graphs." 2003. eprint: `arXiv: math/0306394` (cit. on p. 245).

[VSL15]   Dmitry Vagner, David I. Spivak, and Eugene Lerman. "Algebras of open dynamical systems on the operad of wiring diagrams." *Theory and Applications of Categories* 30 (2015), pp. 1793–1822 (cit. on pp. 211, 219).

[Wal92]   R. F. C. Walters. *Categories and Computer Science*. Cambridge University Press, 1992 (cit. on p. 114).

[Wil07]   Jan C. Willems. "The behavioral approach to open and interconnected systems." *IEEE Control Systems* 27.6 (2007), pp. 46–99 (cit. on p. 179).

[Wis+15] Ryan Wisnesky, David I. Spivak, Patrick Schultz, and Eswaran Subrahmanian. *Functorial Data Migration: From Theory to Practice*. Report G2015-1701. National Institute of Standards and Technology, 2015. arXiv: 1502.05947v2 (cit. on p. 114).

[Zan15] Fabio Zanasi. "Interacting Hopf Algebras – the Theory of Linear Systems." PhD thesis, Ecole normale supérieure de lyon, ENS LYON, 2015. URL: https://tel.archives-ouvertes.fr/tel-01218015 (cit. on p. 179).

Index